Accounting:
Information for Business Decisions

Billie M. Cunningham
University of Missouri

Loren A. Nikolai
University of Missouri

John D. Bazley
University of Denver

Kevin E. Murphy
Oklahoma State University

Mark Higgins
University of Rhode Island

THOMSON

™

SOUTH-WESTERN

Printed in the United States of America

South-Western College Publishing
5191 Natorp Blvd
Mason, Ohio 45040
USA

For information about our products, contact us:
Thomson Learning Academic Resource Center
1-800-423-0563
http://www.swcollege.com

International Headquarters
Thomson Learning
International Division
290 Harbor Drive, 2nd Floor
Stamford, CT 06902-7477
USA

UK/Europe/Middle East/South Africa
Thomson Learning
Berkshire House
168-173 High Holborn
London WCIV 7AA
United Kingdom

Asia
Thomson Learning
60 Albert Street, #15-01
Albert Complex
Singapore 189969

Canada
Nelson Thomson Learning
1120 Birchmount Road
Toronto, Ontario MIK 5G4
Canada

For permission to use material from this text or product, submit a request online at
http://www.thomsonrights.com

Any additional questions about permissions can be submitted by email to
thomsonrights@thomson.com

ISBN 0-324-39156-0

The Adaptable Courseware Program consists of products and additions to existing South-Western College Publishing products that are produced from camera-ready copy. Peer review, class testing, and accuracy are primarily the responsibility of the author(s).

Custom Contents

PART 1

OVERVIEW: BUSINESS, ACCOUNTING, AND THE ROLE OF CREATIVE AND CRITICAL THINKING

CHAPTER 1
INTRODUCTION TO BUSINESS AND ACCOUNTING

CHAPTER 2
CREATIVE AND CRITICAL THINKING, PROBLEM SOLVING, AND THEIR ROLES IN BUSINESS AND ACCOUNTING

This section consists of two chapters which introduce you to business and accounting, and discuss the role of creative and critical thinking in business decisions. After reading these chapters, you will be able to:

- *understand the role of accounting information in business*

- *describe the planning, operating, and evaluating activities of managing a company*

- *know the difference between management accounting and finanacial accounting*

- *identify internal and external accounting reports*

- *explain the meaning of creative and critical thinking*

- *apply creative and critical thinking in business decisions*

INTRODUCTION TO BUSINESS AND ACCOUNTING

"BUSINESS IS
A GAME, THE
GREATEST GAME
IN THE WORLD IF
YOU KNOW HOW TO
PLAY IT."

—THOMAS J.
WATSON SR.

1. Why is it necessary to have an understanding of business before trying to learn about accounting?

2. What is the role of accounting information within the business environment?

3. What is private enterprise, and what forms does it take?

4. What types of regulations do companies face?

5. What activities contribute to the operations of a company?

6. Are there any guidelines for reporting to company managers?

7. Are there any guidelines in the United States for reporting to people outside of a company?

8. What role does ethics play in the business environment?

W hat are you planning to do when you graduate from college—maybe become an accountant or a veterinarian, work your way up to marketing manager for a multinational company, manage the local food bank, or open a sporting goods store? Regardless of your career choice, you will be making business decisions, both in your personal life and at work. We have oriented this book to students like you who are interested in business and the role of accounting in business. You will see that accounting information, used properly, is a powerful tool for making good business decisions. People inside a business use accounting information to help determine and manage costs, set selling prices, and control the operations of the business. People outside the business use accounting information to help make investment and credit decisions about the business. Just what kinds of businesses use accounting? All of them! So let's take a little time to look at what *business* means.

Business affects almost every aspect of our lives. Think for a moment about your normal daily activities. How many businesses do you usually encounter? How many did you directly encounter today? Say you started the day with a quick trip to the local convenience store for milk and eggs. While you were out, you noticed that your car was low on fuel, so you stopped at the corner gas station. On the way to class, you dropped some clothes off at the cleaners. After your first class, you skipped lunch so that you could go to the bookstore and buy the calculator you need; after buying a candy bar for sustenance, you headed to your next class. In just half a day, you already interacted with four businesses: the convenience store, the gas station, the cleaners, and the bookstore.

 Actually, you encountered a fifth business, your school. Why would you describe your school as a business?

Although you were directly involved with four businesses, you were probably *affected* by hundreds of them. For example, two different businesses manufactured the calculator and the candy bar you purchased at the bookstore. Suppose that Unlimited Decadence Corporation manufactured the candy bar that you purchased. As we illustrate in Exhibit 1-1, Unlimited Decadence purchased the candy bar ingredients from many other businesses *(suppliers)*. Each supplier provided Unlimited Decadence with particular ingredients. Shipping businesses *(carriers)* moved the ingredients from the suppliers' warehouses to Unlimited Decadence's factory. Then, after the candy bars were manufactured, a different carrier moved them from Unlimited Decadence to the bookstore. Making and shipping the calculator would follow the same process. You can see that many businesses are involved with manufacturing, shipping, and selling just two products. Now think about all the other products that you used during the morning and all the businesses that were involved with the manufacture and delivery of each product. Before leaving your house, apartment, or dorm this morning, you could easily have been affected by hundreds of businesses.

Products and services affect almost every minute of our lives, and businesses provide these products and services to us. As you will soon see, accounting plays a vital role in both businesses and the business environment by keeping track of a business's economic resources and economic activities, and then by reporting the business's financial position and the results of its activities to people who are interested in how well it is doing. (This is similar to the way statistics are gathered and reported for baseball players and other athletes.)

Accounting focuses on the resources and activities of individual businesses. We will introduce you to accounting by first looking at private enterprise and the environment in which businesses operate. Our discussion will include the types and forms of business, as well as some of the regulatory issues associated with forming and operating a business. Then we will discuss the activities of managers within a business. Next we will introduce the role of accounting information within a business and in the business environment. Finally, we will discuss the importance of ethics in business and accounting.

1 Why is it necessary to have an understanding of business before trying to learn about accounting?

2 What is the role of accounting information within the business environment?

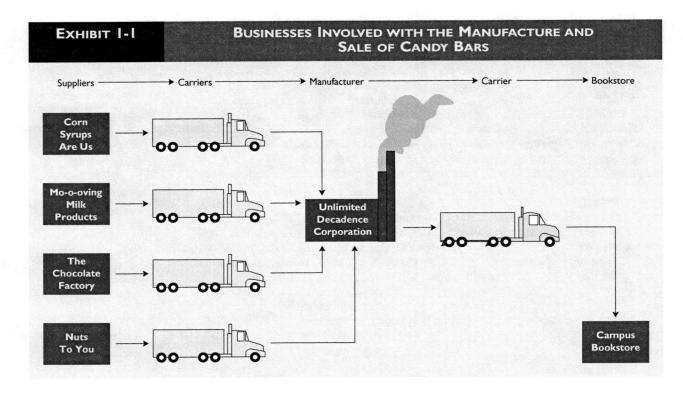

EXHIBIT 1-1 **BUSINESSES INVOLVED WITH THE MANUFACTURE AND SALE OF CANDY BARS**

PRIVATE ENTERPRISE AND CAPITALISM

3 What is private enterprise, and what forms does it take?

Businesses in the United States and most other countries operate in an economic system based on *private enterprise*. In this system, individuals (people like us, rather than public institutions like the government) own *companies* (businesses) that produce and sell services and/or goods for a profit. These companies generally fall into three categories: service companies, merchandising companies, and manufacturing companies.

Service Companies

Service companies perform services or activities that benefit individuals or business customers. The dry cleaning establishment where you dropped off your clothes this morning provides the service of cleaning and pressing your clothes for you. Companies like **A Great Cut, Midas Muffler Shops, Merry Maids,** and **UPS,** and professional practices such as accounting, law, architecture, and medicine, are all service companies. Other companies in the private enterprise system produce or provide goods, or tangible, physical products. These companies can be either *merchandising companies* or *manufacturing companies*.

http://www.agreatcut.com
http://www.midas.com
http://www.merrymaids.com
http://www.ups.com

Merchandising Companies

Merchandising companies purchase goods (sometimes referred to as *merchandise* or *products*) for resale to their customers. Some merchandising companies, such as plumbing supply stores, electrical suppliers, or beverage distributors, are *wholesalers*. Wholesalers primarily sell their goods to retailers or other commercial users, like plumbers or electricians. Some merchandising companies, such as the bookstore where you bought your calculator and candy bar or the convenience store where you bought your milk and eggs, are *retailers*. Retailers sell their goods directly to the final customer or consumer. **JCPenney, Toys 'R' Us, amazon.com,** and **Circuit City** are retailers. Other examples of retailers include shoe stores and grocery stores.

http://www.jcpenney.com
http://www.toysrus.com
http://www.amazon.com
http://www.circuitcity.com

Manufacturing Companies

Manufacturing companies make their products and then sell these products to their customers. Therefore, a basic difference between merchandising companies and manufacturing companies involves the products that they sell. Merchandising companies *buy* products that are physically ready for sale and then sell these products to their customers, whereas manufacturing companies *make* their products first and then sell the products to their customers. For example, the bookstore is a merchandising company that sells the candy bars it purchased from Unlimited Decadence, a manufacturing company. Unlimited Decadence, though, purchases (from suppliers) the chocolate, corn syrup, dairy products, and other ingredients to make the candy bars, which it then sells to the Campus Bookstore and other retail stores. **General Motors, Black & Decker,** and **Dana Corporation** are examples of manufacturing companies. Exhibit 1-2 shows the relationship between manufacturing companies and merchandising companies and how they relate to their customers.

http://www.gm.com
http://www.blackanddecker.com
http://www.dana.com

The line of distinction between service, merchandising, and manufacturing companies is sometimes blurry because a business can be more than one type of company. For example, Dell Computer Corporation manufactures personal computers, sells the computers it manufactures directly to business customers, government agencies, educational institutions, and individuals, and services those computers (through installation, technology transition, and management).

 Do you think a supplier to a manufacturing company is a merchandising company or a manufacturing company? Why?

Whether a company is a service, merchandising, or manufacturing company (or all three), for it to succeed in a private enterprise system, it must be able to obtain cash to begin to operate and then to grow. As we will discuss in the next sections, companies have several sources of cash.

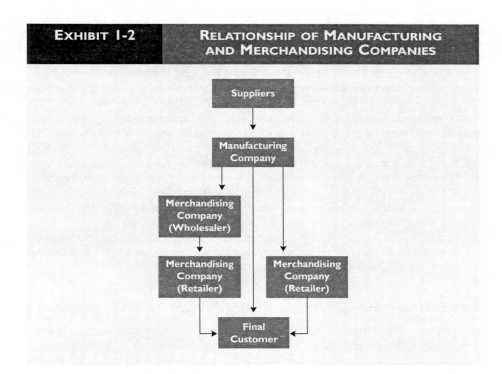

EXHIBIT 1-2   RELATIONSHIP OF MANUFACTURING AND MERCHANDISING COMPANIES

Entrepreneurship and Sources of Capital

Owning a company involves a level of risk, along with a continuing need for **capital**. Although *capital* has several meanings, we use the term here to mean the funds a company needs to operate or to expand operations. In the next two sections we will discuss the risk involved with owning a company and possible sources of capital.

Entrepreneurship

Companies in a private enterprise system produce and sell services and goods for a profit. So, profit is the primary objective of a company. Profit rewards the company's owner or owners for having a business idea and for following through with that idea by investing time, talent, and money in the company. The company's owner hires employees, purchases land and a building (or signs a lease for space in a building), and purchases (or leases) any tools, equipment, machinery, and furniture necessary to produce or sell services or goods, *expecting, but not knowing for sure, that customers will buy what the company provides.* An individual who is willing to risk this uncertainty in exchange for the reward of earning a profit (and the personal reward of seeing the company succeed) is called an **entrepreneur**. Entrepreneurship, then, is a combination of three factors: the company owner's idea, the willingness of the company's owner to take a risk, and the abilities of the company's owner and employees to use capital to produce and sell goods or services. But where does the company get its capital?

Sources of Capital

One source of capital for a company is the entrepreneur's (or company owner's) investment in the company. An entrepreneur invests money "up front" so that the company can get started. The company uses the money to acquire the resources it needs to function. Then, as the company operates, the resources, or capital, of the company increase or decrease through the profits and losses of the company.

When an entrepreneur invests money in a company, he or she hopes to eventually get back the money he or she contributed to the company (a return *of* the contribution). Furthermore, the entrepreneur hopes to periodically receive additional money above the amount he or she originally contributed to the company (a return *on* the contribution). The entrepreneur would like the return *on* the contribution to be higher than the return that could have been earned with that same money on a different investment (such as an interest-bearing checking or savings account).

Borrowing is another source of capital for a company. To acquire the resources necessary to grow or to expand the types of products or services it sells, a company may have to borrow money from institutions like banks (called *creditors*). This occurs when the cash from the company's profits, combined with the company owner's contributions to the company, is not large enough to finance its growth. But borrowing by a company can be risky for the owner or owners. In some cases, if the company is unable to pay back the debt, the owner must personally assume that responsibility.

Borrowing can also be risky for a company. If the company cannot repay its debts, it will be unable to borrow more money and will soon find itself unable to continue operating. In addition to earning a profit, then, another objective of a company is to remain solvent. Remaining **solvent** means that the company can pay off its debts.

The terms *service, merchandising,* and *manufacturing* describe what companies do (perform services, purchase and sell goods, or make and sell products). We next discuss the forms that companies take, or how companies are organized.

THE FORMS THAT COMPANIES TAKE

Several types of organizations use accounting information in their decision-making functions but do not have profit-making as a goal. These organizations are called *not-for-profit organizations* and include many educational institutions, religious institutions, charitable

EXHIBIT 1-3	**TYPES OF BUSINESS ORGANIZATIONS (COMPANIES)**	
Sole Proprietorships • Single owner-manager • Small companies • Most common type of business organization	**Partnerships** • Two or more owners (partners) • Partnership agreement	**Corporations** • Stockholders have separate identity from company • Capital stock • Greatest volume of business

organizations, municipalities, governments, and some hospitals. Since making a profit is not a goal of these organizations, some aspects of accounting for these organizations' activities are unique and beyond the scope of this book.

In this book we emphasize *business* organizations. These business organizations, or *companies,* are a significant aspect of the U.S. and world economies. As Exhibit 1-3 shows, a company may be organized as a (1) sole proprietorship, (2) partnership, or (3) corporation.

Sole Proprietorships

A **sole proprietorship** is a company owned by one individual who is the sole investor of capital into the company. Usually the sole owner also acts as the manager of the company. Small retail stores and service firms often follow this form of organization. The sole proprietorship is the most common type of company because it is the easiest to organize and simplest to operate. In 1999, about 72 percent of all companies were sole proprietorships.[1]

Partnerships

A **partnership** is a company owned by two or more individuals (sometimes hundreds of individuals) who each invest capital, time, and/or talent into the company and share in the profits and losses of the company. These individuals are called *partners*, and their responsibilities, obligations, and benefits are usually described in a contract called a **partnership agreement**. Accounting firms and law firms are examples of partnerships. In 1999, just under 8 percent of all companies were partnerships.[2]

 If you and a friend decide to become business partners, do you think you need a formal partnership agreement? Why or why not?

Corporations

A **corporation** is a company organized as a separate legal entity, or body (separate from its owners), according to the laws of a particular state. In fact, the word *corporation* comes from the Latin word for body *(corpus)*. In 1999, over 20 percent of all companies were corporations.[3]

By being incorporated, a company can enter into contracts, own property, and issue stock. A company issues shares of *capital stock* to owners, called *stockholders,* as evidence of the owners' investment of capital into the corporation. These shares are transferable from stockholder to stockholder, and each share represents part-ownership of the

---

[1] U.S. Treasury Department, Internal Revenue Service, *Statistics of Income Bulletin*, Spring 2002, 294, 295, 297.
[2] Ibid.
[3] Ibid.

http://www.gapinc.com
http://www.intel.com

corporation. A corporation may be owned by one stockholder or by many stockholders (these stockholders are called *investors*). In fact, many large corporations have thousands of stockholders. For example, in their 2001 annual reports, The Gap and Intel Corporation indicated that their stockholders owned 865,726,890 and 6,690,000,000 shares of stock, respectively!

The organization and legal structure of a corporation are more complex than that of a sole proprietorship or a partnership. Although sole proprietorships are the most common type of company, corporations conduct the greatest volume of business in the United States. In 1999, sole proprietorships made over 5 percent, partnerships close to 9 percent, and corporations over 86 percent of all business sales in the United States.[4]

Since most of what we discuss in this text applies to all types of companies, we will use the general term *company* to apply to any company, regardless of structure. If the topic relates only to a specific type of company, we will identify the type of company.

THE REGULATORY ENVIRONMENT OF BUSINESS

4 What types of regulations do companies face?

Companies affect each of us every day, but they also affect each other, the economy, and the environment. Just as individuals must abide by the laws and regulations of the cities, states, and countries in which they live and work, all companies, regardless of type, size, or complexity, must deal with regulatory issues.

Think again about that candy bar you had as a snack today. When Unlimited Decadence Corporation was formed, the company had to do more than build a factory, purchase equipment and ingredients, hire employees, find retail outlets to sell the candy bar, and begin operations. It also had to deal with the regulatory issues involved with opening and operating even the smallest of companies. Furthermore, its managers must continue to address regulatory issues as long as they continue to operate the company.

 Suppose a company is about to open a factory down the street from your house. What concerns do you have? What regulations might help reduce your concerns?

Many different laws and authorities regulate the business environment, covering issues such as consumer protection, environmental protection, employee safety, hiring practices, and taxes. Companies must comply with different sets of regulations depending on where their factories and offices are located. We discuss these sets of regulations next.

Local Regulations

City regulations involve matters such as zoning (parts of the city in which companies may operate), certificates of occupancy, and for some companies, occupational licenses and pollution control. Counties are concerned with issues such as the following: health permits for companies that handle, process, package, and warehouse food; registration of the unique name of each company; and control of pollution to air, land, or water.

State Regulations

States also regulate the activities of companies located within their borders. Most states require corporations to pay some form of state tax, usually an income tax (a tax on profit), a franchise tax (a fee for the privilege of conducting corporate business in the state), or both. New companies (regardless of form) in most states must apply for sales tax numbers and permits. Each state has unemployment taxes that companies operating within that state must pay.

---

[4]Ibid.

Practicing professionals, such as doctors, lawyers, and accountants, must get a license for each state in which they practice. Finally, states regulate companies that conduct certain types of business. For example, in Texas, companies that sell, transport, or store alcoholic beverages must obtain licenses from, and pay fees to, the state of Texas. Massachusetts regulations ban selling fireworks, whereas New Hampshire allows the sale of fireworks.

Federal Regulations

The federal government has a variety of laws and agencies that regulate companies and the business environment. These laws and agencies relate to specific aspects and activities of companies, regardless of the city or state in which the companies are located.

Internal Revenue Service

All companies have some dealings with the Internal Revenue Service (IRS). Each company must withhold taxes from its employees' pay and send these taxes to the IRS. Furthermore, the IRS taxes the profits of the companies themselves. The type of company determines who actually pays the taxes on profits, though. Corporations must pay their own income taxes to the IRS because, from a legal standpoint, they are viewed as being separate from their owners. Sole proprietorships and partnerships, however, do not pay taxes on their profits. Rather, owners of these types of companies include their share of the company profits along with their other taxable income on their personal income tax returns. This is because the tax law does not distinguish the owners of sole proprietorships and partnerships from the companies themselves.

Laws and Other Government Agencies

A variety of laws and government departments and agencies (in addition to the IRS) regulate companies. Federal departments and agencies oversee the administration of laws governing areas such as competition (the Federal Trade Commission and the Department of Justice), fair labor practices (the Department of Labor), safety (the Occupational Safety and Health Administration), employee and customer accessibility (the Department of Justice), workplace discrimination (the Equal Employment Opportunity Commission), control of pollution to air, land, or water (the Environmental Protection Agency), and the like.

International Regulations

When a company conducts business internationally, it also must abide by the laws and regulations of the other countries in which it operates. These laws and regulations address such issues as foreign licensing, export and import documentation requirements, tax laws, multinational production and marketing regulations, domestic ownership of company property, and expatriation of cash (how much of the company's cash can leave the country). Of course, these laws and regulations differ from country to country, so a company operating in several countries must abide by many laws and regulations. Exhibit 1-4

EXHIBIT 1-4	COMMON REGULATORY ISSUES COMPANIES FACE		
City and County Issues	**State Issues**	**Federal Issues**	**International Issues**
zoning	state tax	federal taxes	foreign licensing
certificate of occupancy	sales tax	competition	exports and imports
occupational license	unemployment taxes	labor standards	taxes
environmental regulations	professional licenses	working conditions	multinational production and
health permit	industry-specific regulations	workplace discrimination	marketing
company name and registration			property ownership
			cash restrictions

lists some of the more common regulatory issues facing companies operating in different jurisdictions.

 Suppose that as a manager of a manufacturing company, you have the opportunity to have many parts of your product manufactured in another country where the labor is much cheaper and the environmental regulations are less stringent. What are the pros and cons of taking advantage of this opportunity?

INTEGRATED ACCOUNTING SYSTEM

A company is responsible to many diverse groups of people, both inside and outside the company. For example, its managers and employees depend on the company for their livelihood. Customers expect a dependable product or service at a reasonable cost. The community expects the company to be a good citizen. Owners want returns on their investments, and creditors expect to be paid back. Governmental agencies expect companies to abide by their rules.

People in all of these groups use accounting information about the company to help them assess a company's ability to carry out its responsibilities, and to help them make decisions involving the company. This information comes from the company's integrated accounting system. An **integrated accounting system** is a means by which accounting information about a company's activities is identified, measured, recorded, and summarized so that it can be communicated in an accounting report. A company's integrated accounting system provides much of the information used by the many diverse groups of people outside the company (these are sometimes called **external users**), as well as by the managers and employees within the company (these are sometimes called **internal users**). Two branches of accounting, management accounting and financial accounting, use the information in the integrated accounting system to produce reports for different groups of people. Management accounting provides vital information about a company to internal users; financial accounting gives information about a company to external users.

Management Accounting Information

Management accounting information helps managers plan, operate, and evaluate a company's activities. Managers must operate the company in a changing environment. They need information to help them compete in a world market in which technology and methods of production are constantly changing. Moreover, in a world exploding with new information, managers must manage that information in a way that will let them use it more efficiently and effectively. Accounting is one of the critical tools of information management.

Since management accounting helps managers inside the company, it is free from the restrictions of regulatory bodies interested in how companies report to external users. Therefore, managers can request "tailor made" information in whatever form is useful for their decision making, such as in dollars, units, hours worked, products manufactured, numbers of defective units, or service agreements signed. The integrated accounting system provides information about segments of the company, products, tasks, plants, or individual activities, depending on what information is important for the decisions managers are making.

Financial Accounting Information

Financial accounting information is organized for the use of interested people outside of the company. External users analyze the company's financial reports as one source of useful financial information about the company. For these users to be able to interpret the reports, companies reporting to outsiders follow specific guidelines, or rules, known as *generally accepted accounting principles,* (almost a one-size-fits-all approach to reporting). Since a company's financial reports are not tailored to specific user decisions, external users have to use care to extract from these reports the information that is relevant to their decisions.

Financial accounting information developed by the integrated accounting system is expressed in dollars in the United States and in different currencies (such as yen, euros, and pesos) in other countries. This information emphasizes the whole company and sometimes important segments of the company.

Both internal and external users need accounting information to make decisions about a company. Since external users want to see the reported results of management activities, we discuss these activities next. Then we will discuss how accounting information supports both management activities and external decision making.

MANAGEMENT ACTIVITIES

Managers play a vital role in a company's success—by setting goals, making decisions, committing the resources of the company to achieving these goals, and then by achieving these goals. To help ensure the achievement of these goals and the success of the company, managers use accounting information as they perform the activities of planning the operations of the company, operating the company, and evaluating the operations of the company for future planning and operating decisions. Exhibit 1-5 shows these activities.

5 What activities contribute to the operations of a company?

Planning

Management begins with planning. A clear plan lays out the organization of, and gives direction to, the operating and evaluating activities. **Planning** establishes the company's goals and the means of achieving these goals. Managers use the planning process to identify what resources and employees the company needs in order to achieve its goals. They also use the planning process to set standards, or "benchmarks," against which they later can measure the company's progress toward its goals. Periodically measuring the company's progress against standards or benchmarks helps managers identify whether the company needs to make corrections to keep itself on course. Because the business environment changes so rapidly, plans must be ongoing and flexible enough to deal with change before it occurs or as it is happening.

Managers of companies operating in more than one country have more to consider in their planning process than do those operating only in the United States. Managers of multinational companies must also consider such factors as multiple languages, economic systems, political systems, monetary systems, markets, and legal systems. In such companies, managers must also plan and encourage the communication between and among branches in several countries.

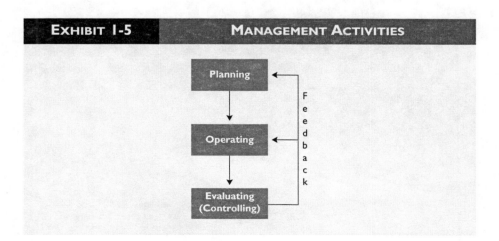

EXHIBIT 1-5   MANAGEMENT ACTIVITIES

Operating

Operating refers to the set of activities that the company engages in to conduct its business according to its plan. For Unlimited Decadence, these are the activities that ensure that candy bars get made and sold. They involve gathering the resources and employees necessary to achieve the goals of the company, establishing organizational relationships among departments and employees, and working toward achieving the goals of the company. In operating the company, managers and work teams must make day-to-day decisions about how best to achieve these goals. For example, accounting information gives them valuable data about a product's performance. With this information, they can decide which products to continue to sell and when to add new products or drop old ones. If the company is a manufacturing company, managers and work teams can decide what products to produce and whether there is a better way to produce them. With accounting information, managers can also make decisions about how to set product selling prices, whether to advertise and how much to spend on advertising, and whether to buy new equipment or expand facilities. These decisions are ongoing and depend on managers' evaluations of the progress being made toward the company's goals and on changes in the company's plans and goals.

Evaluating

Evaluating is the management activity that measures actual operations and progress against standards or benchmarks. It provides feedback for managers to use to correct deviations from those standards or benchmarks, and to plan for the company's future operations. Evaluating is a continuous process that attempts to prevent problems or to detect and correct problems as quickly as possible.

As you might guess, the more countries in which a company operates, the more interesting the evaluating activity becomes. Because of cultural and other differences, evaluation methods and feedback used in some countries may have little meaning in other countries. For example, it would be difficult to convince employees of the importance of high quality if these employees are used to standing in long lines for whatever quality and quantity of merchandise is available in their country. Managers must pay particular attention to the cultural effects of evaluation methods and feedback in order to achieve effective control.

© RYAN MCVAY/GETTY IMAGES

Do you think these people are engaged in planning activities, operating activities, or evaluating activities? Why?

Even coaches of professional sports teams perform the activities of planning, operating, and evaluating. If a team's goal is to win the Super Bowl, how would the head coach implement each of these activities?

Planning, operating, and evaluating all require information about the company. The company's accounting system provides much of the quantitative information managers use.

ACCOUNTING SUPPORT FOR MANAGEMENT ACTIVITIES

Management accounting involves identifying, measuring, recording, summarizing, and then communicating economic information about a company to *internal* users for management decision-making. Internal users include individual employees, work groups or teams, departmental supervisors, divisional and regional managers, and "top management." Management accountants, then, provide information to internal users for planning the operations of the company, for operating the company, and for evaluating the operations of the company. With the help of the management accountant, managers use this information to help them make decisions about the company.

The reports that result from management accounting can help managers *plan* the activities and resources needed to achieve the goals of the company. These reports may provide revenue (amounts charged to customers) estimates and cost estimates of planned activities and resources, and an analysis of these cost estimates. By describing how alternative actions might affect the company's profit and solvency, these estimates and analyses help managers plan.

In *operating* a company, managers use accounting information to make day-to-day decisions about what activities will best achieve the goals of the company. Management accounting helps managers make these decisions by providing timely economic information about how each activity might affect profit and solvency.

Accounting information also plays a vital role in helping managers *evaluate* the operations of the company. Managers use the revenue and cost estimates generated during the planning and decision-making process as a benchmark, and then compare the company's actual revenues and costs against that benchmark to evaluate how well the company is carrying out its plans.

Since managers are making decisions about their own company, and since each company is different, the information the management accountant provides must be "custom fitted" to the information needs of the company. This involves selecting the appropriate information to report, presenting that information in an understandable format (interpreting the information when necessary), and providing the information when it is needed for the decisions being made.

Management accounting responsibilities and activities thus vary widely from company to company. Furthermore, these responsibilities and activities continue to evolve as management accountants respond to the need for new information—a need caused by the changing business environment.

In response to this changing business environment, the Institute of Management Accountants (IMA) publishes guidelines for management accountants called Statements on Management Accounting (SMAs).

Statements on Management Accounting

SMAs serve as guidelines for management accountants to use in fulfilling their responsibilities. The SMAs are nonbinding (they are not rules that must be followed), but because they are developed by professional accountants, as well as leaders in industry and colleges and universities, management accountants turn to SMAs for help when faced with new situations.

6 Are there any guidelines for reporting to company managers?

Framework for Management Accounting

The responsibility for identifying issues to be addressed by SMAs lies with an IMA committee called the Management Accounting Committee. One of the first activities this committee undertook was to develop a framework for the work it was assigned to do. The "Framework for Management Accounting" developed by this committee defines the scope of the SMAs, including a statement of the objectives of management accounting and a description of the activities and responsibilities of management accountants.[5]

Company-specific responsibilities and unique elements of a company's internal reports may change, but the underlying goals of management accounting remain the same for all companies:

- To inform people inside and outside the company about past or future events or circumstances that affect the company
- To interpret information from inside and outside the company and to communicate the implications of this information to various segments of the company
- To establish planning and control systems that ensure that company employees use the company's resources in accordance with company policy
- To develop information systems (manual or computer systems) that contain, process, and manage accounting data
- To implement the use of modern equipment and techniques to aid in identifying, gathering, analyzing, communicating, and protecting information
- To ensure that the accounting system provides accurate and reliable information
- To develop and maintain an effective and efficient management accounting organization

In order to see how a company's accounting information helps managers in their planning, operating, and evaluating activities, briefly consider three key management accounting reports prepared with these goals in mind.

Basic Management Accounting Reports

Budgets, cost analyses, and manufacturing cost reports are examples of management tools the accounting system provides. Exhibit 1-6 illustrates the relationships between management activities and these reports.

 Suppose you are the manager of your company's sales force. What type of information would you want to help you do your job?

Budgets
Budgeting is the process of quantifying managers' plans and showing the impact of these plans on the company's operating activities and financial position. Managers present this information in a report called a *budget* (or *forecast*). Once the planned activities have occurred, managers can evaluate the results of the operating activities against the budget to make sure that the actual operations of the various parts of the company achieved the established plans. For example, Unlimited Decadence might develop a budget showing how many boxes of candy bars it plans to sell during the first three months of 2004. Later, after actual 2004 sales have been made, managers will compare the results of these sales with the budget to determine if their forecasts were "on target" and, if not, to find out why differences occurred. We will discuss budgets further in Chapters 4 and 12.

---

[5]Institute of Management Accountants, *Statements on Management Accounting: Objectives of Management Accounting,* Statement No. 1B, June 17, 1982.

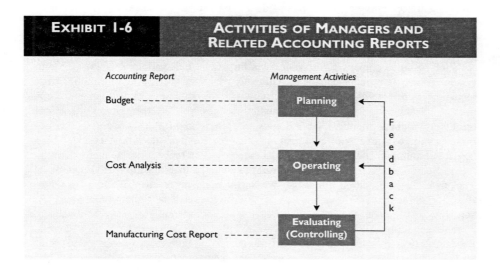

EXHIBIT 1-6 — **ACTIVITIES OF MANAGERS AND RELATED ACCOUNTING REPORTS**

Cost Analyses

Cost analysis, or **cost accounting**, is the process of determining and evaluating the costs of specific products or activities within a company. Managers use cost analyses when making decisions about these products or activities. For example, Unlimited Decadence might use a cost analysis to decide whether to stop or to continue making the Divinely Decadent candy bar. The cost analysis report might show that the candy bar is not profitable because it earns less than it costs to make. If this is the case, the fact that this candy bar does not make a profit will be one factor in the managers' decision. The company's managers also will have to resolve the ethical issue of whether to lay off the employees who produced the candy bar. (Can you think of an alternative to a layoff?)

 Suppose you are a manager of a company that makes a food product thought to create major health problems after long-term use. What facts would you consider in trying to decide whether the company should drop the product or continue producing it?

We will discuss cost analysis reports again in Chapter 15.

Manufacturing Cost Reports

As we mentioned above, managers must monitor and evaluate a company's operations to determine if its plans are being achieved. Accounting information can highlight specific "variances" from plans, indicating where corrections to operations can be made if necessary.

For example, a manufacturing cost report might show that total actual costs for a given month were greater than total budgeted costs. However, it might also show that some actual costs were greater than budgeted costs while others were less than budgeted costs. The more detailed information will be useful for managers as they analyze why these differences occurred, and then make adjustments to the company's operations to help the company achieve its plans. We will discuss manufacturing cost reports again in Chapter 17.

ACCOUNTING SUPPORT FOR EXTERNAL DECISION MAKING

 Say you have been offered a job at Unlimited Decadence. What economic information concerning Unlimited Decadence would you want to know about to help you decide whether to accept the job offer?

Management accounting gives people inside a company vital business information about the company and its performance, but the company also must provide business information about its performance to people outside the company. **Financial accounting** involves identifying, measuring, recording, summarizing, and then communicating economic information about a company to *external* users for use in their various decisions. External users are people and groups outside the company who need accounting information to decide whether or not to engage in some activity with the company. These users include individual investors, stockbrokers and financial analysts who offer investment assistance, consultants, bankers, suppliers, labor unions, customers, and local, state, and federal governments and governments of countries in which the company does business.

The accounting information that helps external users make a decision (for example, a bank's loan officer deciding whether or not to extend a loan to a company) may be different from the information a manager within the company needs. Thus the accounting information prepared for the external user may differ from that prepared for the internal user. However, some of the accounting information the internal user needs also helps the external user and vice versa. For example, Unlimited Decadence may decide to continue to produce and sell a new candy bar if it can borrow enough money to do so. In weighing the likelihood of getting a loan from the bank, company managers will probably want to evaluate the same financial accounting information that the bank evaluates. In deciding whether to make a loan to Unlimited Decadence, the bank will consider the likelihood that Unlimited Decadence will repay the loan. Since this likelihood may depend on current and future sales of the candy bar, the bank also may want to evaluate the company's actual sales, as well as the sales budget that Unlimited Decadence's managers developed as part of the planning process.

Many external users evaluate the accounting information of more than one company, and need comparable information from each company. For example, a bank looks at accounting information from all of its customers who apply for loans, and must use comparable information in order to decide to which customers it will make loans. This need for comparability creates a need for guidelines or rules for companies to follow when preparing accounting information for external users. Over the years, because of the activities of several professional accounting organizations, a set of broad guidelines for financial accounting has evolved in the United States. These guidelines are referred to as *generally accepted accounting principles*.

Generally Accepted Accounting Principles

7 Are there any guidelines in the United States for reporting to people outside of a company?

Generally accepted accounting principles, or **GAAP**, are the currently accepted principles, procedures, and practices that companies use for financial accounting and reporting in the United States. These principles, or "rules," must be followed in the external reports of companies that sell capital stock to the public in the United States and by many other companies as well. GAAP covers such issues as how to account for inventory, buildings, income taxes, and capital stock; how to measure the results of a company's operations; and how to account for the operations of companies in specialized industries, such as the banking industry, the entertainment industry, and the insurance industry. Without these agreed-upon principles, external users of accounting information would not be able to understand the meaning of this information. (Imagine if we all tried to communicate with each other without any agreed-upon rules of spelling and grammar!)

Several organizations contribute to GAAP through their publications (called "pronouncements" or "standards"). The three most important organizations that develop GAAP in the United States are the Financial Accounting Standards Board (FASB), the Securities and Exchange Commission (SEC), and the American Institute of Certified Public Accountants (AICPA). The FASB is a seven-member full-time board of professional accountants and businesspeople; it issues *Statements of Financial Accounting Standards,* which establish new standards or amend previously-established standards. The SEC is a branch of the U.S. government; it issues *Financial Reporting Releases* containing financial accounting guidelines. The AICPA is the professional organization of all certified public accountants (CPAs); it issues *Statements of Position* that also create accounting standards.[6]

Accounting is less standardized worldwide than in the United States because of cultural, legal, economic, and other differences among countries. However, several organizations have made progress in developing worldwide accounting standards. Most notably, the International Accounting Standards Board (IASB) has issued over 40 standards covering issues such as accounting for inventories, property and equipment, and the results of a company's operations. Although compliance with these standards is voluntary, most of the more than 140 accountancy organizations around the world that are represented on this board have agreed to eventually require the *International Accounting Standards* as part of their countries' generally accepted accounting standards.

Many GAAP pronouncements are complex and very technical in nature. In this book, we will introduce only the basic aspects of the generally accepted accounting principles that apply to the issues we discuss. It is important to recognize, however, that these principles do change; they are modified as business practices and decisions change and as better accounting techniques are developed.

Basic Financial Statements

Companies operate to achieve various goals. They may be interested in providing a healthy work environment for their employees, in reaching a high level of pollution control, or in making contributions to civic and social organizations and activities. However, to meet these goals, a company must first achieve its two primary objectives: *earning a satisfactory profit* and *remaining solvent*. If a company fails to meet either of these objectives, it will not be able to achieve its various goals and will not be able to survive in the long run.

Profit (commonly referred to as *net income*) is the difference between the cash and credit sales of a company *(revenues)* and its total costs *(expenses)*. **Solvency** is a company's long-term ability to pay its debts as they come due. As you will see, both internal and external users analyze the *financial statements* of a company to determine how well the company is achieving its two primary objectives.

Financial statements are accounting reports used to summarize and communicate financial information about a company. A company's integrated accounting system produces three major financial statements: the income statement, the balance sheet, and the cash flow statement. It also produces a supporting financial statement: the statement of changes in owner's equity. Each of these statements summarizes specific information that has been identified, measured, and recorded during the accounting process.

Income Statement

A company's **income statement** summarizes the results of its operating activities for *a specific time period* and shows the company's profit for that period. It shows a company's revenues, expenses, and net income (or net loss) for that time period, usually one year. Exhibit 1-7 shows what kind of information appears in a company's income statement. **Revenues** are the prices charged to a company's customers for the goods or services the company provided to them. **Expenses** are the costs of providing the goods or services. These amounts include the costs of the products the company has sold (either the cost of making these products or the cost of purchasing these products), the costs of conducting business (called *operating expenses*), and the costs of income taxes, if any. The **net income** is the excess of revenues over expenses, or the company's profit; a **net loss** arises when expenses are greater than revenues. We will discuss the income statement further in Chapter 6 and throughout the book.

---

[6]Each of these organizations issues other documents that influence and establish GAAP, but they are too numerous to mention here.

EXHIBIT 1-7	WHAT A COMPANY'S INCOME STATEMENT SHOWS

Revenues

Here's where the company shows what it charged customers for the goods or services provided to them during a specific time period.

Expenses

Here's where the company lists the costs of providing the goods and services during that period.

Net Income

This is the difference between revenues and expenses.

Balance Sheet

A company's **balance sheet** summarizes its financial position *on a given date* (usually the last day of the time period covered by the income statement). It is also called a *statement of financial position*. Exhibit 1-8 shows what kind of information appears on a balance sheet. A balance sheet lists the company's assets, liabilities, and owner's equity on the given date. **Assets** are economic resources that a company owns and that it expects will provide future benefits to the company. **Liabilities** are the company's economic obligations (debts) to its creditors—people outside the company such as banks and suppliers— and to its employees. The **owner's equity** of a company is the owner's current investment in the assets of the company, which includes the owner's contributions to the company and any earnings (net income) that the owner leaves in (or invests in) the company. A corporation's owners' equity is called **stockholders' equity**. We will discuss the balance sheet further in Chapter 7 and throughout the book.

Statement of Changes in Owner's Equity

A company's integrated accounting system frequently provides a supporting financial statement, called a **statement of changes in owner's equity**, to explain the amount shown in the owner's equity section of the company's balance sheet. Both the balance sheet and the statement of changes in owner's equity show the owner's investment in the assets of the company on the balance sheet date. However, the statement of changes in owner's equity also summarizes the *changes* that occurred in the owner's investment between the first day and the last day of the time period covered by the company's income statement. Exhibit 1-9 shows the kind of changes in owner's equity that appear on this statement.

EXHIBIT 1-8	WHAT A COMPANY'S BALANCE SHEET SHOWS

Assets

Here's where the company lists its economic resources, such as cash, inventories of its products, and equipment it owns.

Liabilities

Here's where the company lists its obligations to creditors, such as banks and suppliers, and to employees.

Owner's Equity

Here's where the company lists the owner's current investment in the assets of the company.

EXHIBIT 1-9	WHAT A COMPANY'S STATEMENT OF CHANGES IN OWNER'S EQUITY SHOWS

Beginning Owner's Equity

Here's where the company shows the Owner's Equity amount at the beginning of the income statement period (the last day of the previous income statement period). This amount also appears on the balance sheet on the last day of the previous income statement period.

+ Net Income

Here's where the company adds the net income from the period's Income Statement (the profit that the company earned during the income statement period).

+ Owner's Contributions

Here's where the company adds any additional contributions to the company that the company's owner made during the income statement period.

– Withdrawals

Here's where the company subtracts any withdrawals of cash from the company that the company's owner made during the income statement period.

Ending Owner's Equity

Here's where the company shows the resulting Owner's Equity amount that also appears on the company's balance sheet on the last day of the income statement period.

Net income earned during the period increases the owner's investment in the company's assets (and the assets themselves) as the owner "reinvests" the profit of the company back into the company. Similarly, additional contributions of money by the owner to the company during the time period also increase the owner's investment in the company's assets (and the assets themselves). On the other hand, a net loss, rather than a net income, decreases the owner's investment in the company (and the company's assets), as does the owner's choice to remove (or withdraw) money from the company ("disinvesting" the profit from the company). We will discuss the statement of changes in owner's equity further in Chapter 6 and throughout the book.

Cash Flow Statement

A company's **cash flow statement** summarizes its cash receipts, cash payments, and net change in cash for a specific time period. Exhibit 1-10 shows what kind of information appears in a cash flow statement. The cash receipts and cash payments for operating activities, such as products sold or services performed and the costs of producing the products or services, are summarized in the *cash flows from operating activities* section of the statement. The cash receipts and cash payments for investing activities are summarized in the *cash flows from investing activities* section of the statement. Investing activities include the purchases and sales of assets such as buildings and equipment. The cash receipts and cash payments for financing activities, such as money borrowed from and repaid to banks, are summarized in the *cash flows from financing activities* section of the statement. We will discuss the cash flow statement further in Chapter 8, and throughout the book.

A company may publish its income statement, balance sheet, and cash flow statement (and statement of changes in owner's equity), along with other related financial accounting information, in its **annual report**. Many companies (mostly corporations) do so. We will discuss the content of an annual report in Chapter 10.

EXHIBIT 1-10	WHAT A COMPANY'S CASH FLOW STATEMENT SHOWS

Cash Flows from Operating Activities

Here's where the company lists the cash it received and paid in selling products or performing services for a specific time period.

Cash Flows from Investing Activities

Here's where the company lists the cash it received and paid in buying and selling assets such as equipment and buildings.

Cash Flows from Financing Activities

Here's where the company lists the cash it received and paid in obtaining and repaying bank loans and from contributions and withdrawals of cash made by the company's owners.

ETHICS IN BUSINESS AND ACCOUNTING

8 What role does ethics play in the business environment?

A company's financial statements are meant to convey information about the company to internal and external users in order to help them make decisions about the company. But if the information in the financial statements does not convey a realistic picture of the results of the company's operations or its financial position, the decisions based on this information can have disastrous consequences.

Consider the fallout from Enron Corporation's 2001 financial statements.[7] On October 1, 2001, Enron Corporation was the seventh-largest company in the United States, employing 21,000 people in more than 40 countries. It was also the largest energy trading company in the United States. *Fortune* magazine had ranked Enron 24th in its "100 Best Companies to Work for in America" in the year 2000.[8] It's stock was trading for about $83 per share. Two weeks later, after reporting incredible profits for its first two quarters (January through June) of 2001, Enron reported a third-quarter (July through September) loss, in part because of adjustments caused by previously misstated profits. But by November 1, JP Morgan Chase and Citigroup's Salomon Smith Barney had attempted to rescue Enron by extending the company an opportunity to borrow $1 billion (above what Enron already owed them). On November 19, Enron publicly acknowledged that its financial statements did not comply with GAAP in at least two areas. This failure resulted in huge misstatements on Enron's financial statements: assets and profits were overstated, and liabilities were understated. On December 2, 2001, Enron declared bankruptcy.

The rapid demise of one of the largest, and what appeared to be one of the most successful, companies in the world to the largest corporate failure in the United States created a wave of economic and emotional effects around the world. Before Enron reported a third-quarter loss, its stock was selling for around $83 per share. After Enron reported its loss, its stock dropped to $0.70 per share. Most of those who had purchased shares of Enron stock lost money. Many lost hundreds of thousands of dollars! The Enron employees' pension plan, 62 percent of which was Enron stock, lost $1 billion dollars or more, virtually wiping out the retirement savings of most of Enron's employees, many of whom were nearing retirement age. Many Enron employees were laid off from their jobs. Enron left behind $15 billion in debts, with JP Morgan owed $900 million and Citigroup up to $800 million. Many banks around the world also were affected by having lent money to Enron.

---

[7]http://specials.ft.com/enron and http://news.bbc.co.uk
[8]*Fortune* magazine, January 10, 2001, pp. 82–110.

Ethical behavior on the part of all of Enron's managers would not have guaranteed the company's success. However, it could have prevented much of the damage suffered by those inside and outside the company who depended on Enron's financial statements to provide them with dependable information about the company.

Do you think JP Morgan or Citigroup would have lent Enron as much money if Enron had not overstated its net income and assets, and understated its liabilities? Why or why not? What might Enron's employees have done differently if Enron's financial statements had been properly stated?

Enron was not the first, nor (unfortunately) will it be the last, company to get into trouble for misleading financial reporting. While it seems clear that some of what Enron's managers, and those of some other companies, disclosed on their financial statements was wrong, many business and accounting issues and events in the business environment cannot be interpreted as absolutely right or wrong. Every decision or choice has pros and cons, costs and benefits, and people or institutions who will be affected positively or negatively by the decision. Even in a setting where many issues and events fall between the extremes of right and wrong, it is very important for accountants and businesspeople to maintain high ethical standards. Several groups have established codes of ethics addressing ethical behavior to help accountants and their business associates work their way through the complicated ethical issues associated with business issues and events. These groups include the American Institute of Certified Public Accountants (AICPA), the Institute of Management Accountants (IMA), the International Federation of Accountants (IFAC), and most large companies.

Professional Organizations' Codes of Ethics

The members of the AICPA adopted a code of professional conduct that guides them in their professional work.[9] It addresses such issues as self-discipline, honorable behavior, moral judgments, the public interest, professionalism, integrity, and technical and ethical standards. The IMA has a code of conduct that is similar to the AICPA's code.[10] It addresses competence, confidentiality, integrity, objectivity, and resolution of ethical conflict.

The IFAC is an independent, worldwide organization. Its stated purpose is to "develop and enhance a coordinated worldwide accountancy profession with harmonized standards." As part of its efforts, it has developed a code of ethics for accountants in each country to use as the basis for founding their own codes of ethics.[11] Because of the wide cultural, language, legal, and social diversity of the nations of the world, the IFAC expects professional accountants in each country to add their own national ethical standards to the code to reflect their national differences, or even to delete some items of the code at their national level. The code addresses objectivity, resolution of ethical conflicts, professional competence, confidentiality, tax practice, cross-border activities, and publicity. It also addresses independence, fees and commissions, activities incompatible with the practice of accountancy, clients' money, relations with other professional accountants, and advertising and solicitation.

Ethics at the Company Level

Many companies have codes or statements of company and business ethics. For example, **Texas Instruments Incorporated (TI)**, which manufactures microchips, calculators, and other electronic equipment, has several documents containing guidelines for ethical

http://www.TI.com

---

[9]American Institute of Certified Public Accountants, *Code of Professional Conduct,* http://www.aicpa.org.

[10]Institute of Management Accountants, *Statements on Management Accounting: Standards of Ethical Conduct for Management Accountants,* http://www.imanet.org.

[11]International Federation of Accountants, *Code of Ethics for Professional Accountants,* http://www.ifac.org.

decision making. It even has an ethics office and a director of ethics! The most important ethics document at TI is called *Ethics in the Business of TI*. It was originally published in 1961 and has been periodically revised since then.

The spirit of TI's code of ethics is described by the president and chief executive officer on the inside cover of the publication. "Texas Instruments will conduct its business in accordance with the highest ethical and legal standards. . . . We will always place integrity before shipping, before billings, before profits, before anything. If it comes down to a choice between making a desired profit and doing it right, we don't have a choice. We'll do it right."[12] The code addresses the marketplace, gifts and entertainment, improper use of corporate assets, political contributions, payments in connection with business transactions, conflict of interest, investment in TI stock, TI proprietary information, trade secrets and software of others, transactions with governmental agencies, and disciplinary action, among other subjects.

In our society, we expect people to behave within a range of civilized standards. This expectation allows our society to function with minimal confusion and misunderstanding. In both our personal and our business lives, ethics and integrity are our "social glue."

FRAMEWORK OF THE BOOK

Now that you have been introduced to business and accounting, it is almost time to begin a more in-depth study of the use of accounting information in the business environment. But first we will take a chapter to discuss the types of thinking necessary for one to prosper in this environment. Chapter 2 describes creative and critical thinking, the types of thinking done by people who succeed in accounting and business. In each subsequent chapter you will see examples of creative and critical thinking and will be given the opportunity to practice them.

Beginning in Chapter 3 we will discuss, in more depth, accounting and its use in the management activities of planning, operating, and evaluating, starting with a simple company. Then, in later chapters, we will progress through more complex companies. We will also discuss the use of accounting by decision makers outside the company.

As you read through the book, you will begin to notice the same topics reemerging; but note that each time, a topic will be refined or enhanced by a different company structure, a different type of business, or a different user perspective. You will also notice that we continue to discuss ethical considerations. That's because ethical considerations exist in all aspects of business and accounting.

You will also notice that international issues appear again and again. Many companies operating in the United States have home offices, branches, and subsidiaries in other countries or simply trade with companies in foreign countries. Managers must know the implications of conducting business in foreign countries and with foreign companies. External users of accounting information also must know the effects of these business connections.

SUMMARY

At the beginning of the chapter we asked you several questions. During the chapter, we asked you to STOP and answer some additional questions to build your knowledge about specific issues. Be sure you answered these additional questions. Below are the questions from the beginning of the chapter, with a brief summary of the key points relating to the answers. Use your thinking skills to expand on these key points to develop more complete answers to the questions and to determine what other questions you have that might lead you to learn more about the issues.

---

[12]Texas Instruments Incorporated, *Ethics in the Business of TI* (1990).

1 **Why is it necessary to have an understanding of business before trying to learn about accounting?**

Accounting involves identifying, measuring, recording, summarizing, and communicating economic information about a company for decision making. It focuses on the resources and activities of companies. Therefore, you need to understand companies and the business environment in which they exist, before trying to learn how to account for their resources and activities.

2 **What is the role of accounting information within the business environment?**

Accounting information helps people inside and outside companies make decisions. It supports management activities by providing managers with quantitative information about their company to aid them in planning, operating, and evaluating the company's activities. Accounting information supports external decision making by providing people outside of the company—such as investors, creditors, stockbrokers, financial analysts, bankers, suppliers, labor unions, customers, and governments—with financial statements containing economic information about the performance of the company.

3 **What is private enterprise, and what forms does it take?**

Companies in the private enterprise system produce goods and services for a profit. These companies can be service, merchandising, or manufacturing companies. Entrepreneurs, or individuals, invest money in companies so that the companies can acquire resources, such as inventory, buildings, and equipment. The companies then use these resources to earn a profit. The three types of business organization are (1) the sole proprietorship, owned by one individual, (2) the partnership, owned by two or more individuals (partners), and (3) the corporation, incorporated as a separate legal entity and owned by numerous stockholders who hold capital stock in the corporation.

4 **What types of regulations do companies face?**

The activities of companies must be regulated because these activities affect us, other companies, the economy, and the environment. All companies, regardless of type, size, or complexity, must contend with regulatory issues. Numerous laws and authorities regulate companies on issues ranging from environmental protection to taxes. Each city, county, state, and country has its own regulations. Owners of companies must learn and comply with the regulations issued by the governments where the companies are located and in the areas in which the companies conduct business.

5 **What activities contribute to the operations of a company?**

Managers strive to make their company successful through setting and achieving the goals of their company, making decisions, and committing the resources of the company to the achievement of these goals. Planning provides the organization and direction for the other activities. Operating involves gathering the necessary resources and employees and implementing the plans. Evaluating measures the actual progress against standards or benchmarks so that problems can be corrected.

6 **Are there any guidelines for reporting to company managers?**

The Institute of Management Accountants publishes a broad set of nonbinding guidelines for management accountants to use in fulfilling their responsibilities. These guidelines provide help for management accountants when they are faced with new situations.

7 **Are there any guidelines in the United States for reporting to people outside of a company?**

So that external users can understand the meaning of accounting information, companies follow agreed-upon principles in their external reports. The FASB, the SEC, and the AICPA contribute to the development of generally accepted accounting principles, the standards or "rules" that many companies must follow.

8 **What role does ethics play in the business environment?**

Since the world is a complex place, where issues are not always clear, decisions must be made in an ethical context with the best available information. Accounting information can be relied on only if it is generated in an ethical environment. Many groups have established codes of ethics.

KEY TERMS

annual report *(p. 19)*
assets *(p. 18)*
balance sheet *(p. 18)*
budgeting *(p. 14)*
capital *(p. 6)*
cash flow statement *(p. 19)*
corporation *(p. 7)*
cost analysis or cost accounting *(p. 15)*
entrepreneur *(p. 6)*
evaluating *(p. 12)*
expenses *(p. 17)*
external users *(p. 10)*
financial accounting *(p. 16)*
financial statements *(p. 17)*
generally accepted accounting principles (GAAP) *(p. 16)*
income statement *(p. 17)*
integrated accounting system *(p. 10)*
internal users *(p. 10)*
liabilities *(p. 18)*

management accounting *(p. 13)*
manufacturing companies *(p. 5)*
merchandising companies *(p. 4)*
net income *(p. 17)*
net loss *(p. 17)*
operating *(p. 12)*
owner's equity *(p. 18)*
partnership *(p. 7)*
partnership agreement *(p. 7)*
planning *(p. 11)*
profit *(p. 17)*
revenues *(p. 17)*
service companies *(p. 4)*
sole proprietorship *(p. 7)*
solvency *(p. 17)*
solvent *(p. 6)*
statement of changes in owner's equity *(p. 18)*
stockholders' equity *(p. 18)*

SUMMARY SURFING

Here is an opportunity to gather information on the Internet about real-world issues related to the topics in this chapter. Go to http://www.cunningham.swlearning.com and click on the Interactive Study Center. Click on this chapter number then click on Summary Surfing and answer the following questions.

- Click on **FEI** (Financial Executives Institute). Click on *About the FEI.* Then click on *Code of Ethics.* The FEI's Code of Ethics lists responsibilities that its members have to several groups of people. To what groups is the Code referring? Give an example of a responsibility that members have to each group.

- Click on **TI** (Texas Instruments). In the *Search TI box,* type "ethics," and click *Go.* Then click on *Texas Instruments Ethics.* What are the three primary functions of the TI Ethics Office? Now click on *Ethics Organization at TI.* What does the TI Ethics Committee do?

APPENDIX

The Profession of Accountancy

Perhaps you are wondering what the profession of accountancy is all about. Accountancy has emerged as a profession, alongside other professions such as medicine, law, and architecture. Two characteristics distinguish a profession from an occupation. One is that its members have exclusive technical competence in their field requiring extensive training and specialized study. These members usually demonstrate their initial competence by taking a standardized exam. The second distinguishing characteristic of a profession is that its members adhere to a service ideal (they render a specialized service to the public) and its supporting standards of conduct and ethics. As we discuss in this appendix, the profession of accountancy meets both of these criteria. First of all, the study and the practice of accountancy require a broad understanding of concepts in such areas as business, economics, sociology, psychology, and public administration, as well as an in-depth technical knowledge of specialized accounting areas. Accountants demonstrate their understand-

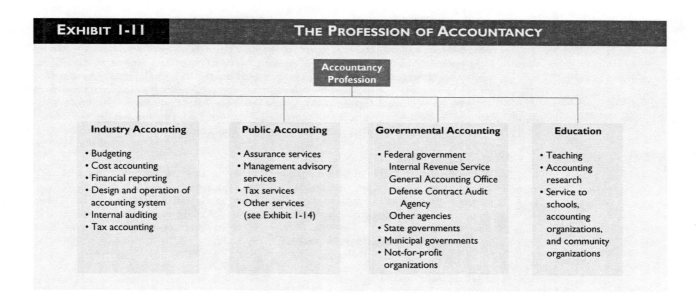

EXHIBIT 1-11 — **THE PROFESSION OF ACCOUNTANCY**

ing of these concepts and their technical accounting knowledge by taking standardized exams such as the CPA Exam and the CMA Exam, which we will discuss later. Secondly, accountants render specialized accounting services in four general areas (or fields) and must comply with specific standards of conduct and ethics (as you read about in the main part of this chapter).

The four general fields of accountancy include (1) industry accounting, (2) public accounting, (3) governmental accounting, and (4) education, each of which has several accounting specialty areas. We summarize them in Exhibit 1-11 and discuss them briefly here.

Industry Accounting

A company employs industry, or management, accountants to perform its internal (management) accounting activities and to prepare its financial reports. A high-level manager, such as the company's **controller**, usually coordinates these activities. This manager frequently reports directly to a top manager of the organization, such as the **chief financial officer**—an indication of how important the management accounting functions are to the company's operations.

Another indication of the importance of management accounting is the Certificate in Management Accounting (CMA). The CMA is granted to those who meet specific educational and professional standards and who pass a uniform CMA examination, administered twice yearly by the **Institute of Management Accountants**. Although the CMA is not required as a license to practice, accountants holding the CMA are recognized as professional experts in the area of management accounting.

http://www.imanet.org

Management accounting activities encompass several areas: budgeting, cost accounting, and financial reporting (which we discussed briefly in this chapter), as well as designing and operating accounting systems, internal auditing, and tax accounting. We discuss the last three of these areas next.

Design and Operation of an Integrated Accounting System
One duty of the management accountant is to design and operate a company's integrated accounting system, which may be a part of its bigger enterprise resource planning (ERP) system (which we discuss in Chapter 10). This function is sometimes referred to as **general accounting** because of the wide variety of activities involved. These activities include, among others, deciding how much of the accounting system will be computer or manually operated, determining the information needs of different managers and departments, integrating the accounting activities for those departments, and designing accounting procedures, forms, and reports.

Internal Auditing
One part of the design of an accounting system is establishing good internal control. As we discuss in Chapter 9, **internal control** involves the procedures needed to control or minimize a

company's risks (of losses, earnings drops, fraud, fines, scandal, and so forth), to safeguard a company's economic resources, and to promote the efficient and effective operation of its accounting system. Internal auditing is a part of a company's internal control procedures and has as its purpose the review of the company's operations to ensure that all company employees are following these procedures. Internal auditing is becoming increasingly important because, as you will see shortly, the procedures for a company's *external* audit depend, to a great degree, on the quality of its internal control. As evidence of professionalism in internal auditing, an accountant may earn a Certificate of Internal Auditing (CIA), awarded by the **Institute of Internal Auditors, Inc.** Although not a license to practice, this certificate states that the holder has met specified educational and practical experience requirements and has passed the uniform CIA examination.

http://www.theiia.org

Tax Accounting

Although companies often assign their tax work to the tax services department of a public accounting firm, many of them maintain their own tax departments as well. Accountants with expertise in the tax laws that apply to the company make up the staff of these departments. These accountants handle income tax planning and the preparation of the company's state, federal, and foreign income tax returns. They also work on real estate taxes, personal property taxes (such as taxes on inventories), and other taxes.

Public Accounting

A public accountant is an independent professional who provides accounting services to clients for a fee. Many accountants practice public accounting as individual practitioners or work in local or regional public accounting firms. Others work in large public accounting firms that have offices in most major U.S. and international cities. These firms provide accounting services to large corporations, some of which span the United States as well as the world in their activities.

Most public accountants are **certified public accountants** (CPAs). A CPA has met the requirements of the state in which he or she works, and holds a license to practice accounting in that state. States use licensing as a means of protecting the public interest by helping to ensure that public accountants provide high-quality, professional services. Although the licensing requirements vary from state to state, all CPAs must pass the Uniform CPA Examination. The "CPA exam," which tests the skills and knowledge of entry-level accountants, is a national examination given by the **American Institute of Certified Public Accountants (AICPA)**. Exhibit 1-12 lists the skills and knowledge tested on the CPA exam. In addition to passing the exam, a CPA must have met a state's minimum educational and practical experience requirements to be licensed in that state.

http://www.aicpa.com

But what do public accountants *do*? Next, we discuss several services that public accountants provide to their client.

Assurance Services

As a result of the information age, the volume of all types of information has grown, and much of that information is readily available to companies, governments, organizations, and individuals. Since decision makers increasingly rely on such information, they need assurance that the information is valid for the purposes for which they intend to use it. In this regard, "an assurance involves expression of a written or oral conclusion on the reliability and/or relevance of information and/or information systems."[13] Auditing evolved from a need for a specific type of assurance service.

Auditing   Accounting information is one type of information for which decision makers need assurance. One way a company communicates accounting information is by issuing financial statements. Managers of the company issuing the statements are responsible for preparing these statements. But because of the potential bias of these managers, external users of the financial statements need objective assurance that the statements *fairly* represent the results of the activities of the company. Therefore, both the **New York Stock Exchange** and the **American Stock Exchange**, as well as the **Securities and Exchange Commission**, require that the set of financial statements that certain corporations (those offering equity securities for public sale) issue every year be audited. This audited set of financial statements is called an **annual report**. Similarly, a bank may require a company to provide audited financial statements when the company applies for

http://www.nyse.com
http://www.amex.com
http://www.sec.gov

---

[13]AICPA-sponsored meeting of representatives of small and medium-sized firms, regulators, and scholars.

EXHIBIT 1-12	**REVISED CPA EXAM**[a]

Content (Knowledge) Areas

Skills Needed to Apply Knowledge

Auditing & Attestation—tests knowledge of auditing procedures, generally accepted auditing standards, and other standards related to attest engagements, and the skills needed to apply that knowledge in those engagements

Ability to communicate

Ability to perform research

Financial Accounting & Reporting—tests knowledge of generally accepted accounting principles for business enterprises, not-for-profit organizations, and governmental entities, and the skills needed to apply that knowledge

Ability to analyze and organize information

Regulation—tests knowledge of federal taxation, ethics, professional and legal responsibilities, and business law, and the skills needed to apply that knowledge

Ability to understand and apply knowledge from diverse areas

Business Environment & Concepts—tests knowledge of general business environment and business concepts that candidates need to know in order to understand the underlying business reasons for and accounting implications of business transactions, and the skills needed to apply that knowledge

Ability to use judgment

[a]"Computerizing the Uniform CPA Exam," Briefing Paper #2, March 2001, Joint AICPA/NASBA Computerization Implementation Committee, pp. 11, 12.

a loan. For the same reason, other types of economic entities, such as universities and charitable organizations, also issue audited financial statements. But what does it mean to be audited?

Auditing involves the examination, by an *independent* CPA, of a company's accounting records and financial statements. Based on sample evidence gathered in the auditing process, including evidence about the quality of the company's internal controls, the CPA forms and expresses a professional, unbiased opinion about (or *attests* to) the fairness of the accounting information in the company's financial statements. Auditing plays an important role in society because many external users rely on CPAs' opinions when making decisions about whether to engage in activities with companies, universities, charitable organizations, and other economic entities.

Other Assurance Services

Recently, because of their clients' needs for assurance about other types of information, public accounting firms have begun to expand their assurance services. In the interest of helping public accounting firms serve their customers, the AICPA, through one of its committees,[14] identified some specific opportunities for these firms to provide assurance services. Exhibit 1-13 summarizes some of these opportunities.

EXHIBIT 1-13	**OPPORTUNITIES FOR PROVIDING ASSURANCE SERVICES**

- Assessing whether an entity has identified all its risks and is effectively managing them
- Evaluating whether an entity's performance measurement system contains relevant and reliable measures of its progress toward its goals and objectives
- Assessing whether an entity's integrated information system (or its ERP system) provides reliable information for decision making
- Assessing whether systems used in e-commerce provide appropriate data integrity, security, privacy, and reliability
- Assessing the effectiveness of health care services provided by HMOs, hospitals, doctors, and other providers
- Assessing whether various caregivers are meeting specified goals regarding care for the elderly

---

[14]AICPA Special Committee on Assurance Services.

Management Advisory Services

In addition to auditing departments, many public accounting firms have separate management advisory services departments to conduct special studies to advise non-audit client companies about improving their internal operations and to aid client managers in their various activities. These departments, in part, help to provide some of the assurance services we discussed earlier.

Management advisory services in public accounting firms include the design or improvement of a company's financial accounting system for identifying, measuring, recording, summarizing, and reporting accounting information. These services also may include assistance in areas such as developing cost-control systems, planning manufacturing facilities, and installing computer operations. To provide these services, public accounting firms also must have employees who have a strong understanding of the industries in which their clients operate. Therefore, in addition to hiring accountants, public accounting firms hire people with other specialties—people such as lawyers, industrial engineers, and systems analysts.

Tax Services

The federal government, governments of other countries, and most state governments require corporations and individuals to file income tax returns and to pay taxes. Because of the high tax rates, complex tax regulations, and special tax incentives today, most companies (and individuals) can benefit from carefully planning their activities to minimize or postpone their tax payments. This is called **tax planning**. Many public accounting firms have separate tax services departments that employ tax professionals who are experts in the various federal, foreign government, and state tax regulations. These tax professionals assist companies and individuals in tax planning. In addition to tax planning, the tax services departments of public accounting firms frequently prepare client corporation or individual income tax returns that reflect the results of these tax-planning activities.

Other Services

When clients hire an accountant or an accounting firm, they really are not as interested in reports that look good as they are in good advice, sound thinking, and creative answers to difficult problems. Computers, the Internet, and other recent high-tech developments allow accountants to do less "number crunching" and more advising, thinking, and creating. As a result of client demand, both large and small public accounting firms have expanded the types of services they offer their individual, small and medium-sized companies, and large corporate clients. These services can range from asset valuation to creative financing.

For example, one Dallas CPA, a sole practitioner, worked on one of the largest divorce cases in Texas history. He was hired to assign a value to the feuding couple's family-owned baseball card company. Texas is a community property state, and the company had to be sold in the divorce proceedings so the unhappy couple could split the cash received from the sale of the company. By analyzing the company's assets and what they would be worth when sold, and by looking at the company's potential future cash flows and net income, the accountant helped the couple determine the value of the company, which eventually sold for $87.5 million![15]

Accountants who focus on small companies can help their clients find sources of creative financing. Most new companies have difficulty finding capital to pay for their continued growth. Without a long business history, many of these companies don't qualify for traditional bank loans. Accountants can help these companies locate alternative financing sources, such as asset-based lending, leasing, and loans guaranteed by the Small Business Administration. Exhibit 1-14 lists some additional services that public accountants provide to their clients.

Governmental and Quasi-Governmental Accounting

Certain governmental and quasi-governmental agencies also employ accountants. The Internal Revenue Service, for example, is responsible for the collection of federal income taxes. State revenue agencies also perform similar functions. Administrators of other federal, state, and local government agencies are responsible for the control of both tax revenues and tax expenditures. These agencies hire accountants to provide accounting information for use in the administration of these activities.

Several other governmental organizations also employ accountants. As we mention in this chapter, the **Securities and Exchange Commission** (SEC) is responsible for overseeing the

http://www.sec.gov

---

[15]"CPAs Enter New Era," *The Dallas Morning News*, June 12, 2001, 8F.

EXHIBIT 1-14	OTHER SERVICES THAT PUBLIC ACCOUNTANTS PROVIDE FOR THEIR CLIENTS
Estate planning	Fraud detection
Forensic accounting	Business succession planning
Real estate advisory services	Debt restructuring and bankruptcy advising
Technology consulting	Business planning
Business valuation	Personal financial planning
Merger and acquisition assistance	E-commerce advising
International accounting	Environmental accounting

reported financial statements of certain corporations and has the legal authority to establish accounting regulations for them. The SEC employs accountants to identify appropriate accounting standards and to verify that corporations are following existing regulations. The **General Accounting Office (GAO)** is responsible for cooperating with various agencies of the federal government in the development and operation of their accounting systems to improve the management of these agencies. It also oversees the administration of government contracts and the spending of federal funds. The **Defense Contract Audit Agency (DCAA)** audits all federally funded defense contracts. Its work resembles the audit services of public accounting firms. Other federal and state agencies, such as the **Federal Bureau of Investigation**, the **Environmental Protection Agency**, and the **Federal Communications Commission**, also employ accountants to prepare and use accounting information.

http://www.gao.gov

http://www.dtic.mil/dcaa

http://www.fbi.gov
http://www.epa.gov
http://www.fcc.gov

Administrators of federal, state, municipal, and other not-for-profit organizations such as colleges and universities, hospitals, and mental health agencies are responsible for their organizations' efficient and effective operations. The accounting information needed by these organizations is similar to that needed by companies. But because they are not-for-profit organizations financed in part by public funds, these organizations are required to use somewhat different accounting procedures (sometimes called *fund accounting*). These organizations hire accountants to design and operate their accounting systems.

As evidence of professionalism in governmental accounting, an accountant may become a Certified Government Financial Manager (CGFM). A CGFM must have met specified educational and practical experience requirements and must have passed a uniform CGFM exam.

The Accountant of the 21st Century

In Chapter 2 we discuss the broad skills (communication, interpersonal, and intellectual skills) needed by businesspeople to do business effectively in a changing environment. These same skills, as well as others, also apply to accountants, and make accountants more effective in dealing with their clients. In 1999, the AICPA reframed these broad skills into a set of core competencies that all college graduates entering the *profession of accountancy* should possess, in addition to the traditional technical accounting skills they studied in college. It divided these competencies into three categories: *functional competencies* (technical competencies most closely aligned with the value that accounting professionals add to the business environment), *personal competencies* (individual attributes and values), and *broad business perspective competencies* (relating to an understanding of the internal and external business environment). Exhibit 1-15 lists and describes the competencies in each category.

In order to help accountants acquire the core competencies (or know if they already have these competencies), the **AICPA** also identified the elements (or "sub-competencies") that comprise the competencies, and lists both the competencies and their elements on its web site. Beginning in 2004, candidates taking the CPA exam will be required to demonstrate their ability to apply many of these competencies and elements in all four sections of the exam (in addition to, and in the context of, the traditional accounting skills typically tested on the exam). This will ensure that new accountants "have what it takes" to deliver the best services to their clients and to adapt to the ever-changing business environment.

http://www.aicpa.org/edu/corecomp.htm

EXHIBIT 1-15	AICPA CORE COMPETENCIES*

FUNCTIONAL COMPETENCIES relate to the technical competencies, which are most closely aligned with the value contributed by accounting professionals. Functional competencies include:

- Decision Modeling
- Risk Analysis
- Measurement
- Reporting
- Research
- Leverage Technology to Develop and Enhance Functional Competencies

PERSONAL COMPETENCIES relate to the attitudes and behaviors of individuals preparing to enter the accounting profession. Developing these personal competencies will enhance the way professional relationships are handled and facilitate individual learning and personal improvement. Personal competencies include:

- Professional Demeanor
- Problem Solving and Decision Making
- Interaction
- Leadership
- Communication
- Project Management
- Leverage Technology to Develop and Enhance Personal Competencies

BROAD BUSINESS PERSPECTIVE COMPETENCIES relate to the context in which accounting professionals perform their services. Individuals preparing to enter the accounting profession should consider both the internal and external business environments and how their interactions determine success or failure. They must be conversant with the overall realities of the business environment. Broad business perspective competencies include:

- Strategic/Critical Thinking
- Industry/Sector Perspective
- International/Global Perspective
- Resource Management
- Legal/Regulatory Perspective
- Marketing/Client Focus
- Leverage Technology to Develop and Enhance a Broad Business Perspective

* http://www.aicpa.org/edu/corecomp.htm

Professional Organizations

A number of *professional* organizations (composed of accounting professionals) exist to facilitate communication among members of the profession, provide professional development opportunities, alert their members to emerging accounting and management issues, and promote ethical conduct. In addition to the Securities and Exchange Commission and the Financial Accounting Standards Board that we mentioned earlier in the chapter, these organizations also influence generally accepted accounting principles (GAAP). Exhibit 1-16 provides a summary of these organizations.

EXHIBIT 1-16	PROFESSIONAL ORGANIZATIONS THAT INFLUENCE GAAP

Web Site	Organization	Description
http://www.aicpa.org	American Institute of Certified Public Accountants (AICPA)	National professional organization of CPAs. In addition to influencing accounting principles, the AICPA influences auditing standards. The Auditing Standards Board of the AICPA develops auditing standards that govern the way CPAs perform audits. The AICPA also prepares and grades the CPA examination and dispenses the results to the individual states, which then issue licenses to those who have passed the examination and who meet the other qualifications of the state.
http://www.feiaz.org	Financial Executives Institute (FEI)	Organization of financial executives of major corporations and accounting professors in academia. Examples of member executives include chief financial officers, financial vice-presidents, controllers, treasurers, and tax executives.

(continued)

EXHIBIT 1-16		CONTINUED

Web Site	Organization	Description
http://www.imanet.org	Institute of Management Accountants (IMA)	Organization of management accountants and others interested in management accounting. Besides influencing the practice of management accounting, the IMA prepares and grades the CMA examination.
http://www.accounting.rutgers.edu/raw/aaa	American Accounting Association (AAA)	National professional organization of academic and practicing accountants interested in both the academic and research aspects of accounting. Members of the AAA, many of whom are accounting practitioners, influence accounting standard setting through their research on accounting issues.

INTEGRATED BUSINESS AND ACCOUNTING SITUATIONS

Answer the Following Questions in Your Own Words.

Testing Your Knowledge

1-1   How would you describe private enterprise?

1-2   What distinguishes a service company from a merchandising or manufacturing company?

1-3   How is a merchandising company different from a manufacturing company? How are the two types of company the same?

1-4   What is entrepreneurship?

1-5   Suppose you were an entrepreneur. Where might you go for business capital?

1-6   What distinguishes a corporation from a partnership and a sole proprietorship?

1-7   What types of regulations must companies comply with in different jurisdictions?

1-8   What is the purpose of an integrated accounting system?

1-9   Given what you have learned from this chapter, how would you define *accounting*?

1-10   How would you describe the similarities and differences between management accounting and financial accounting? Why are they different and why are they similar?

1-11   How do management accounting reports help managers with their activities?

1-12   What is the purpose of Statements on Management Accounting (SMAs)?

1-13   What are generally accepted accounting principles?

1-14   How do financial accounting reports help external users?

1-15   Why have various business groups found it necessary to establish codes of ethics?

1-16   (Appendix) What does a company's controller do?

1-17   (Appendix) What do you know about an accountant who holds a Certificate in Management Accounting (CMA)?

1-18   (Appendix) What is internal control?

1-19   (Appendix) What is the purpose of internal auditing?

1-20   (Appendix) What do you know about an accountant who holds a Certificate of Internal Auditing (CIA)?

1-21   (Appendix) What are the responsibilities of the accountants who work in a company's tax department?

1-22   (Appendix) What do you know about an accountant who is a certified public accountant (CPA)?

1-23   (Appendix) In addition to knowledge of accounting, what other skills and knowledge prepare a college graduate to enter the profession of accountancy?

1-24   (Appendix) What is an assurance?

1-25   (Appendix) What is auditing?

1-26   (Appendix) What do management advisory services include?

1-27   (Appendix) What tax services does a public accounting firm's tax department perform?

1-28   (Appendix) In addition to assurance services, tax services, and traditional management advisory services, what other services do accountants perform for their clients?

1-29   (Appendix) What different types of jobs might a governmental accountant hold?

1-30   (Appendix) What do you know about an accountant who is a certified government financial manager (CGFM)?

1-31   (Appendix) What are four professional organizations of accountants and who are their members?

Applying Your Knowledge

http://www.aa.com
http://www.toyota.com

1-32   How is **American Airlines** an example of a service company? How is **Toyota Motor Corporation** an example of a manufacturing company?

1-33   How might knowledge of a company's cash receipts and payments affect a bank's decision about whether to loan the company money? What financial statement would the loan officer want to look at to begin to understand the company's cash receipts and payments?

1-34   What factors would you consider in deciding whether to operate your company as a sole proprietorship, a partnership, or a corporation?

1-35   Suppose you are Ichabod Cook, CEO of Unlimited Decadence Corporation, maker of candy bars. Unlimited Decadence currently operates in the northeastern United States, and you are considering opening a factory and sales office in California. What questions do you want answered before you proceed with this idea?

1-36   Refer to 1-35. Suppose, instead, that you are considering opening a factory and sales office in Tokyo. What questions do you want answered before proceeding with *this* idea? How do you explain the similarities and differences in these two sets of questions?

1-37   What are some examples of company information in which both internal and external users have an interest?

1-38   Suppose you are a manager of The Foot Note, a small retail store that sells socks. Give an example of information that would help you in each of the management activities of planning, operating, and evaluating the operations of the store.

1-39   What are generally accepted accounting principles, and how do they affect the accounting reports of companies in the United States? Why might the owner or owners of a company be concerned about a proposed new accounting principle?

1-40   A friend of yours, Timorous ("Tim," for short) Ghostly, who has never taken an accounting course, has been assigned a short speech in his speech class. In this speech, Tim must describe the financial statements of a company. Tim has come to you for help (with his professor's permission). He says, "Please describe what financial statements are, what the major financial statements are, and what each financial statement includes." Prepare a written response to Tim's request.

1-41 How do codes of ethics help businesspeople make decisions?

Making Evaluations

1-42 Your friend, Vito Guarino (an incredible cook!), plans to open a restaurant when he graduates from college. One evening, while extolling the virtues of linguini to you and some of your other friends, he glances down at your accounting textbook, which is open to Exhibit 1-2. "What kind of a company is a restaurant?" he asks. "How would a restaurant fit into this exhibit?" Everyone in the room waits with great anticipation for your answer and the rationale behind your answer. What are you going to say?

1-43 You and your cousin, Harvey, have decided to form a partnership and open a landscaping company in town. But before you do, you and Harvey would like to "iron out" a few details about how to handle various aspects of the partnership and then write a partnership agreement outlining these details. What specific issues would you like to see addressed in the partnership agreement before you begin your partnership with Harvey?

1-44 Suppose you are thinking about whether presidents of companies should be allowed to serve on the FASB. What do you think are the potential benefits of allowing them to serve? What do you think are the potential problems?

1-45 Read a daily newspaper for the next week. What evidence do you find that supports the need for business codes of ethics?

1-46 You just nabbed a plum job joining a team of consultants writing an advice column, "Dear Dr. Decisive," for the local newspaper. Yesterday, you received your first letter:

DR. DECISIVE

Dear Dr. Decisive:

Yesterday, my boyfriend and I got into a high-spirited "discussion" about lucky people in business. I say that most successful businesspeople are just plain lucky. They've been in the right place at the right time. He says that these successful people have worked hard preparing themselves for the time when they will be in the right place at the right time. OK, I think we're saying the same thing. He says there is an important difference. Now he won't call me unless I admit I'm wrong (which I'm not) or until you say I'm right.

I'm right, right?

Call me "Lucky."

Required: Meet with your Dr. Decisive team and write a response to "Lucky."

CHAPTER 3

DEVELOPING A BUSINESS PLAN: COST-VOLUME-PROFIT ANALYSIS

"HE WHO EVERY MORNING PLANS THE TRANSACTION OF THE DAY, AND FOLLOWS OUT THAT PLAN, CARRIES A THREAD THAT WILL GUIDE HIM THROUGH THE MAZE OF THE MOST BUSY LIFE. BUT WHERE NO PLAN IS LAID, WHERE THE DISPOSAL OF TIME IS SURRENDERED MERELY TO THE CHANCE OF INCIDENCE, CHAOS WILL SOON REIGN."

—VICTOR HUGO

61-68

⋆ 68

1 Since the future is uncertain and circumstances are likely to change, why should a company bother to plan?

2 What should a company include in its business plan?

3 How does accounting information contribute to the planning process?

4 What must decision makers be able to predict in order to estimate profit at a given sales volume?

5 How can decision makers predict the sales volume necessary for estimated revenues to cover estimated costs?

6 How can decision makers predict the sales volume necessary to achieve a target profit?

7 How can decision makers use accounting information to evaluate alternative plans?

uppose your sister Anna has hired you, as an employee-advisor, to help her open and run a candy store. Anna, who earned her degree last year with a major in marketing, has an insatiable sweet tooth and has always "hungered" to own a candy store. After long and lively discussions with you about the name of the company, Anna decides to name it "Sweet Temptations." You and Anna arrange to obtain retail space, to purchase display fixtures, supplies, and candy, to hire an employee to sell candy, and to advertise in the newspaper. Now you are ready to open for business. But whoa! Not so fast. Have you thought of everything? If you and Anna want Sweet Temptations to succeed, there are other issues that you must consider before you open your company. Instead of rushing into business when the idea is fresh, first you would be smart to develop a detailed business plan that addresses these issues.

1 Since the future is uncertain and circumstances are likely to change, why should a company bother to plan?

PLANNING IN A NEW COMPANY

Planning is an ongoing process for successful companies. It begins before a company opens for operations and continues throughout the life of the company. A **business plan** is an evolving report that describes a company's goals and its current plans for achieving those goals. The business plan is used by both internal and external users. A business plan typically includes

2 What should a company include in its business plan?

1. a description of the company,
2. a marketing plan,
3. a description of the operations of the company, and
4. a financial plan.

We will discuss each of these parts in later sections.

A business plan has three main purposes. First, it helps an entrepreneur to visualize and organize the company and its operations. Remember from Chapter 2 how Basil tested the strengths and weaknesses of the proposal to make the Empty Decadence candy bar? Similarly, thinking critically about your hopes for the business and putting a plan on paper will help you and Anna imagine how the plan will work and will help you evaluate the plan, develop new ideas, and refine the plan. By looking at the plan from different points of view, such as those of managers who have responsibility for marketing the company's products or purchasing its inventory of products, you can discover and correct flaws before implementing the plan. Then "paper mistakes" won't become real mistakes!

Second, a business plan serves as a "benchmark," or standard, against which the entrepreneur can later measure the actual performance of a company. You and Anna will be able to evaluate differences between the planned performance of Sweet Temptations, as outlined in its business plan, and its actual performance. Then you will be able to use the results of your evaluation to adjust Sweet Temptations' future activities. For instance, suppose in its first month of business, sales are higher than you and Anna predicted. If you decide that sales will continue at this level, you can use this information to increase Sweet Temptations' future candy purchases.

Third, a business plan helps an entrepreneur obtain the financing that new and growing companies often need. When Anna starts looking for additional funding for Sweet Temptations, potential investors and creditors may request a copy of the company's business plan to help them decide whether or not to invest in Sweet Temptations or to loan it money. For example, as part of its loan-making decisions, **Central Bank** in Jefferson City, Missouri, routinely evaluates the business plans of companies that apply for business loans at the bank.

http://www.centralbank.net

Investors and creditors, such as Central Bank, have two related concerns when they are making investment and credit decisions. One concern is the level of risk involved with their decisions. **Risk** usually refers to how much uncertainty exists about the future operations of the company. The other concern is the **return**, or money back, that they will receive from their investment and credit decisions. A thorough business plan will provide useful information for helping investors and creditors evaluate their risk and potential return. Now let's look at the parts of a business plan.

Description of the Company

A business plan usually begins with a description of the company and its basic activities. Details of this description include information about the organization of the company, its product or service, its current and potential customers, its objectives, where it is located, and where it conducts its business.

For example, Sweet Temptations is a new retailing company located in a "high-growth" suburb north of a major metropolitan area. Initially, Sweet Temptations will sell only one kind of candy—boxes of chocolates. You and your sister Anna will expand the "product line" to include other kinds of candy as the company grows. After the sale of chocolates is "up and running," and after you graduate, you plan to join Anna full-time as a partner in the company. You and Anna are eager to begin marketing and operating the company but are waiting to do so until after you finish writing the company's business plan and obtain financing. You realize that writing the plan is helping you to think through the various aspects of the business so that you don't "miss something" important in planning your activities. Exhibit 3-1 illustrates how you might describe Sweet Temptations in its business plan.

The organization of a company and its personnel can have a major influence on the success of a company. Therefore, the description of the company also includes a listing of the important people and the major roles they will play in the company. This listing can include the individuals responsible for starting the company, significant investors who

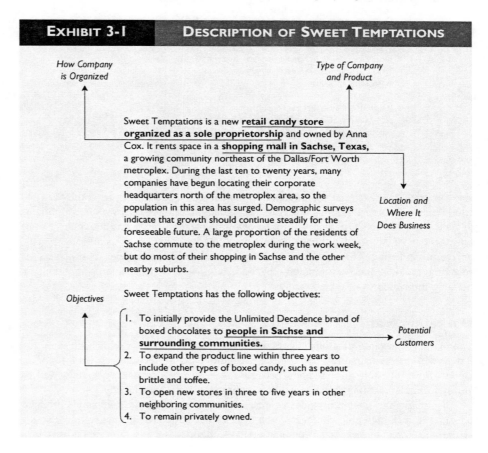

EXHIBIT 3-1   **DESCRIPTION OF SWEET TEMPTATIONS**

How Company is Organized

Type of Company and Product

Sweet Temptations is a new **retail candy store organized as a sole proprietorship** and owned by Anna Cox. It rents space in a **shopping mall in Sachse, Texas,** a growing community northeast of the Dallas/Fort Worth metroplex. During the last ten to twenty years, many companies have begun locating their corporate headquarters north of the metroplex area, so the population in this area has surged. Demographic surveys indicate that growth should continue steadily for the foreseeable future. A large proportion of the residents of Sachse commute to the metroplex during the work week, but do most of their shopping in Sachse and the other nearby suburbs.

Location and Where It Does Business

Objectives

Sweet Temptations has the following objectives:

1. To initially provide the Unlimited Decadence brand of boxed chocolates to **people in Sachse and surrounding communities.**
2. To expand the product line within three years to include other types of boxed candy, such as peanut brittle and toffee.
3. To open new stores in three to five years in other neighboring communities.
4. To remain privately owned.

Potential Customers

EXHIBIT 3-2	ORGANIZATION OF SWEET TEMPTATIONS

The team at Sweet Temptations is composed of four people, one of whom is a financial consultant. The members of this team are as follows:

Anna Cox	Owner	*Listing of Important People and the Roles They Will Play*
(Your name)	Employee/adviser	
Jaime Gonzales	Employee	
Joe Smiley	Consultant	

Each of these individuals brings special skills to Sweet Temptations. Anna Cox graduated last year with a B.B.A. in marketing from State U. She has already earned a reputation for her marketing and business skills. While in school, she won the National Student Marketing Association's prestigious Student Marketer of the Year Award and the coveted Small Business Institute's Rising Star Award. She graduated with highest honors. While in school, Anna worked for three retail stores, two of which were start-up companies. One of the start-up companies was a candy store.

(Your name) is an honors business student at State U. and will be graduating in two years. (Your first name) has worked twenty hours per week "keeping the books" at a local candy store for the past two years. Prior to that, (your first name) worked summers and part time during the school year doing miscellaneous jobs at the same candy store.

Qualifications of Important People

Jaime Gonzales is an honors business student at State U. Jaime has worked summers at several restaurants in Sachse.

Joe Smiley is a partner in the management advisory services area of (name of company), a large public accounting firm in Dallas. His firm specializes in consulting with start-up companies.

also are providing expertise and direction to the company, and influential employees and consultants who have a strong impact on the company. Exhibit 3-2 shows some highlights of how you might discuss Sweet Temptations' organization in its business plan. Notice how this part of the plan highlights the combination of your major in business and Anna's degree in marketing. This part may also contain the company's policies or strategies for selecting, training, and rewarding employees. These issues are particularly important for the long-term success of the company.

Marketing Plan

The marketing section of a business plan shows how the company will make sales and how it will influence and respond to market conditions. This section receives a lot of attention from investors and creditors because the company's marketing strategy and its ability to implement that strategy can be very important to the company's success.

The marketing section provides evidence of the demand for the company's products or services, including any market research that has been conducted. This section also describes the current and expected competition in the market, as well as relevant government regulations. The marketing section describes how the company will promote, price, and distribute its products (the company's "marketing strategy"), as well as the predicted growth, market share, and sales of its products (its "sales forecast") by period. This information is helpful to the entrepreneur as a starting point for thinking about the company's other activities related to sales, such as timing the purchase of its inventories. The marketing section is also helpful to people outside the company, such as bank loan officers, because it shows how well the entrepreneur has thought through the company's sales potential and how the company will attract and sell to customers.

Sweet Temptations' business plan may be an inch thick! We don't have room to show each part of its plan, so in the next sections we will ask you to think about what to include. The following is a brief description of Sweet Temptations' market conditions. Initially, Sweet Temptations will have a temporary marketing advantage. Currently, community members must drive at least 30 miles to purchase boxes of Unlimited Decadence chocolates (and they actually make the drive!). After evaluating the community's available retail space (and plans for building retail space), you and Anna believe that there will be very little competition during the next several years. However, you eventually expect competing stores to open in the community. In the meantime, part of your marketing plan is to build a reputation for friendly service and quality products. Your advertising will focus on the quality ingredients used in the chocolates. Furthermore, your initial advertising "punch" will include the fact that Unlimited Decadence now produces, and Sweet Temptations sells, mini-versions of "everyone's favorite candy bars" in boxed form. You believe Sweet Temptations has a distinct advantage in selling Unlimited Decadence chocolates because of the already-established good reputation and popularity of the Unlimited Decadence candy bars.

 What information about market conditions facing Sweet Temptations would you include in the marketing section of its business plan?

Description of Company Operations

Since a company is organized to deliver a product or service to a market, the business plan must address how the company will develop and enhance its products or services. The company operations section of a business plan includes a description of the relationships between the company, its suppliers, and its customers, as well as a description of how the company will develop, service, protect, and support its products or services. This section also includes any other influences on the operations of the company. The company operations section of the business plan is important because it helps the entrepreneur think through the details of making the idea work. Also, it helps outside users evaluate the entrepreneur's ability to successfully carry out the idea.

Here is a brief description of Sweet Temptations' operations. Sweet Temptations has a ready supply of chocolates. Unlimited Decadence has no sales agreements with any other candy stores within a 30-mile radius of Sweet Temptations. Furthermore, you know of other potential suppliers—candy manufacturers who have high production standards, quality ingredients, and good reputations in the candy industry. In fact, Anna is now talking with representatives of these companies and visiting their kitchens so that she will have identified and selected other suppliers by the time Sweet Temptations is ready to sell other types of candy.

 What information about Sweet Temptations' operations would you include in its business plan?

Other influences on the operations of the company might also be described in this section. These other influences might include the availability of employees, concerns of special-interest groups, regulations, the impact of international trade, and the need for patents, trademarks, and licensing agreements.

If Sweet Temptations' major supplier of chocolates was a company in Brussels, Belgium, rather than Unlimited Decadence Corporation, what additional issues do you think should be included in this section of the business plan? What else do you think managers, owners, creditors, and investors would like to know?

Financial Plan

Since Sweet Temptations is a new company, it has no credit history or recent financial statements. Therefore, Anna should also provide a detailed, realistic financial plan in Sweet Temptations' business plan. The purpose of the financial plan section is to identify the company's capital requirements and sources of capital, as well as to describe the company's projected financial performance. For a new company, this section also highlights the company's beginning financial activities, or "start-up" costs.

Here is some information about Sweet Temptations' start-up costs:

> Anna has decided that she will invest $15,000 of her own money as capital to run Sweet Temptations. Based on the rent charged for space in the shopping mall, she has determined that it will cost $1,000 per month to rent store space in the mall. When Sweet Temptations signs a rental contract for the store in December 2003, it will pay six months' rent in advance, totaling $6,000. Based on a supplier's cost quotation, Anna has determined that Sweet Temptations can buy store equipment for $1,800. The supplier will allow Sweet Temptations to make a $1,000 down payment and to sign a note (a legal document, referred to as a *note payable*) for the remaining amount, to be paid later. Based on the purchases budget (which we will discuss in Chapter 4), Sweet Temptations will purchase 360 boxes of chocolates for "inventory" in December 2003 at a cost of $1,620 from Unlimited Decadence. Unlimited Decadence has agreed to allow Sweet Temptations to pay for this inventory in January 2004. Sweet Temptations will also purchase $700 of supplies in December 2003, paying for the supplies at that time.

What information about Sweet Temptations' start-up costs would you include in the financial section of its business plan?

Identifying Capital Requirements

Most companies eventually need additional funding, or **capital**. The financial section of a business plan should include a discussion of the company's capital requirements and potential sources of that capital. For new companies and small companies, this discussion can be the most important part of the business plan. As you may have noticed while reading the business section of your local newspaper, if a company does not have enough capital and sources of capital, it will have a difficult time surviving.

An entrepreneur can determine a company's capital requirements by analyzing two major issues. First, the entrepreneur should decide what resources the company needs, such as buildings, equipment, and furniture. Then, the entrepreneur can estimate how much capital the business will need in order to acquire those resources. Cost quotations, appraisals, and sales agreements are a good starting point for this estimate. Next, the entrepreneur should analyze the company's projected cash receipts and payments to determine whether it will have enough cash to buy the resources and, if not, how much cash the company will need to borrow. Planning capital requirements involves projections, not guarantees, so the entrepreneur must expect and provide for reasonable deviations from plans. Suppose, for example, that cash sales for the month turn out to be less than expected. For "surprises" like this, the entrepreneur should plan to have a "cash buffer," which is extra cash on hand above the projected short-run cash payments of the company. One purpose of this buffer is to protect the company from differences between actual cash

Significant "start-up" costs for a company.

flows and projected cash flows, and also from unanticipated problems such as having to replace a refrigerated display case sooner than expected. A cash buffer lets the company operate normally through downturns without having to look for financing. It also lets the company take advantage of unexpected opportunities that require cash.

 Can you think of an example of an unexpected opportunity for which an entrepreneur or manager might find a cash buffer to be handy?

Sources of Capital

Once the entrepreneur knows the company's capital requirements, potential sources of capital can be identified. Here, the entrepreneur must know both the length of time that the company plans to use the capital before paying it back to creditors or returning it to investors, and the availability of short- and long-term sources of capital. The entrepreneur can determine how long the company will need to use the capital by analyzing the company's projected cash receipts and payments. We will discuss the tools of this analysis more thoroughly in Chapter 4.

Short-term capital will be repaid within a year or less. Short-term capital can come from two sources. First, suppliers provide short-term capital to some of their customers through what is called "trade credit." Trade credit involves allowing a customer to purchase inventory "on credit" if the customer agrees to pay soon, usually within 30 days. You and Anna have an arrangement with Unlimited Decadence that will allow Sweet Temptations to buy boxed chocolates on credit and pay Unlimited Decadence 30 days later.

 If Sweet Temptations took longer than 30 days, on the average, to sell its inventory of chocolates, do you think its arrangement with Unlimited Decadence would be valuable? What other questions would you like answered to help you determine the answer to this question?

http://www.sba.gov

Second, financial institutions, such as commercial banks, provide loans to companies, many of which are guaranteed by government agencies such as the **U.S. Small Business Administration**. These institutions require a more formal agreement with a company do issuers of trade credit. Also, they charge interest on these short-term loans. point, Anna may talk with her banker to arrange a small line of credit for Sw tations. A **line of credit** allows a company to borrow money "as needed," arranged, agreed-upon interest rate and a specific payback schedule. We disc capital more in Chapter 9.

Long-term capital will be repaid to creditors or returned to investors after more than a year. Initially, as we mentioned in Chapter 1, companies obtain capital from the owner and from bankers. Sweet Temptations obtained its initial capital from Anna, who invested money from her savings account. Other sources of long-term capital can include friends and relatives, commercial banks, and leasing companies. Many loans are guaranteed by the Small Business Administration or the state's economic-development agency. For example, after being turned down for a bank loan, the owner of **Frosty Factory of America**, a Ruston, Louisiana company that manufactures slush machines for making sorbets and frozen drinks, took his company's business plan to the Louisiana State Economic Development Corporation. After reviewing the plan, the state agency agreed to guarantee 50 percent of the loan. The bank reconsidered and loaned the company $325,000.[1]

http://www.frostyfactory.com

All institutions require a formal agreement with the company about payment dates and interest rates. But suppose Sweet Temptations borrows money from Anna's and your friends and relatives. Do you think it is necessary to have a formal written agreement between these friends and relatives and Sweet Temptations? Why or why not?

Eventually, as a company grows too large to be financed by the owner and these other sources, it may offer private placements or public offerings. Private placements are securities that are sold directly to private individuals or groups (called *investors*). Public offerings involve issuing bonds or stocks to the public (investors) through securities firms or investment bankers. We will discuss bonds and stocks as a source of long-term capital in Chapters 22 and 24.

For the near future, several of Anna's and your friends and relatives have agreed to lend Sweet Temptations specific amounts of money, as needed. Anna and these friends and relatives have agreed that the interest rate on these loans will match the market interest rate at the time of each loan. Sweet Temptations includes this information in its financial plan.

Projected Financial Performance

This section of the financial plan projects the company's financial performance. Suppose Anna has assigned you the responsibility of preparing this section of Sweet Temptations' financial plan. Although projecting a company's financial performance involves uncertainty, if you follow some guidelines, the financial performance information will be more dependable.

First, the data that you use should be as reliable as possible. Since Sweet Temptations is a new company, you don't have historical data to use for planning purposes. When you have sketchy data (or no data at all), industry averages found in such sources as *Moody's, Standard & Poor's,* and *Robert Morse Associates* can serve as a guide.

If you use Moody's, Standard & Poor's, or Robert Morse Associates for industry information, you must be able to identify the industry in which Sweet Temptations is operating. What are some key words that you could use to identify the industry?

Second, because predicting a company's financial performance is uncertain, you should consider several scenarios. "What if" questions are useful for this type of planning. What if we sell only 800 boxes of chocolates? What if we sell 1,300 boxes of chocolates? The scenarios should be realistic and perhaps should consider the best case, the worst case, and the most probable case.

Third, you should revise your projection as more facts become available. Finally, it is important that the financial plan is consistent with the information in the other sections of the business plan. For example, since the marketing section of Sweet Temptations' business plan refers to the advertising that you plan to do, the financial plan section must show advertising costs.

---

[1]Pete Weaver, "Need a Loan? See Your State," *National Business,* April 1995, 51R.

The financial performance section of the financial plan includes projected financial statements,[2] supported by cost-volume-profit analysis and budgets. Budgets include reports on such items as estimated sales, purchases of inventory, and expenses, as well as estimated cash receipts and payments. In the remainder of this chapter we will discuss cost-volume-profit analysis and its relationship to the projected income statement. In Chapter 4, we will discuss budgets and how they fit into a company's financial plan.

In summary, you have just learned that the business plan shows the direction a company will be taking during the next year. You have also learned that the business plan includes a description of the company, a marketing plan, a description of company operations, and a financial plan. Accountants are most involved with the financial plan, which includes an analysis of predicted costs, sales volumes, and profits. We thus will spend the remainder of this chapter discussing cost-volume-profit analysis and its use in planning.

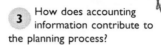

COST-VOLUME-PROFIT (C-V-P) PLANNING

3 How does accounting information contribute to the planning process?

Determining if a company will be profitable is difficult before it begins operations. This uncertainty is part of the risk that the entrepreneur takes in starting a business. Although it can be scary, it is also part of the fun. Uncertain profit does not mean that the entrepreneur should disregard any type of analysis before beginning the operations of a company, however. It is possible to take educated risks based on estimations of costs, sales volumes, and profits. The financial plan should include an analysis of these factors. One type of analysis that uses these three factors is called *cost-volume-profit analysis*.

Cost-Volume-Profit Analysis

Cost-volume-profit (C-V-P) analysis shows how profit will be affected by alternative sales volumes, selling prices of products, and various costs of the company. C-V-P analysis sometimes is called "break-even analysis." Entrepreneurs use C-V-P analysis to help them understand how the plans they make will affect profits. This understanding can produce more-informed decisions during the ongoing planning process.

C-V-P analysis is based on a simple profit computation involving revenues and costs. This computation can be shown in an equation or in a graph. Although equations provide precise numbers, C-V-P graphs provide a convenient visual form for presenting the analysis to decision-makers. However, to understand a C-V-P equation or graph, decision makers also must understand how costs behave.

Cost Behavior

A careful cost analysis considers the activity level of the operation that causes the cost. For example, Unlimited Decadence, a manufacturing company, might measure its activity by using the number of cases of chocolate bars produced or the number of hours worked in producing these cases of chocolate bars. On the other hand, Sweet Temptations, a retail company, might measure its activity by using the number of boxes of chocolates *sold*. The activity level (the number of boxes of candy bars sold) is often referred to as **volume**. The relationship between an activity's cost and its volume helps us determine the cost's behavior pattern.

To understand what C-V-P equations and graphs reveal about a company's potential profitability, let's first look at two cost behavior patterns that describe how have. These are called *fixed costs* and *variable costs*.

---

[2]The financial plan usually includes a projected balance sheet, but to simplify the discussion in this chapter the projected balance sheet until Chapter 12.

Fixed Costs

Fixed costs are constant in total for a specific time period; they are *not* affected by differences in volume during that same time period. Managers' annual salaries are usually fixed costs, for instance. For another example, think about the $1,000 monthly rent that Sweet Temptations will pay for its retail space. Sweet Temptations' activity level is its sales volume—the number of boxes of chocolates sold. The rent cost of the retail space will not change as a result of a change in the sales volume, assuming you have planned carefully and have leased enough retail space. Sweet Temptations will pay its monthly rent of $1,000 no matter how many candy bars it sells that month. Since the rent cost does not change as volume changes, it is a fixed cost. The graph in Exhibit 3-3 illustrates the relationship between the rent cost and the sales volume. As you can see, the rent cost will be $1,000 whether Sweet Temptations sells 500 boxes of chocolates or 1,000 boxes.

Note in Exhibit 3-3 that we show a fixed cost as a horizontal straight line on the graph, indicating that the cost will be the same (fixed) over different volume levels. It is important not to be misled about fixed costs. Saying that a cost is "fixed" does not mean that it cannot change from one time period to the next. In the next period, Sweet Temptations could rent more retail space if needed or the landlord could raise the rent when the lease is renewed, causing the rent cost to be higher. To be fixed, a cost must remain constant for a time period in relation to the volume attained *in that same time period.* For example, most companies consider the costs of using their buildings, factories, office equipment, and furniture—called *depreciation*[3]—to be fixed. That is, depreciation costs within a specific time period will not change even if volume changes within that time period.

You have estimated that Sweet Temptations' monthly fixed costs will include the $1,000 rent cost plus $2,050 total salaries for you and Jaime Gonzales (the employee Anna hired to sell candy), $200 consulting costs, $305 advertising costs, $30 supplies costs, $15 depreciation of the store equipment, and $250 telephone and utilities costs.[4] Sweet Temptations' total fixed costs will be the sum of the individual fixed costs, or $3,850.

 What would the graph look like for Sweet Temptations' $3,850 total fixed costs? Why?

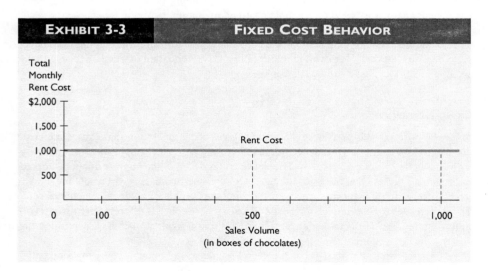

EXHIBIT 3-3 — FIXED COST BEHAVIOR

---

[3]We will discuss in Chapters 5 and 21 how a company determines its depreciation cost. We include a brief discussion here because most companies have some depreciation costs to consider in evaluating their operations.

[4]Some supplies, telephone, and utility costs may have minimum charges, but their total costs are affected by changes in the volume of usage. These costs are called *mixed costs,* which we will discuss in Chapter 11. For simplicity, here we assume they are fixed costs.

If this box of chocolates contained more pieces of candy, would the company's total fixed costs decrease?

© GETTY IMAGES, INC./PHOTODISC

Decision makers sometimes state fixed costs as a dollar amount *per unit,* computed by dividing total fixed costs by the volume in units. This can be misleading and should be avoided. For instance, at a sales volume of 500 boxes of chocolates, Sweet Temptations' fixed cost per box of chocolates will be $7.70 ($3,850 fixed costs ÷ 500 boxes of chocolates). At a sales volume of 1,000 boxes of chocolates, the fixed cost per box of chocolates will only be $3.85 ($3,850 fixed costs ÷ 1,000 boxes of chocolates). Comparing $7.70 with $3.85, you might think that total fixed costs decrease as sales volume increases. This is not true! Sweet Temptations' total fixed costs will be $3,850 regardless of the sales volume.

Variable Costs

A **variable cost** is constant *per unit* of volume, and changes in total in a time period in direct proportion to the change in volume. For instance, consider Sweet Temptations' cost of purchasing chocolates from Unlimited Decadence to resell to its customers. You have estimated that it will cost Sweet Temptations $4.50 for each box of chocolates that it purchases. The *total cost* of boxes of chocolates sold varies in proportion to the *number* of boxes sold. If Sweet Temptations sells 500 boxes of chocolates in January, the total variable cost of these boxes of chocolates sold will be $2,250 (500 boxes of chocolates × $4.50 per box). If the volume doubles to 1,000 boxes of chocolates, the total variable cost of boxes of chocolates sold will also double to $4,500 (1,000 boxes of chocolates × $4.50 per box). It is important to remember that the total variable cost for a time period increases in proportion to volume in that same time period because each unit has the same variable cost.

Exhibit 3-4 shows the estimated total variable costs of boxes of chocolates sold by Sweet Temptations at different sales volumes. Note that total variable costs are shown by a straight line sloping upward from the origin of the graph. This line shows that the total variable cost increases as volume increases. If no boxes of chocolates are sold, the total variable cost will be $0. If 500 boxes of chocolates are sold, the total variable cost will be $2,250. The slope of the line is the rate at which the total variable cost will increase each time Sweet Temptations sells another box of chocolates. This rate is the variable cost per unit of volume, or $4.50 per each additional box of chocolates sold.

 How could rent be a variable cost? If it were a variable cost, how do you think it would affect Sweet Temptations' variable costs line in Exhibit 3-4?

Because graphs are easy to see, we used them to show Sweet Temptations' fixed and variable costs in Exhibits 3-3 and 3-4. For C-V-P analysis, however, it is often better to use equations because they show more precise numbers. For instance, the equation for the total amount of a variable cost is

Total variable cost $= vX$
where:
v = variable cost per unit sold, and
X = sales volume.

The equation for the variable cost line in Exhibit 3-4 is

Total variable cost of boxes of chocolates sold $= \$4.50X$
where:
X = sales volume.

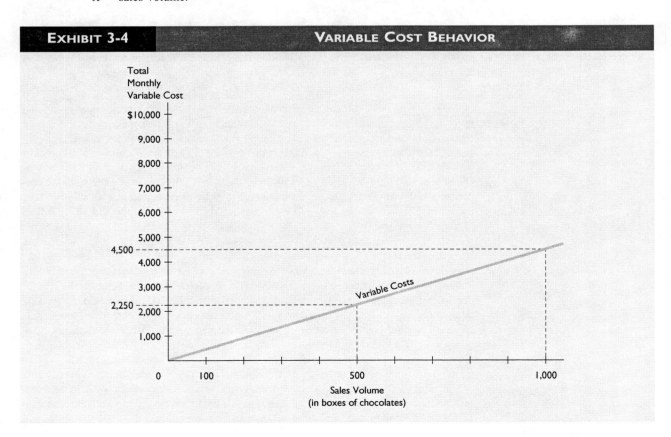

EXHIBIT 3-4   **VARIABLE COST BEHAVIOR**

Total Costs

Total costs at any volume are the sum of the fixed costs and the variable costs at that volume. For example, at a sales volume of 500 boxes of chocolates, Sweet Temptations' estimated fixed costs are $3,850 and its estimated variable costs are $2,250 (500 × $4.50), for an estimated total cost of $6,100 at that volume. At a sales volume of 1,000 boxes of chocolates, estimated fixed costs are $3,850, estimated variable costs are $4,500 (1,000 × $4.50), and the estimated total cost is $8,350. Exhibit 3-5 illustrates the total cost in relation to sales volume. Notice that if no boxes of chocolates are sold, the total cost will be equal to the fixed costs of $3,850. As sales increase, the total cost will increase by $4.50 per box, the amount of the variable cost per box.

The equation for the total cost is

$$\text{Total cost} = f + \nu X$$
where:
f = total fixed costs,
ν = variable cost per unit sold, and
X = sales volume.

The equation for the total cost line in Exhibit 3-5 is

$$\text{Total cost of boxes of chocolates sold} = \$3,850 + \$4.50X$$
where:
X = sales volume.

Now that you understand the relationships of volume, fixed costs, and variable costs to the total cost, we can use C-V-P analysis to estimate profit.

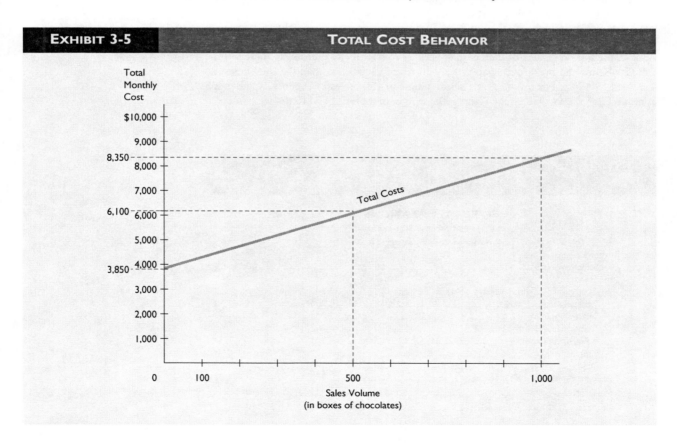

EXHIBIT 3-5 — TOTAL COST BEHAVIOR

Profit Computation

According to the marketing plan, Sweet Temptations expects to sell 720 boxes of chocolates at $10 each in January. Exhibit 3-6 shows Sweet Temptations' projected income statement in the format that is presented to external users. This is the same format that we discussed in Chapter 1 and illustrated in Exhibit 1-7 on page 18. External decision makers find this format understandable and use this form of income statement for their investment and credit decisions. This income statement results from the following equation:

$$\text{Net Income (Profit)} = \text{Revenues} - \text{Expenses}$$

In this equation, revenues (the selling prices of all the boxes of chocolates sold to customers) include cash and credit sales, and expenses (the costs of providing the boxes of chocolates to customers) include the cost of boxes of chocolates sold and the expenses to operate the business.

Profit Graph

One way of graphing a company's net income (profit) is to show both its revenues and its costs (expenses) on the same graph. Recall that the graph of a company's total costs includes its fixed costs and its variable costs, as we illustrated in Exhibit 3-5 for Sweet Temptations. The graph of a company's revenues is shown by a straight line sloping upward from the origin of the graph. The slope of the line is the rate (selling price per unit) at which the total revenues increase each time the company sells another unit.

The graph in Exhibit 3-7 shows the estimated total revenue line and the estimated total cost line for Sweet Temptations. Note that the total revenue line crosses the total cost line at 700 boxes of chocolates. At this point, the total revenues will be $7,000, and the total costs will be $7,000, so there will be zero profit. The unit sales volume at which a company earns zero profit is called the **break-even point**. Above the break-even unit sales volume, the total revenues of the company are more than its total costs, so there will be a profit. Below the break-even point, the total revenues are less than the total costs, so there will be a loss. For instance, at a sales volume of 720 boxes of chocolates, the graph in Exhibit 3-7 shows that Sweet Temptations will earn a profit of $110 (as we computed

EXHIBIT 3-6	PROJECTED INCOME STATEMENT FOR EXTERNAL USERS

SWEET TEMPTATIONS
Projected Income Statement
For the Month Ended January 31, 2004

Revenues:		
Sales revenues		$7,200
Expenses:		
Cost of boxes of chocolates sold	$3,240	
Rent expense	1,000	
Salaries expense	2,050	
Consulting expense	200	
Advertising expense	305	
Supplies expense	30	
Depreciation expense: display cases	15	
Telephone and utilities expense	250	
Total expenses		(7,090)
Net income		$ 110

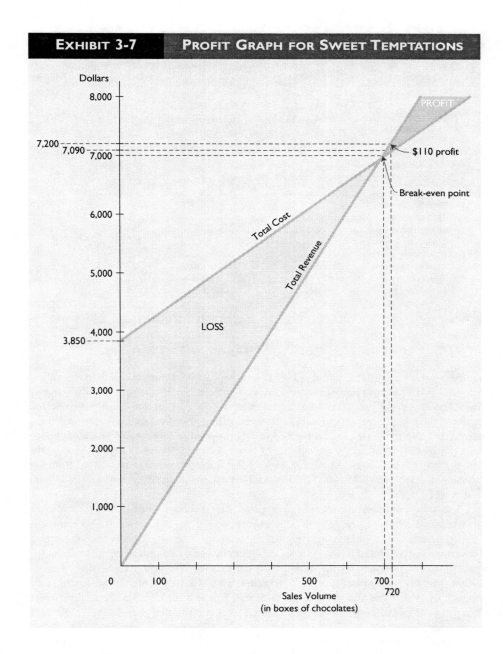

EXHIBIT 3-7   PROFIT GRAPH FOR SWEET TEMPTATIONS

in the income statement in Exhibit 3-6), the difference between the $7,200 estimated total revenue and $7,090 estimated total cost at this volume. Although some decision makers use this type of graph, many others prefer to use a different graph that shows a company's contribution margin, as we discuss next.

Contribution Margin

To estimate profit at different volume levels, the entrepreneur needs C-V-P information in a form that relates the estimated revenues and estimated variable costs to the estimated fixed costs. Exhibit 3-8 shows the same information as Exhibit 3-6, but in a format that is more useful for the internal decision makers in performing C-V-P analysis because it

EXHIBIT 3-8	PROJECTED INCOME STATEMENT FOR INTERNAL USERS (CONTRIBUTION MARGIN APPROACH)

SWEET TEMPTATIONS
Projected Income Statement
For the Month Ended January 31, 2004

Total sales revenues ($10 × 720 boxes of chocolates)...		$7,200
Less total variable costs:		
Cost of boxes of chocolate sold ($4.50 × 720 boxes)..		(3,240)
Total contribution margin		$3,960
Less total fixed costs:		
Rent expense...	$1,000	
Salaries expense...	2,050	
Consulting expense...	200	
Advertising expense..	305	
Supplies expense..	30	
Depreciation expense: display cases	15	
Telephone and utilities expense..	250	
Total fixed costs..		(3,850)
Profit...		$ 110

shows expenses as variable and fixed. This income statement format is sometimes called the *contribution margin approach*. Notice that, on this income statement, Sweet Temptations first calculates its estimated sales revenue ($7,200) by multiplying the number of boxes of chocolates it expects to sell (720) by the selling price per box ($10). Sweet Temptations next determines the total estimated variable costs of selling the 720 boxes of chocolates ($3,240) by multiplying the number of boxes it expects to sell (720) by the variable cost per box of chocolates ($4.50). These total variable costs are then subtracted from total sales revenue. The $3,960 ($7,200 − $3,240) difference is called the *total contribution margin*.

The **total contribution margin**, at a given sales volume, is the difference between the estimated total sales revenue and the estimated total variable costs. It is the amount of revenue remaining, after subtracting out the total variable costs, that will contribute to "covering" the estimated fixed costs. To compute the estimated profit, we subtract the total estimated fixed costs for the month from the total contribution margin. If the contribution margin is more than the total fixed costs, there will be a profit. If the contribution margin is less than the total fixed costs, there will be a loss. Exhibit 3-8 shows that Sweet Temptations' estimated profit is $110 ($3,960 total contribution margin − $3,850 total fixed costs).

The contribution margin may also be shown on a per-unit basis. The **contribution margin per unit** is the difference between the estimated sales revenue per unit and the estimated variable costs per unit. For Sweet Temptations, the contribution margin per unit is $5.50 ($10 sales revenue − $4.50 variable costs). At 720 units, the total contribution margin will be $3,960 (720 × $5.50), which is the same as shown in Exhibit 3-8. Later, you will see that computing the total contribution margin (by either method described above) is the key to understanding the relationship between profit and sales volume.

Exhibit 3-9 shows what the total contribution margin will be at different unit sales this graph, since the contribution margin of one box of chocolates is $5.50, ntribution margin increases at a rate of $5.50 per box of chocolates sold. For a volume of 500 boxes of chocolates, the contribution margin will be $2,750

EXHIBIT 3-9	RELATIONSHIP BETWEEN THE TOTAL CONTRIBUTION MARGIN AND THE UNIT SALES VOLUME

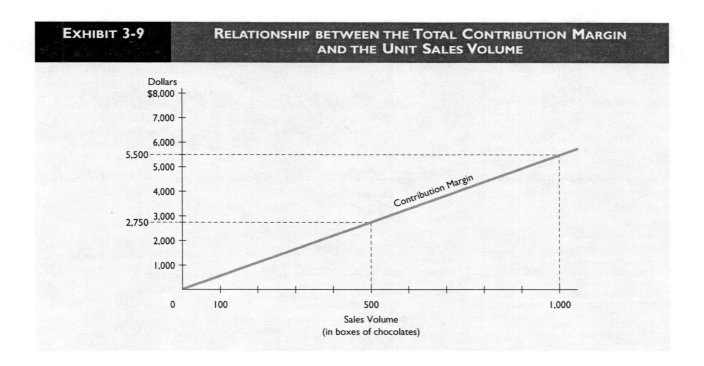

(500 boxes × $5.50). At a volume of 1,000 boxes of chocolates, the contribution margin will be $5,500 (1,000 boxes × $5.50).

 If variable costs were higher per unit, would you expect the contribution margin line in Exhibit 3-9 to be steeper or flatter than it is? Why?

Showing C-V-P Relationships

Now that you understand the contribution margin and fixed costs, we can show the estimated profit or loss at different sales volumes in a graph. Exhibit 3-10 shows how sales volume affects the estimated profit (or loss) for Sweet Temptations. Two lines are drawn on this graph. One line shows the estimated total contribution margin at different sales volumes. It is the same line as shown in Exhibit 3-9. The other line shows the $3,850 total estimated fixed costs. The vertical distance between these lines is the estimated profit or loss at the different sales volumes. Remember, estimated profit is the total contribution margin minus the estimated total fixed costs. Note that this graph shows that Sweet Temptations will earn $0 profit if it sells 700 boxes of chocolates; this is its break-even point. Above the break-even unit sales volume (such as at a volume of 1,000 boxes), the total contribution margin ($5,500) is more than the total estimated fixed costs ($3,850), so there would be a profit ($5,500 − $3,850 = $1,650). Below the break-even point (such as at a volume of 500 boxes), the total contribution margin ($2,750) is less than the total estimated fixed costs, so there would be a loss ($2,750 − $3,850 = −$1,100).

 If fixed costs were greater, would you expect Sweet Temptations to break even at a lower sales volume or a higher sales volume? Why?

Profit Computation (Equation Form)

In Exhibit 3-10, we show a graph of the C-V-P relationships for Sweet Temptations. Graphs are usually a helpful tool for an entrepreneur (and students!) to see a "picture" of these relationships. Sometimes, however, an entrepreneur (or student) does not need a pic-

EXHIBIT 3-10	COST-VOLUME-PROFIT RELATIONSHIPS FOR SWEET TEMPTATIONS

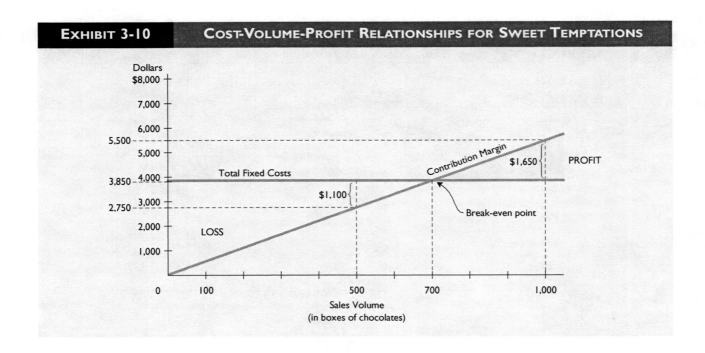

ture to understand the relationships. In this case, using equations may be a better and faster way to understand C-V-P relationships. (You may have already thought of these equations as you studied Exhibit 3-10.) In this section, we look at how to use equations for C-V-P analysis to answer the following questions:

1. How much profit will the company earn at a given unit sales volume?
2. How many units must the company sell to break even?
3. How many units must the company sell to earn a given amount of profit? (The given amount is usually a desired profit that the company uses as a goal.)

In the following discussion, we use Sweet Temptations' revenue and cost information from the projected income statement for internal decision makers, in Exhibit 3-8. We determined the total sales revenue by multiplying the selling price per unit by the estimated sales volume. We determined the total estimated variable costs by multiplying the variable cost per unit by the same sales volume. And we subtracted the total estimated variable and fixed costs from the total estimated sales revenue to determine the estimated profit. We can show this in a "profit equation" as follows:

$$\text{Profit (for a given sales volume)} = \left[\begin{array}{ccc} \text{Selling} \\ \text{price} & \times & \text{Unit sales} \\ \text{per unit} & & \text{volume} \end{array} \right] - \left[\begin{array}{ccc} \text{Variable} \\ \text{cost} & \times & \text{Unit sales} \\ \text{per unit} & & \text{volume} \end{array} \right] - \begin{array}{c} \text{Total} \\ \text{fixed} \\ \text{costs} \end{array}$$

The profit equation can be used in C-V-P analysis. For Sweet Temptations, for instance, if we use X to stand for a given sales volume of boxes of chocolates, and if we include the estimated selling price and variable cost per unit, the equation is written as follows:

$$\text{Profit} = \$10X - \$4.50X - \$3,850$$
$$= (\$10 - \$4.50)X - \$3,850$$
$$= \$5.50X - \$3,850$$

This equation, then, can be used in solving various C-V-P questions for Sweet Temptations.[5]

USING C-V-P ANALYSIS

4 What must decision makers be able to predict in order to estimate profit at a given sales volume?

C-V-P analysis is useful in planning because it shows the impact of alternative plans on profit. This analysis can help the entrepreneur make planning decisions and can help investors and creditors evaluate the risk associated with their investment and credit decisions. For instance, suppose Anna has asked you to answer, for Sweet Temptations, the three questions we mentioned earlier. In this section we describe how to do so.

Estimating Profit at Given Unit Sales Volume

Suppose Anna wants you to estimate Sweet Temptations' monthly profit if it sells 750 boxes of chocolates (i.e., a unit sales volume of 750 boxes) a month. Remember that Sweet Temptations' selling price is $10 per unit and its variable cost is $4.50 per unit. You can estimate monthly profit when 750 boxes of chocolates are sold in a month by using the profit equation as follows:

$$\text{Profit} = \left[\begin{array}{c} \text{Selling} \\ \text{price} \\ \text{per unit} \end{array} \times \begin{array}{c} \text{Unit} \\ \text{sales} \\ \text{volume} \end{array} \right] - \left[\begin{array}{c} \text{Variable} \\ \text{cost} \\ \text{per unit} \end{array} \times \begin{array}{c} \text{Unit} \\ \text{sales} \\ \text{volume} \end{array} \right] - \begin{array}{c} \text{Total} \\ \text{fixed} \\ \text{costs} \end{array}$$

$$= (\$10 \times 750) - (\$4.50 \times 750) - \$3,850$$
$$= \$7,500 - \$3,375 - \$3,850$$
$$= \underline{\$275}$$

Thus, you can tell Anna that Sweet Temptations will make a monthly profit of $275 if it sells 750 boxes of chocolates a month.

Finding the Break-Even Point

5 How can decision makers predict the sales volume necessary for estimated revenues to cover estimated costs?

Suppose Anna wants you to estimate how many boxes of chocolates Sweet Temptations must sell to break even each month. Recall that the break-even point is the unit sales volume that results in zero profit. This occurs when total sales revenue equals total costs (total variable costs plus total fixed costs). To find the break-even point, we start with the profit equation. Remember that the contribution margin per unit is the difference between the sales revenue per unit and the variable costs per unit. With this in mind, we can rearrange the profit equation[6] into a break-even equation as follows:

---

[5]Note in the last line of the equation that the $5.50 is the contribution margin per unit. This can come in handy as a "shortcut" when using the profit equation, so that the equation becomes:

$$\begin{array}{c} \text{Profit} \\ \text{(for a given} \\ \text{sales volume)} \end{array} = \left[\begin{array}{c} \text{Contribution} \\ \text{margin} \\ \text{per unit} \end{array} \times \begin{array}{c} \text{Unit} \\ \text{sales} \\ \text{volume} \end{array} \right] - \begin{array}{c} \text{Total} \\ \text{fixed} \\ \text{costs} \end{array}$$

[6]For those of you who want "proof" of this break-even equation, since the contribution margin per unit is the selling price per unit minus the variable cost per unit, we can substitute the total contribution margin per unit into the profit equation as follows:

$$\text{Profit} = \left[\begin{array}{c} \text{Contribution} \\ \text{margin} \\ \text{per unit} \end{array} \times \begin{array}{c} \text{Unit} \\ \text{sales} \\ \text{volume} \end{array} \right] - \begin{array}{c} \text{Total} \\ \text{fixed} \\ \text{costs} \end{array}$$

Since break-even occurs when profit is zero, we can omit the profit, move the total fixed costs to the other side of the equation, and rewrite the equation as follows:

$$\text{Total fixed costs} = \left[\begin{array}{c} \text{Contribution} \\ \text{margin} \\ \text{per unit} \end{array} \times \begin{array}{c} \text{Unit} \\ \text{sales} \\ \text{volume} \end{array} \right]$$

Finally, we can divide both sides of the equation by the contribution margin per unit to derive the break-even equation:

$$\frac{\text{Total fixed costs}}{\text{Contribution margin per unit}} = \begin{array}{c} \text{Unit sales volume} \\ \text{(to earn zero profit)} \end{array}$$

$$\text{Unit sales volume (to earn zero profit)} = \frac{\text{Total fixed costs}}{\text{Contribution margin per unit}}$$

So for Sweet Temptations, you can tell Anna that the break-even point is 700 boxes of chocolates, computed using the break-even equation as follows (letting X stand for the unit sales volume):

$$\text{Unit sales volume (to earn zero profit)} = \frac{\$3,850 \text{ total fixed costs}}{(\$10 \text{ selling price} - \$4.50 \text{ variable cost}) \text{ per unit}}$$

$$X = \frac{\$3,850}{\$5.50}$$

$$X = \underline{700} \text{ boxes of chocolates}$$

You can verify the break-even sales volume of 700 boxes of chocolates with the following schedule:

Total sales revenue (700 boxes of chocolates @ $10.00 per box)	$7,000
Less: Total variable costs (700 boxes of chocolates @ $4.50 per box)	(3,150)
Total contribution margin (700 boxes of chocolates @ $5.50 per box)	$3,850
Less: Total fixed costs	(3,850)
Profit	$ 0

Finding the Unit Sales Volume to Achieve a Target Profit

Finding the break-even point gives the entrepreneur useful information. However, most entrepreneurs are interested in earning a profit that is high enough to satisfy their goals and the company's goals. A company often states its profit goals at amounts that result in a satisfactory return on the average total assets used in its operations. Since this is an introduction to C-V-P analysis, we will wait to discuss what is meant by "satisfactory return" and "average total assets" until Chapter 11. Here we will assume an amount of profit that is satisfactory. Suppose Anna's goal is that Sweet Temptations earn a profit of $110 per month. How many boxes of chocolates must Sweet Temptations sell per month to earn $110 profit? To answer this question, we slightly modify the break-even equation.

6 How can decision makers predict the sales volume necessary to achieve a target profit?

The break-even point is the sales volume at which the total contribution margin is equal to, or "covers," the total fixed costs. Therefore, each additional unit sold above the break-even sales volume increases profit by the contribution margin per unit. Hence, to find the sales volume at which the total contribution margin "covers" both total fixed costs *and* the desired profit, we can modify the break-even equation simply by adding the desired profit to fixed costs, as follows:

$$\text{Unit sales volume (to earn zero profit)} = \frac{\text{Total fixed costs} + \text{Desired profit}}{\text{Contribution margin per unit}}$$

So, if we let X stand for the unit sales volume, Sweet Temptations needs to sell 720 boxes of chocolates to earn a profit of $110 a month, computed as follows:

$$X = \frac{\$3,850 + \$110}{\$5.50 \text{ per box of chocolates}}$$

$$X = \underline{720} \text{ boxes of chocolates}$$

an verify the $110 profit with the schedule on the following page.

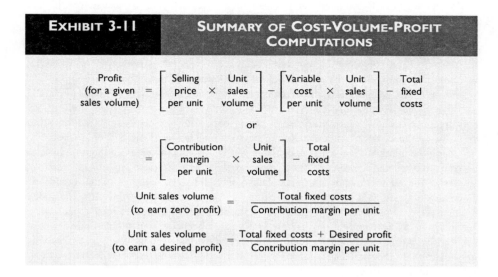

EXHIBIT 3-11	SUMMARY OF COST-VOLUME-PROFIT COMPUTATIONS

$$\text{Profit (for a given sales volume)} = \left[\begin{array}{c}\text{Selling} \\ \text{price} \\ \text{per unit}\end{array} \times \begin{array}{c}\text{Unit} \\ \text{sales} \\ \text{volume}\end{array}\right] - \left[\begin{array}{c}\text{Variable} \\ \text{cost} \\ \text{per unit}\end{array} \times \begin{array}{c}\text{Unit} \\ \text{sales} \\ \text{volume}\end{array}\right] - \begin{array}{c}\text{Total} \\ \text{fixed} \\ \text{costs}\end{array}$$

or

$$= \left[\begin{array}{c}\text{Contribution} \\ \text{margin} \\ \text{per unit}\end{array} \times \begin{array}{c}\text{Unit} \\ \text{sales} \\ \text{volume}\end{array}\right] - \begin{array}{c}\text{Total} \\ \text{fixed} \\ \text{costs}\end{array}$$

$$\begin{array}{c}\text{Unit sales volume} \\ \text{(to earn zero profit)}\end{array} = \frac{\text{Total fixed costs}}{\text{Contribution margin per unit}}$$

$$\begin{array}{c}\text{Unit sales volume} \\ \text{(to earn a desired profit)}\end{array} = \frac{\text{Total fixed costs} + \text{Desired profit}}{\text{Contribution margin per unit}}$$

Total sales revenue (720 boxes of chocolates @ $10.00 per box)	$7,200
Less: Total variable costs (720 boxes of chocolates @ $4.50 per box)	(3,240)
Total contribution margin (720 boxes of chocolates @ $5.50 per box)	$3,960
Less: Total fixed costs	(3,850)
Profit	$ 110

Since Anna had included the desired profit of $110 per month in Sweet Temptations' business plan, the income statement for internal decision makers shown in Exhibit 3-8 is an expanded version of the preceding schedule.

Summary of the C-V-P Analysis Computations

Exhibit 3-11 summarizes the equations that we used in our discussion of C-V-P analysis. Although it may be tempting to try to memorize them, you should strive to understand how these equations relate to one another.

OTHER PLANNING ISSUES

7 How can decision makers use accounting information to evaluate alternative plans?

Providing answers to the previous three questions showed how C-V-P analysis is useful in planning. There are many other planning issues for which C-V-P analysis provides useful information. For instance, suppose you and Anna are considering alternative plans for Sweet Temptations to raise its monthly profit. These plans include:

1. Raising the selling price of the boxes of chocolates to $11 per box. With this alternative, the variable costs per box of chocolates and the total fixed costs do not change.

2. Purchasing a premium line of chocolates rather than the superior line, thus increasing the variable costs to $4.60 per box. You and Anna are considering this alternative because the improvement in the quality of the chocolate may cause the sales volume of boxes of chocolates to increase. With this change, neither the selling price per unit nor the total fixed costs change.

3. Increasing the total fixed costs by spending $110 more on advertising. With this alternative, the selling price per unit and the variable costs per unit do not change, but the additional advertising may cause an increase in sales volume.

How would you modify the graph in Exhibit 3-10 to provide information for Plan #1?

We do not show C-V-P analysis for these three issues at this time because we will discuss similar issues in Chapter 11. We raise these issues here to get you to think about how to use the C-V-P equations or graphs to provide helpful information. The C-V-P analysis for these three alternative plans, however, does not provide all the information you need to make a decision. It is a helpful tool, but it is most effective when used with critical thinking. You must think about the effects each of the alternatives has on your customers.

For instance, each of the alternatives is likely to affect the number of boxes of chocolates that Sweet Temptations can sell. A change in selling price would certainly affect your customers' decisions to purchase boxes of chocolates. A decrease in selling price would bring the boxes of chocolates into the spending range of more people (probably increasing the number of boxes of chocolates you could sell), whereas an increase in selling price may make the boxes of chocolates too expensive for some customers (possibly decreasing the number of boxes of chocolates you could sell). Selling a higher quality of chocolates may attract a different, or additional, group of customers, thus affecting sales volume. Increasing advertising may make more people aware of, and may attract more customers to, Sweet Temptations. Before you make a decision, you should consider how it will affect customers' interest in your product and estimate the probable unit sales volume for each alternative. Then, for whatever sales volume you expect, the analysis can provide a more realistic profit estimate.

WHAT CAN HAPPEN IF A COMPANY DOESN'T HAVE A BUSINESS PLAN[7]

As the last century ended, many Internet companies were in such a rush to join their apparently wildly successful dot-com peers that they forgot one small detail: a sound business plan. By the end of 2001, more than 519 dot-com companies had failed. Some went bankrupt; others had to make radical adjustments to the way they conducted business, including massive layoffs (98,522 employees in 2000). Some such as Art.com and Wine.com began again with new owners and business plans, and still others simply shut down their web sites.

A look back on that period reveals that many of those companies needed huge revenue growths just to break even. For example, the revenue of Tickets.com had increased by 38.77% during 2000—but in order to break even, the company's revenue would have had to grow 606.7%! Nineteen companies had an even more grim situation. The worst was E-Loan, whose revenue had increased by 85.24% during 2000—but whose revenue would have had to grow 5,065.2% in order for the company to break even! Both of these companies had failed by the end of 2000.

What happened? Many company owners, in an effort to compete and attract customers, thought they could start out selling their products for less than what they paid for them and then, after increasing their volume of customers, make up the difference later by raising the selling prices of their products. This means that they started out with negative contribution margins, causing them to lose money right from the start on every sale that they made. (And for some companies, even the low selling prices didn't attract enough customers.) Unfortunately, there was no "later," because these companies ran out of capital or attracted an insufficient number of customers to "make a go of it." Additionally, some owners didn't consider the extremely high costs necessary to run and advertise a web site (particularly the marketing and salary costs), as well as the costs of storing and distributing their companies' products.

A business plan, along with its C-V-P analysis, could have helped the owners of these companies discover the "flaws" in their thinking before their companies got into trouble. With C-V-P analysis it would have been easy for them to confirm that the planned selling

[7]"What detonated dot-bombs?" *USA Today,* December 4, 2000, 2A,2B; "Dot-coms without plans die," *USA Today,* December 4, 2000, 2B; "Dead and (Mostly) Gone," *Fortune,* December 24, 2001, 46,47.

prices of their products initially would not have been high enough, *at any volume,* for the companies to break even. Furthermore, a business plan would have focused the owners' attention on the high marketing and salary costs, and the product storage and distribution costs, thereby helping the owners determine the selling prices that would most likely help their companies break even *and* earn desired profits. A business plan also could have helped the owners see the possible effects on their companies' profits of sales predictions that were too optimistic or too pessimistic.

BUSINESS ISSUES AND VALUES: WASTE NOT, WANT NOT

C-V-P accounting information is one factor that influences business decisions, but entrepreneurs also need to consider the nonfinancial effects of their decisions. For example, if the managers of a company are thinking about lowering the company's total costs by omitting toxic waste cleanup around the factory, they must ask questions such as the following: What will be the impact on the environment? What health effects might the employees suffer later? What might be the health impact on the company's neighbors? Legally, can we even consider not cleaning up the toxic waste? Although omitting toxic waste cleanup may reduce total costs dramatically, these managers might consider the other, more-difficult-to-measure costs to be too high. Therefore, after weighing all the factors surrounding the alternatives, the managers may choose a more socially acceptable alternative that results in a less favorable profit.

SUMMARY

At the beginning of the chapter we asked you several questions. During the chapter, we asked you to STOP and answer some additional questions to build your knowledge about specific issues. Be sure you answered these additional questions. Below are the questions from the beginning of the chapter, with a brief summary of the key points relating to the answers. Use your creative and critical thinking skills to expand on these key points to develop more complete answers to the questions and to determine what other questions you have that might lead you to learn more about the issues.

1 Since the future is uncertain and circumstances are likely to change, why should a company bother to plan?

A business plan helps the owners or managers of a company organize the company, serves as a benchmark against which they can evaluate actual company performance, and helps the company obtain financing. The business plan consists of a description of the company, a marketing plan, a description of the company's operations, and a financial plan. Accounting information contributes to the planning process by providing information for C-V-P analysis and by including in the financial plan the effects that estimated revenues, variable costs, and fixed costs have on the company's profits.

2 What should a company include in its business plan?

A business plan should include a description of the company, a marketing plan, a description of the operations of the company, and a financial plan. The description should include information about the organization of the company, its products or services, its current and potential customers, its objectives, where it is located, and where it conducts business. The marketing plan shows how the company will make sales and how it will influence and respond to market conditions. The company operations section includes a description of the relationships between the company, its suppliers, and its customers, as well as a description of how the company will develop, service, protect, and support its products or services. The financial plan identifies the company's capital requirements and sources of capital, and describes the company's projected financial performance.

3 How does accounting information contribute to the planning process?

Accountants determine how revenues, variable costs, and fixed costs affect profits based on their observations of how costs "behave" and on their estimates of future revenues and costs. By observing cost behavior patterns, accountants are able to classify the costs as fixed or variable, and then to use this classification to predict the amounts of the costs at different activity levels. Accounting information, then, can help decision-makers evaluate alternative plans by using C-V-P analysis to show the profit effect of each plan. C-V-P analysis is a tool that helps managers think critically about the different aspects of each plan.

4 What must decision makers be able to predict in order to estimate profit at a given sales volume?

To estimate profit at a given sales volume, decision makers must be able to predict the product's selling price, the costs that the company will incur, and the behavior of those costs (whether they are fixed or variable costs). The fixed costs will not change because of sales volume, but the variable costs will change directly with changes in sales volume.

5 How can decision makers predict the sales volume necessary for estimated revenues to cover estimated costs?

To predict the sales volume necessary for estimated revenues to cover estimated costs, decision makers must rearrange the profit equation into the break-even equation. Using what they know about the product's selling price and the behavior of the company's costs, the decision makers can determine the contribution margin per unit of product by subtracting the estimated variable costs per unit from the product's estimated selling price. Then they can substitute the contribution margin and the estimated fixed costs into the equation and solve for the necessary sales volume.

6 How can decision makers predict the sales volume necessary to achieve a target profit?

Predicting the sales volume necessary to achieve a target profit is not very different from predicting the sales volume necessary for estimated revenues to cover estimated costs. The only difference is that the decision makers must modify the break-even equation by adding the desired profit to the estimated fixed costs. Then, after substituting the contribution margin and the estimated fixed costs plus the desired profit into the equation, they can solve for the necessary sales volume.

7 How can decision makers use accounting information to evaluate alternative plans?

Decision makers can determine how changes in costs and revenues affect the company's profit. Based on accounting information alone, the alternative that leads to the highest profit will be the best solution. However, decision makers should also consider the nonfinancial effects that their decisions may have.

KEY TERMS

break-even point *(p. 73)*
business plan *(p. 61)*
capital *(p. 65)*
contribution margin per unit *(p. 75)*
cost-volume-profit (C-V-P) analysis
  (p. 68)
fixed costs *(p. 69)*
line of credit *(p. 66)*

long-term capital *(p. 67)*
return *(p. 61)*
risk *(p. 61)*
short-term capital *(p. 66)*
total contribution margin *(p. 75)*
total costs *(p. 72)*
variable cost *(p. 70)*
volume *(p. 68)*

SUMMARY SURFING

Here is an opportunity to gather information on the Internet about real-world issues related to the topics in this chapter. Go to http://www.cunningham.swlearning.com and click on the Interactive Study Center. Click on this chapter number then click on Summary Surfing and answer the following questions.

- Click on **SBA** (U.S. Small Business Administration). Click on *Learn about SBA*. Scroll to the bottom of the page and click on *Search all of SBA*. Type "Strategic Plan" in the search box (be sure to include quotation marks) and click on *search*. Click on *SBA FY2001-FY2006 Strategic Plan*. Look under "How We Make A Difference." What is the mission of the SBA? Go back to "Learn about SBA." What percent of all employers are represented by the SBA? What percent of the gross national product do small businesses generate? How much technological innovation do small businesses invent? Click on *Profile of SBS,* and then look under "Financing." What percent of a loan can the SBA guarantee? What is the maximum interest rate allowed?

- Click on **SBA** (U.S. Small Business Administration). Scroll down, and click on *Starting Your Business;* then, on the left side, click on *Business Plans*. In "Business Plan Outline," what are the three elements (identified by roman numerals) of a business plan? Identify a few components under each section. In "The Business Plan: What It Includes," what are the four distinct sections? How do these compare with what we discussed in the chapter?

INTEGRATED BUSINESS AND ACCOUNTING SITUATIONS

Answer the Following Questions in Your Own Words.

Testing Your Knowledge

3-1   Since the future is uncertain and circumstances are likely to change, why should the managers and owners of a company bother to plan?

3-2   Describe the three main functions of a business plan.

3-3   Describe the components of a business plan. How does each of these components help an investor, a creditor, and a manager or owner make decisions about a company?

3-4   Why is it important for a company to have a cash buffer on hand?

3-5   How can an entrepreneur determine a company's capital requirements?

3-6   What is the difference between short-term and long-term capital?

3-7   Explain what cost-volume-profit analysis is.

3-8   How does cost-volume-profit analysis help entrepreneurs develop their companies' business plans?

3-9   How can you tell whether a cost is a variable cost or a fixed cost?

3-10 What is a contribution margin?

3-11 Explain what it means when a company breaks even.

3-12 Indicate the effect (increase, decrease, no change, or not enough information) that each of the following situations has on break-even unit sales. If you answer "not enough information," list the information that you need in order to be able to determine the effect.
      (a) A retail company purchases price tags to use in place of the stickers it has used in the past.
      (b) An athletic equipment store leases more retail space.

(c) A bakery increases its advertising expense.

(d) A merchandiser plans to increase the selling price of its product. To counter potential decreases in sales, the merchandiser also plans to increase the amount of per-product commission that the sales staff earns.

(e) An accounting firm plans to increase its billing rate per hour.

(f) A retail company has found a supplier that will provide the same merchandise its old supplier provided, but at a lower price.

(g) A private college in the Northwest installs air conditioning in its dormitories.

(h) A retail company reduces advertising expenses and increases the commissions of its sales force.

(i) Instead of having its office building cleaned by a cleaning service, a company plans to hire its own cleaning crew.

3-13 If the total variable cost per unit increases while the selling price per unit, the fixed costs, and the sales volume remain the same, how would you expect the change in variable costs to affect profit? the break-even point?

3-14 If total fixed costs increase while the selling price per unit, the variable costs per unit, and the sales volume remain the same, how would you expect the change in fixed costs to affect profit? the break-even point?

3-15 How does the income statement shown in Exhibit 3-8 help internal decision makers perform cost-volume-profit analysis?

Applying Your Knowledge

3-16 Imagine that you are going to start your own company. Think about the concept for a minute.

Required: What will you call your company? What kind of product or service will you sell? What price will you charge for your product or service? Why? What variable costs and what fixed costs do you think you will incur?

3-17 Suppose you want to start a company that sells sports equipment.

Required: Go to the reference section of your library. What type of information can you find in *Moody's* or *Standard & Poor's* to help you prepare projected financial statements for your company?

3-18 TLC Company sells a single product, a food basket (containing fruit, cheese, nuts, and other items) that friends and family can purchase for college students who need a little extra TLC. This product, called the Exam-O-Rama, sells for $10 per basket. The variable cost is $6 per basket, and the total fixed cost is $24,000 per year.

Required: (1) Draw one graph showing TLC's (a) total revenues and (b) total costs as volume varies. Locate the break-even point on the graph.
(2) What is TLC's profit equation in terms of units sold?
(3) What is TLC's break-even point in units?

3-19 Bathtub Rings Company sells shower-curtain rings for $1.60 per box. The variable cost is $1.20 per box, and the fixed cost totals $20,000 per year.

Required: (1) What is Bathtub Rings' profit equation in terms of boxes of shower-curtain rings sold?
(2) Draw a graph of Bathtub Rings' total contribution margin and total fixed cost as volume varies. Locate the break-even point on this graph.
(3) What is Bathtub Rings' break-even point in units?
(4) What would total profits be if Bathtub Rings sold 500,000 boxes of shower-curtain rings?
(5) How many boxes of shower-curtain rings would Bathtub Rings have to sell to earn $50,000 of profit?

3-20 Go Figure Company sells small calculators for $12 each. This year, Go Figure's fixed cost totals $110,000. The variable cost per calculator is $7.

Required: (1) Compute the break-even point in number of calculators.
(2) Compute the number of calculators required to earn a profit of $70,000.
(3) If the total fixed cost increases to $160,000 next year,
    (a) what will Go Figure's break-even point be in number of calculators?
    (b) what profit (or loss) will Go Figure have if it sells 30,000 calculators?
    (c) how many calculators will Go Figure have to sell to earn a profit of $70,000?

3-21 Silencer Company sells a single product, mufflers for leaf blowers. The company's profit computation for last year is shown here:

Sales revenue (2,000 units @ $25)	$50,000
Less variable costs	(20,000)
Contribution margin	$30,000
Less fixed costs	(22,000)
Profit	$ 8,000

Silencer has decided to increase the price of its product to $30 per muffler. The company believes that if it increases its fixed advertising (selling) cost by $3,400, sales volume next year will be 1,800 mufflers. Variable cost per muffler will be unchanged.

Required: (1) Using the above income statement format, show the computation of expected profit for Silencer's operations next year.
(2) How many mufflers would Silencer have to sell to earn as much profit next year as it did last year?
(3) Do you agree with Silencer's decision? Explain why or why not.

3-22 Rapunzel Company currently sells a single product, shampoo, for $4 per bottle. The variable cost per bottle is $3. Rapunzel's fixed cost totals $6,000.

Required: (1) Compute the following amounts for Rapunzel Company:
    (a) Contribution margin per bottle of shampoo
    (b) Break-even point in bottles of shampoo
    (c) The profit that Rapunzel will earn at a sales volume of 25,000 bottles of shampoo
    (d) The number of bottles of shampoo that Rapunzel must sell to earn a profit of $16,000
(2) Rapunzel is considering increasing its total fixed cost to $8,000 and then also increasing the selling price of its product to $5. The variable cost per bottle of shampoo would remain unchanged. Repeat the computations from (1), using this new information. Will this decision be a good one for Rapunzel? Why or why not?
(3) Draw a graph with four lines to show the following:
    (a) Total contribution margin earned when Rapunzel sells from 0 to 10,000 bottles of shampoo at a selling price of $4 per bottle
    (b) Total contribution margin earned when Rapunzel sells from 0 to 10,000 bottles of shampoo at a selling price of $5 per bottle
    (c) Rapunzel's fixed cost total of $6,000
    (d) Rapunzel's fixed cost total of $8,000
    (e) Rapunzel's break-even point in bottles of shampoo before and after the selling price and fixed cost changes
(4) Does the graph support your conclusion in (2) above? If so, how does it support your conclusion? If not, what new or different information did you get from the graph?

3-23 The Body Shop Equipment Company sells a small, relatively lightweight multipurpose exercise machine. This machine sells for $700. A recent cost analysis shows that The Body Shop's cost structure for the coming year is as follows:

Variable cost per unit	$ 325
Total annual fixed costs	125,000

Required: (1) Draw a graph that clearly shows (a) total fixed cost, (b) total cost, (c) total sales revenue, and (d) total contribution margin as the sales volume of exercise machines increases. Locate the break-even point on the graph.

(2) Compute the break-even point in number of machines.

(3) How many machines must the Body Shop sell to earn $30,000 of profit per year?

(4) How much profit would be earned at a sales volume of $420,000?

(5) Sean McLean, the owner of the Body Shop Equipment Company, is considering traveling a circuit of gyms and fitness centers around the United States each year to demonstrate the exercise machine, distribute information, and obtain sales contracts. He estimates that this will cost about $6,000 per year. How many additional exercise machines must the company sell per year to cover the cost of this effort?

3-24 Lady MacBeth Company sells bottles of dry cleaning solvent (spot remover) for $10 each. The variable cost for each bottle is $4. Lady MacBeth's total fixed cost for the year is $3,600.

Required: (1) Answer the following questions about the company's break-even point.

(a) How many bottles of spot remover must Lady MacBeth sell to break even?

(b) How would your answer to (1a) change if Lady MacBeth lowered the selling price per bottle by $2? What if, instead, it raised the selling price by $2?

(c) How would your answer to (1a) change if Lady MacBeth raised the variable cost per bottle by $2? What if, instead, it lowered the variable cost by $2?

(d) How would your answer to (1a) change if Lady MacBeth increased the total fixed cost by $60? What if, instead, Lady MacBeth decreased the total fixed cost by $60?

(2) Answer the following questions about the company's profit.

(a) How many bottles must Lady MacBeth sell to earn $4,800 profit?

(b) How would your answer to (2a) change if Lady MacBeth lowered the selling price per bottle by $2?

(c) Suppose that for every $1 the selling price per bottle decreases below its current selling price of $10 per bottle, Lady MacBeth predicts sales volume will increase by 325 bottles. Assume that before lowering the selling price, Lady MacBeth predicts that it can sell exactly 1,400 bottles. Can Lady MacBeth earn $4,800 profit by lowering the selling price per bottle by $2? Explain why or why not.

(d) Suppose that for every $1 the selling price per bottle increases above its current selling price of $10 per bottle, Lady MacBeth predicts sales volume will decrease by 200 bottles. Assume that before raising the selling price, Lady MacBeth predicts that it can sell exactly 1,400 bottles. Can Lady MacBeth earn $4,800 profit by raising the selling price per bottle by $2? Explain why or why not.

(e) How would your answer to (2a) change if Lady MacBeth raised the variable cost per bottle by $2? What if, instead, it lowered the variable cost per bottle by $2?

(f) How would your answer to (2a) change if Lady MacBeth raised the total fixed cost by $60? What if, instead, Lady MacBeth lowered the total fixed cost by $60?

3-25 The Brickhouse Company is planning to lease a delivery van for its northern sales territory. The leasing company is willing to lease the van under three alternative plans:

Plan A—Brickhouse would pay $0.34 per mile and buy its own gas.

Plan B—Brickhouse would pay $320 per month plus $0.10 per mile and buy its own gas.

Plan C—Brickhouse would pay $960 per month, and the leasing company would pay for all gas.

The leasing company will pay for all repairs and maintenance, insurance, license fees, and so on. Gas should cost $0.06 per mile.

Required: Using miles driven as the units of volume, do the following:
   (1) Write out the cost equation for the cost of operating the delivery van under each of the three plans.
   (2) Graph the three cost equations on the same graph (put cost on the vertical axis and miles driven per month on the horizontal axis).
   (3) Determine at what mileage per month the cost of Plan A would equal the cost of Plan B.
   (4) Determine at what mileage per month the cost of Plan B would equal the cost of Plan C.
   (5) Compute the cost, under each of the three plans, of driving 3,500 miles per month.

3-26 The Mallory Motors Company sells small electric motors for $2 per motor. Variable costs are $1.20 per unit, and fixed costs total $60,000 per year.

Required: (1) Write out Mallory's profit equation in terms of motors sold.
   (2) Draw a graph of Mallory's total contribution margin and total fixed cost as volume varies. Locate the break-even point on this graph.
   (3) Compute Mallory's break-even point in units.
   (4) What total profit would Mallory expect if it sold 500,000 motors?
   (5) How many motors would Mallory have to sell to earn $40,000 profit?

3-27 The Campcraft Company is a small manufacturer of camping trailers. The company manufactures only one model and sells the units for $2,500 each. The variable costs of manufacturing and selling each trailer are $1,900. The total fixed cost amounts to $180,000 per year.

Required: (1) Compute Campcraft's contribution margin per trailer.
   (2) Compute Campcraft's profit (or loss) at a sales volume of 160 trailers.
   (3) Compute the number of units that Campcraft must sell for it to break even.
   (4) Compute the number of units that Campcraft must sell for it to earn a profit of $30,000.

3-28 This year Babco's fixed costs total $110,000. The company sells babushkas for $13 each. The variable cost per babushka is $8.

Required: (1) Compute the break-even point in number of babushkas.
   (2) Compute the number of babushkas that Babco must sell to earn a profit of $70,000.
   (3) If the total fixed cost increases to $150,000 next year,
      (a) what will be Babco's break-even point in babushkas?
      (b) what profit (or loss) will Babco have if it sells 28,000 babushkas?
      (c) how many babushkas will Babco have to sell to earn a profit of $70,000?

3-29 The Cardiff Company sells a single product for $40 per unit. Its total fixed cost amounts to $360,000 per year, and its variable cost per unit is $34.

Required: (1) Compute the following amounts for the Cardiff Company:
      (a) Contribution margin per unit
      (b) Break-even point in units
      (c) The number of units that must be sold to earn $30,000 profit
   (2) Repeat all computations in (1), assuming Cardiff decides to increase its selling price per unit to $44. Assume that the total fixed cost and the variable cost per unit remain the same.

Making Evaluations

3-30 Suppose your wealthy Aunt Gert gave you and your cousins $10,000 each. Assume for a moment that you are not associated with Sweet Temptations and that you are considering loaning the $10,000 to Sweet Temptations.

Required: From the information included in Sweet Temptations' business plan so far, do you think this would be a wise investment on your part? Why or why not? What else would you like to know before making a decision (you don't have to limit your thinking to Sweet Temptations)?

3-31 Refer to 3-30. What if Aunt Gert instead gave you $100,000 and you were interested in investing it in Sweet Temptations?

Required: Would this change your answers to 3-30? Why or why not?

3-32 Joe Billy Ray Bob's Country and Western Company sells a single product—cowboy hats—for $24 per hat. The total fixed cost is $180,000 per year, and the variable cost per hat is $15.

Required: (1) Compute the following amounts for Joe Billy Ray Bob's Country and Western Company:
  (a) Contribution margin per hat
  (b) Break-even point in hats
  (c) The numbers of hats that must be sold to earn $27,000 of profit
(2) Repeat all computations in (1), assuming Joe Billy Ray Bob's decides to increase its selling price per hat to $25. Assume that the total fixed cost and the variable cost per hat remain the same.
(3) Do you agree with Joe Billy Ray Bob's decision to increase its selling price per hat? What other factors should the managers consider in making this decision?

3-33 The Vend-O-Bait Company operates and services bait vending machines placed in gas stations, motels, and restaurants surrounding a large lake. Vend-O-Bait rents 200 machines from the manufacturer. It also rents the space occupied by the machines at each location where it places the machines. Arnie Bass, the company's owner, has two employees who service the machines. Monthly fixed costs for the company are as follows:

Machine rental:
  200 machines × $100 per month................................. $20,000
Space rental:
  200 locations × $60 per month.................................... 12,000
Employee wages:
  2 employees × $800 per month.................................... 1,600
Other fixed costs... 2,400
Total $36,000

Currently, Vend-O-Bait's only variable costs are the costs of the night crawlers, which it purchases for $1.20 per pack. Vend-O-Bait sells these night crawlers for $1.80 per pack.

Required: (1) Answer the following questions:
  (a) What is the monthly break-even point (in packs sold)?
  (b) Compute Vend-O-Bait's monthly profit at monthly sales volumes of 52,000, 56,000, 64,000, and 68,000 packs, respectively.
(2) Suppose that instead of paying $60 fixed rent per month, Arnie Bass could arrange to pay $0.20 for each pack of night crawlers sold at each location to rent the space occupied by the machines. Repeat all computations in (1).
(3) Would it be desirable for Arnie Bass to try to change his space rental from a fixed cost ($60 per location) to a variable cost ($0.20 per pack sold)? Why or why not?

3-34 Refer to 3-24. Suppose your boss at Lady MacBeth is considering some alternative plans and would like your input on the following three independent alternatives:
(a) Increase the selling price per bottle by $3
(b) Decrease the variable cost per bottle by $2 by purchasing an equally effective, but less "environmentally friendly," solvent from your supplier
(c) Decrease the total fixed cost by $1,260.

Assume again that Lady MacBeth currently sells bottles of dry cleaning solvent for $10 each, the variable cost for each bottle is $4, and the total fixed cost for the year is $3,600.

Required: (1) How many bottles would Lady MacBeth have to sell to break even under each of the alternatives? Using this accounting information alone, write your boss a memo in which you recommend an alternative.

(2) Maybe your boss would like to earn a profit of $4,320. How many bottles would Lady MacBeth have to sell to earn a profit of $4,320 under each of the alternatives? Which of the three alternatives would you recommend to your boss? Is this consistent with your recommendation in (1)? Why or why not? What other issues did you consider when making your recommendation?

http://www.isuzu.co.jp/world/
index.htm

3-35   Japan's **Isuzu Motors Ltd.**, manufacturer of trucks and sports utility vehicles (SUVs), announced in May 2001 that it planned to eliminate 9,700 jobs (26% of its worldwide work force) over the next three years. Isuzu planned to achieve these job cuts through normal attrition, a freeze on hiring, and an early-retirement program. The company had experienced years of losses but predicted that it would earn a profit as early as the following year.[8]

Required: (1) What effect would you expect the decision to have on Isuzu's break-even point? on the number of trucks and SUVs Isuzu would have to sell to earn a desired profit?

(2) What nonfinancial issues do you think Isuzu's owners had to resolve in order to make this decision?

(3) What questions do you think the owners had to answer in order to resolve these issues?

3-36   Suppose you work for the Miniola Hills Bus Company. The company's 10 buses made a total of 80 trips per day on 310 days last year, for a total of 350,000 miles. Another year like last year will put the company out of business (and you out of a job!). Your boss has come to you for help. Last year, instead of earning a profit, the company lost $102,000, as shown here:

Revenue from riders (496,000 @ $0.50)		$248,000
Less operating costs:		
Depreciation on buses	$100,000	
Garage rent	20,000	
Licenses, fees, and insurance	40,000	
Maintenance	15,000	
Drivers' salaries	65,000	
Tires	20,000	
Gasoline and oil	90,000	(350,000)
Loss		($102,000)

Your boss is considering the following two plans for improving the company's profitability:

(a) Plan A—change the bus routes and reduce the number of trips to 60 per day in order to reduce the number of miles driven

(b) Plan B—sell bus tokens (five for $1.00) and student passes ($2.50 to ride all week) in order to increase the number of riders

Required: (1) Write your boss a memo discussing the effect that each of these plans might have on the costs and revenues of the bus company. Identify in your memo any assumptions you have made.

(2) If you were making this decision, what questions would you like answered before making the decision?

---

[8]cnnfn.cnn.com/news/specials/layoffs, May 28, 2001.

3-37   Yesterday, you received the following letter for your advice column in the local paper:

DR. DECISIVE

Dear Dr. Decisive:

What do you think about this situation? My boyfriend refuses to meet me for lunch until I admit I am wrong about this, which I'm NOT. The other day, when we went to lunch at Subs and Floats on campus, he noticed that they had raised the price of BLT subs. He got mad because he thinks the only reason they raised the price was to increase their profit. I told him that, first of all, their profit might not increase and that, second, he was basing his conclusion on some assumptions that might not be true and that if he would just *open up his mind,* he might see how those assumptions are affecting his conclusion. Well, *then* he got mad at *me.* I'm really upset because I know I'm right and because now I have to buy my own lunch. Will you please explain why I'm right? I know he'll listen to you (he reads your column daily). Until you answer, I'll be

"Starving."

Required: Meet with your Dr. Decisive team and write a response to "Starving."

THE ACCOUNTING SYSTEM: CONCEPTS AND APPLICATIONS

Read All

"PROFITS ARE THE MECHANISM BY WHICH SOCIETY DECIDES WHAT IT WANTS TO SEE PRODUCED."

—HENRY C. WALLICH

1. Why do managers, investors, creditors, and others need information about a company's operations?

2. What are the basic concepts and terms that help identify the activities that a company's accounting system records?

3. What do users need to know about the accounting equation for a company?

4. Why are at least two effects of each transaction recorded in a company's accounting system?

5. What are revenues and expenses, and how is the accounting equation expanded to record these items?

6. What are the accounting principles and concepts related to net income?

7. Why are end-of-period adjustments necessary?

D
o you have a "system" for keeping track of your financial activities? Do you plan your monthly cash receipts and payments by using a budget? When you get your paycheck, do you always review it to be sure you have been paid correctly for the hours that you worked? Do you record every check that you write, and keep a running total of the amount you have in your checkbook? Do you balance your checkbook at the end of each month by comparing it to your bank statement? When you "charge" something on your credit card, do you always check the amount of the receipt before you sign it? Do you keep your credit card receipts and compare them to the charges on your monthly credit card statement before you pay your bill? When you pay your landlord, do you always pay your rent at the beginning of the month? Do you have your bank automatically deduct your car payments from your savings account? Do you pay for your car insurance soon after you get the bill? At the end of each month, do you compare your actual receipts and payments to what you budgeted to see how you stand?

After you graduate, you may want to become a manager of a company. As we discussed in Chapters 3 and 4, accounting methods, such as C-V-P analysis and budgeting, help managers carry out the planning, operating, and evaluating activities of a company. However, managers must also keep track of the company's operations in order to evaluate its performance (and also their own performance as managers). Managers develop and use a company's accounting system for this purpose. For example, a company's accounting system shows managers whether the company sold as many goods as it expected and whether it stayed within its budgets.

Managers are not the only people interested in the operations of a company. External users need information about the company's operations to help them decide whether to do business with a company. In this chapter we discuss the role of financial accounting in decision making and explain the basics of the financial accounting process.

FINANCIAL ACCOUNTING INFORMATION AND DECISION MAKING

Let's return to our discussion of Unlimited Decadence Corporation, the candy bar manufacturer, to see why external users need accounting information. As you can imagine, it takes lots of sugar to make candy bars. Suppose for a moment that you are the president of Sugar Supply Company and that Unlimited Decadence Corporation is considering a purchase of sugar. Unlimited Decadence wants to make bulk purchases on credit and pay for them 30 days later when it has collected money from its candy bar customers.

 As president of Sugar Supply Company, how would you initially react to this request? Why? What facts may change your mind?

Although your immediate response may be to sell the sugar to Unlimited Decadence, you should think carefully before agreeing to the credit arrangement. Certainly, companies like to make sales. However, increasing sales by extending credit is a good decision only if a company is reasonably sure that its credit customers will pay their bills. If Unlimited Decadence doesn't pay its bills, Sugar Supply Company will have given up some of its resources and have nothing to show for it.

The four-step problem-solving process we discussed in Chapter 2 provides an excellent framework for analyzing this credit decision. You already did the first step—recognizing that the problem is to decide whether to sell sugar to Unlimited Decadence Corporation on credit. You now can take the second step—identifying your company's alternatives. You might decide not to extend credit to Unlimited Decadence, to extend credit under more strict or more lenient terms, or to agree to the original request.

The third step, evaluating each alternative by weighing its advantages and disadvantages, helps you decide which alternative best helps your company meet its goals of remaining solvent and earning a satisfactory profit. The alternative that you choose will depend, in part, on your company's ability to extend credit and on its existing credit policies. When

1 Why do managers, investors, creditors, and others need information about a company's operations?

you perform this step, financial accounting information about Unlimited Decadence plays a big role, helping you determine how good a customer Unlimited Decadence will be.

Exhibit 5-1 shows a simplified income statement and a simplified balance sheet for Unlimited Decadence for the first quarter of 2004.[1] When you analyze these financial statements, you learn from the income statement that during the quarter, Unlimited Decadence earned $18,100,000 of revenues from selling candy bars and made $720,000 net income. From the balance sheet, you learn that on March 31, 2004, Unlimited Decadence had $1,200,000 cash in the bank, inventories of $1,300,000 , and other assets (e.g., trucks, factory, etc.) totaling $16,800,000, and that it owed $3,000,000 to suppliers and $2,000,000 to the bank. Each of these items should affect the specific credit terms, if any, that you are willing to offer. After evaluating the alternatives, you are ready to make a decision about Unlimited Decadence's credit request.

This is just one example of how financial accounting information can help external decision makers choose whether or not to do business with a company. Another example is when a banker studies a company's financial statements to decide the conditions for granting a loan. Businesspeople routinely make decisions like these. In each case, financial statements provide information that is important in solving business problems.

Making good decisions based on information in financial statements assumes that there is agreement about what is included in those statements and how the amounts are

EXHIBIT 5-1	INCOME STATEMENT AND BALANCE SHEET

UNLIMITED DECADENCE
Income Statement
For Quarter Ended March 31, 2004
(in thousands of dollars)

Revenues:		
Sales revenue		$18,100
Expenses:		
Cost of candy bars sold	$11,500	
Selling expenses	3,460	
General and administrative expenses	1,940	
Total expenses		(16,900)
Income before income taxes		$ 1,200
Income tax expense		(480)
Net Income		$ 720

UNLIMITED DECADENCE
Balance Sheet
March 31, 2004
(in thousands of dollars)

Assets		Liabilities	
Cash	$ 1,200	Accounts payable (suppliers)	$ 3,000
Inventories	1,300	Notes payable (bank)	2,000
Other assets	16,800	Total Liabilities	$ 5,000
		Stockholders' Equity	
		Total Stockholders' Equity	$14,300
		Total Liabilities and	
Total Assets	$19,300	Stockholders' Equity	$19,300

---

[1]For simplicity, we assume here that Unlimited Decadence sells only one type of candy bar. We will relax this assumption in later chapters. Furthermore, Unlimited Decadence Corporation's actual financial statements have many more items, which we don't show here because you have not studied them yet. We will show more complete financial statements in later chapters.

measured. Without agreement on what accounting information the balance sheet, income statement, and cash flow statement should contain, the statements would be essentially useless. If every company defined and measured financial statement items such as assets, liabilities, revenues, and expenses differently, there would be no way to compare one company's information with another's.

 What difficulties do you think would be caused if each state defined traffic laws differently (e.g., laws stipulating on which side of the road to drive) and used different traffic signs?

The "generally accepted accounting principles" (GAAP) that we mentioned in Chapter 1 were developed to overcome this problem by setting rules for companies to follow when they prepare financial statements. Thus, if you know that a company's financial statements are prepared according to GAAP and you know what rules are included in GAAP, you can confidently use the information in its financial statements for your decision making. The rest of this chapter will give you a basic understanding of the financial accounting process. This process provides the information a company needs for preparing financial statements according to GAAP. Once you have learned some fundamental concepts, we will discuss how the accounting process accumulates and reports information about a company's activities.

BASIC CONCEPTS AND TERMS USED IN ACCOUNTING

Several basic concepts and terms help us identify the activities that a company's accounting process records:

2 What are the basic concepts and terms that help identify the activities that a company's accounting system records?

1. Entity concept
2. Transactions
3. Source documents
4. Monetary unit concept
5. Historical cost concept

Each of these items is important for understanding the process of accumulating and reporting information about a company's activities.

Entity Concept

As you saw in Chapter 1, there are three broad forms of companies—sole proprietorships, partnerships, and corporations. Regardless of a company's form, its accounting records must remain separate from those of its owner, or owners. Even though Anna Cox is the sole proprietor of Sweet Temptations, she doesn't consider her personal assets as belonging to Sweet Temptations, nor does she consider Sweet Temptations' assets to be hers. If Anna owns all or part of several companies, she will keep separate records for each of them. This separation is the basis of the entity concept. An **entity** is considered to be separate from its owners and from any other company. Thus, each company is an entity and has its own accounting system and accounting records. An owner's personal financial activities are *not* included in the accounting records of the company unless this activity has a *direct* effect on the company. For instance, if Anna Cox buys a car only for personal use, its purchase would *not* affect Sweet Temptations' accounting records. On other hand, if Anna uses personal funds to buy a delivery van to be used in the company, the purchase *would* affect the company's records.

Why do you think it is important to treat each entity separately?

Combining company-related items and personal items would make it hard to tell which items are intended for business purposes and which items are for personal use. External users interested in a company's activities would gain little information if you gave them financial statements that included the combined items. With a separate accounting system for a company, it is much easier to identify, measure, and record company activities and to prepare financial statements for the company. Therefore, the financial statements provide more useful information to managers and external users for evaluating the effectiveness of the company's operations.

Transactions

Recall from Chapter 1 that accounting involves the identification, measurement, recording, accumulation, and communication of a company's economic information for use in decision making. The accounting process usually begins with a business transaction. A **transaction** is an exchange of property or service by a company with another entity. For example, when Unlimited Decadence Corporation purchases sugar, it exchanges cash (or the promise to pay cash) for the ingredients needed to make candy. Many events or activities of a company may be described as transactions. Someone, such as the company's accountant or owner, initially records these transactions based on information from source documents.

Source Documents

A **source document** is a business record used as evidence that a transaction has occurred. A source document may be a company check, a sales receipt, a bill from a supplier, a bill sent to a customer, a payroll time card, or a log of the miles driven in the company's delivery truck. Although a company's accounting process begins when a transaction occurs, the identification, measurement, and recording of information are based on an analysis of the related source document. For instance, the check that Unlimited Decadence writes to pay for a bulk purchase of sugar shows the date of the transaction, the dollar amount, the name of the company to whom the check was written (called the *payee*), and possibly the reason for the check. Several source documents may be used as evidence of a single transaction. In addition to the canceled check, source documents for Unlimited Decadence's sugar purchase include the sales invoice from the sugar supplier and the report from the loading dock stating that the sugar arrived at Unlimited Decadence's factory.

Monetary Unit Concept

The source documents for transactions show the value of the exchange in terms that both internal and external users agree on and understand. Since the purpose of recording and analyzing a company's transactions is to understand the company's financial activities, it makes sense to record transactions in terms of money. This idea is known as the **monetary unit concept**. In the United States the monetary unit is the dollar, and therefore U.S.-based companies show their financial statements in dollars. The monetary unit used depends on the national currency of the company's country. For example, Sony uses the Japanese yen, while Volkswagen and Benetton use the European Union euro.

 If a company does business in several countries, what currency do you think it uses to prepare its financial statements? Why? Since Jaguar is owned by Ford, do you think its financial statements are prepared in British pounds or U.S. dollars?

Historical Cost Concept

As we all know, the value of every country's currency changes as a result of inflation. Also, the values of particular goods and services change in the marketplace as supply and demand change. So a company has to decide whether to adjust the recorded amounts to

If Unlimited Decadence purchases a Volkswagen from this dealership in Germany, do you think this candy maker should record the purchase in dollars or euros? Why?

include these types of changes. Under generally accepted accounting principles in the United States, companies generally do *not* record the change in the value of either the currency or the individual goods and services. Instead, they use the historical cost concept or, simply, the cost concept. The **historical cost concept** states that a company records its transactions based on the dollars exchanged (the cost) at the time the transaction occurred. The related source documents show this cost, and the company's accounting records continue to show the *cost* involved in each transaction regardless of whether the *value* of the property or service owned increases (or decreases) or whether the *value* of the currency changes over time. For instance, suppose that a company acquires land for $100,000 and that a year later the value of the land has increased to $130,000. Under the historical cost concept, the company continues to show the land in its accounting records at $100,000, the acquisition cost. However, later in the book you will see that companies do adjust some assets for changes in their values.

 Why do you think most accountants wouldn't want to change the recorded value for the land from $100,000 to $130,000? Why might some want to change?

In Exhibit 5-2 we combine the entity concept, the monetary unit concept, and the historical cost concept to develop Unlimited Decadence Corporation's balance sheet that we showed in Exhibit 5-1. In this balance sheet, we (a) separate company items from personal items according to the *entity concept*, and (b) use the *monetary unit concept* and the *historical cost concept* to show dollar values for each company-related item.

These three concepts are the foundation of what the accounting process shows. With this in mind, you can see how that process functions. The accountant, or the owner, uses the *entity concept* to separate the activities of a company from the owner's activities, which are not related to the company. The company's *transactions* are identified by analyzing *source documents*. The accountant or owner then enters the transactions into the company's accounting records using *monetary units* based on the *costs* involved in its activities. Every time a company activity occurs, the accountant or owner uses these concepts to help decide the proper way to record that activity. After the economic information about a company's activities is recorded and accumulated, the ultimate goal of the accounting process is to communicate this information in the company's balance sheet, its income statement, and its cash flow statement, each prepared according to GAAP.

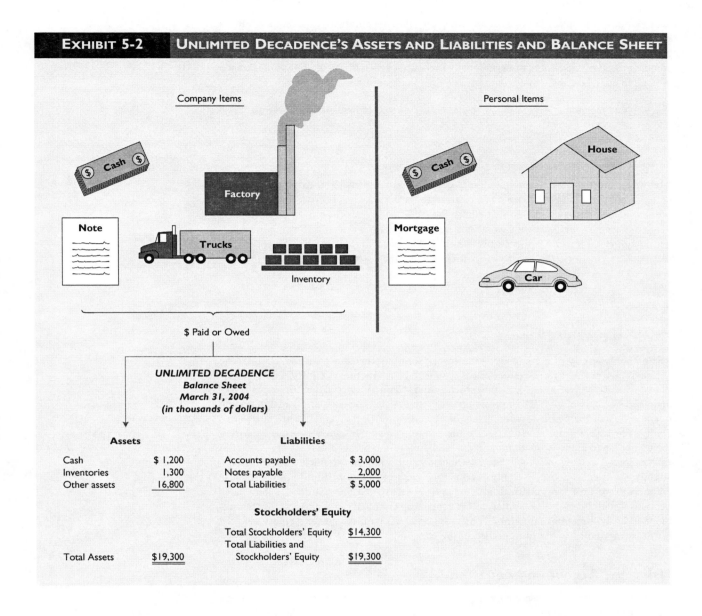

EXHIBIT 5-2   UNLIMITED DECADENCE'S ASSETS AND LIABILITIES AND BALANCE SHEET

Company Items

Personal Items

Cash

Factory

Note

Trucks

Inventory

Cash

House

Mortgage

Car

$ Paid or Owed

UNLIMITED DECADENCE
Balance Sheet
March 31, 2004
(in thousands of dollars)

Assets		Liabilities	
Cash	$ 1,200	Accounts payable	$ 3,000
Inventories	1,300	Notes payable	2,000
Other assets	16,800	Total Liabilities	$ 5,000
		Stockholders' Equity	
		Total Stockholders' Equity	$14,300
		Total Liabilities and	
Total Assets	$19,300	Stockholders' Equity	$19,300

COMPONENTS OF THE ACCOUNTING EQUATION

We can now begin to discuss how the **accounting system** works—the process used to identify, measure, record, and retain information about a company's activities so that the company can prepare its financial statements. This process is based on the three sections of a balance sheet: the asset, liability, and owner's equity sections. Every time a company records the exchange of property or services with another party, the transaction affects at least one of the sections of the balance sheet. So before moving on, consider the following expanded definitions of assets, liabilities, and owner's equity.

Assets

Assets are a company's economic resources that will provide future benefits to the company. A company may own many assets, some of which are physical in nature—such as land, buildings, supplies to be used, and inventory that the company expects to sell to its customers. Other assets do not have physical characteristics but are economic resources because of the legal rights (benefits) they convey to the company. These assets include amounts owed by customers to the company (**accounts receivable**), the right to insurance protection (**prepaid insurance**), and investments made in other companies.

 Can you think of more examples of assets? How do each of these examples meet the definition of assets?

Liabilities

Liabilities are the economic obligations (debts) of a company. The external parties to whom a company owes the debts are referred to as the **creditors** of the company. Liabilities include amounts owed to suppliers for credit purchases (**accounts payable**) and amounts owed to employees for work they have done (**wages and salaries payable**). Legal documents are often evidence of liabilities. These documents establish a claim (**equity**) by the creditors (the **creditors' equity**) against the assets of a company.

 Can you think of more examples of liabilities? How do each of these examples meet the definition of liabilities?

Owner's Equity

The **owner's equity** of a company is the owner's current investment in the assets of the company. (A partnership's balance sheet would refer to **partners' equity**, and a corporation's balance sheet would call this **stockholders' equity**, as you saw in Exhibits 5-1 and 5-2.) The capital invested in the company by the owner, the company's earnings from operations, and the owner's withdrawals of capital from the company all affect owner's equity. For a sole proprietorship, the balance sheet shows the owner's equity by listing the owner's name, the word *capital,* and the amount of the owner's current investment in the company. As you will see later, partners' equity and stockholders' equity appear slightly differently. Owner's equity is sometimes referred to as **residual equity** because creditors have first legal claim to a company's assets. Once the creditors' claims have been satisfied, the owner is entitled to the remainder (residual) of the assets. Sometimes the total of the liabilities (creditors' equity) is combined with the owner's equity, and the result is referred to as the **total equity** of the company.

Using the Accounting Equation

In summary, accountants use the term *assets* to refer to a company's economic resources, and they use the terms *liabilities* and *owner's equity* to describe claims on those resources. All of a company's economic resources are claimed by either creditors or owners. Therefore, the financial accounting system is built on a simple equation:

3 What do users need to know about the accounting equation for a company?

Economic Resources = Claims on Economic Resources

Using the accounting terms you have learned, we can restate the equation:

Assets = Liabilities + Owner's Equity

ematical expression is known as the basic **accounting equation**. The equal- to the liabilities plus owner's equity is the reason a company's statement ition is called a *balance* sheet: the monetary total for the economic re- of the company must always be *in balance* with the monetary total for ies + owner's equity) on the economic resources. Like the components ion, the components of this equation may be transposed. Another way ation is shown on the next page.

Assets − Liabilities = Owner's Equity

In this form of the equation, the left-hand side (i.e., assets minus liabilities) is referred to as **net assets**. This form of the equation also stresses that owner's equity may be thought of as a residual amount. Regardless of what form the equation takes, it must always balance. Because a transaction normally begins the accounting process, a company must record each transaction in a way that maintains this equality. Keeping this equality in mind will help you understand other aspects of the accounting process.

The Dual Effect of Transactions

4 Why are at least two effects of each transaction recorded in a company's accounting system?

To keep the accounting equation in balance, *a company must make at least two changes in its assets, liabilities, or owner's equity* when it records each transaction. This is called the **dual effect of transactions**. For instance, when an owner invests $20,000 in a company, assets (cash) are increased by $20,000 and owner's equity (owner's capital) is increased by $20,000. This transaction causes two changes—one change in the asset section of the company's balance sheet and one change in the owner's equity section of its balance sheet. Because the left-hand side *and* the right-hand side both increase by the same amount, the accounting equation (assets = liabilities + owner's equity) stays in balance.

The fact that transactions always have a dual effect does not mean that every transaction will affect both sides of the equation—or even two components of the equation. A transaction may affect only one side, by increasing one asset and decreasing another asset by the same amount. For example, assume a company buys office equipment by paying $400 cash. In this case, the asset Office Equipment increases by $400 and the asset Cash decreases by $400. The accounting equation still balances after the company records this transaction because the transaction does not affect the right side of the equation and because the *total* for the asset (left) side of the equation is not changed.

To understand how the accounting equation and the dual effect of transactions provide structure to a company's accounting system, think about these concepts as describing a company's "transaction scales." Exhibit 5-3 shows a set of "transaction scales." Instead

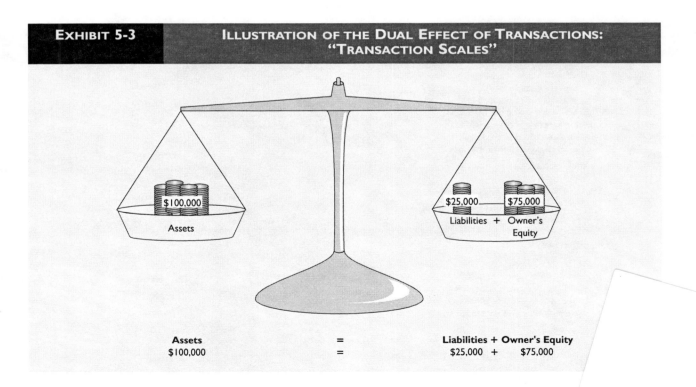

EXHIBIT 5-3   **ILLUSTRATION OF THE DUAL EFFECT OF TRANSACTIONS: "TRANSACTION SCALES"**

$100,000
Assets

$25,000 _____ $75,000
Liabilities + Owner's Equity

Assets	=	Liabilities + Owner's Equity
$100,000	=	$25,000 + $75,000

of measuring the weight of various objects, using ounces or pounds as measuring units, these scales measure transactions, using dollars (historical cost monetary units). Suppose a company currently has assets of $100,000, liabilities of $25,000, and owner's equity of $75,000. Assume that the company's accountant or owner "places" the company's current economic resources (assets of $100,000) on the left side of the scales and "places" current claims on those resources (liabilities of $25,000 + owner's equity of $75,000) on the right side of the scales. Remember, after each transaction the scales must balance. The dual effect of transactions provides a way to keep the scales in balance as company activities are placed (recorded) on the scales. Note that in Exhibit 5-3 the left side of the scales holds $100,000 in total assets and the right side holds $25,000 in liabilities and $75,000 in owner's equity. The scales balance according to the accounting equation:

Assets   = Liabilities + Owner's Equity
$100,000 = $25,000   + $75,000

As we stated earlier, regardless of the type of transaction that occurs, the accounting equation, like our set of transaction scales, must always balance. Exhibits 5-4 and 5-5 use

---

EXHIBIT 5-4   "TRANSACTION SCALES": INCREASE IN ASSETS AND OWNER'S EQUITY

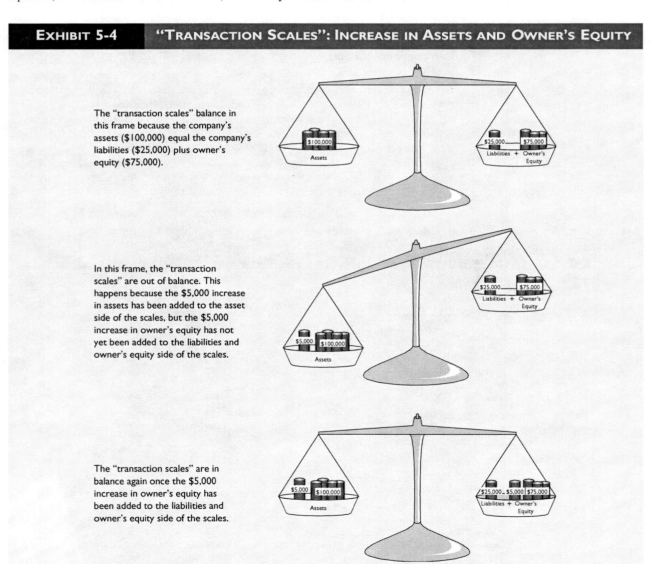

The "transaction scales" balance in this frame because the company's assets ($100,000) equal the company's liabilities ($25,000) plus owner's equity ($75,000).

In this frame, the "transaction scales" are out of balance. This happens because the $5,000 increase in assets has been added to the asset side of the scales, but the $5,000 increase in owner's equity has not yet been added to the liabilities and owner's equity side of the scales.

The "transaction scales" are in balance again once the $5,000 increase in owner's equity has been added to the liabilities and owner's equity side of the scales.

EXHIBIT 5-5	"TRANSACTION SCALES": DECREASE IN ASSETS AND LIABILITIES

The "transaction scales" balance in this frame because the company's assets ($105,000) equal the company's liabilities ($25,000) plus owner's equity ($80,000).

The "transaction scales" do not balance in this frame because, although the $20,000 decrease in assets has been taken off the scales, the liabilities have not yet been changed.

The "transaction scales" are in balance again once the $20,000 decrease in liabilities is removed from the liabilities and owner's equity side of the scales.

the scales to illustrate two more transactions. In Exhibit 5-4, you can see what happens when the company's owner deposits $5,000 *from personal funds* into the company's bank account. The first frame shows the accounting equation in balance before the owner's deposit. The second frame shows the equation *out of balance* because only one change, the $5,000 increase in company assets on the left side of the equation, has been recorded. In the last frame, the scales again balance, showing that the $5,000 owner's equity increase on the right side of the equation has been recorded. After this transaction, the accounting equation is as follows:

$$\textbf{Assets} \quad \textbf{= Liabilities + Owner's Equity}$$
$$\$105,000 = \quad \$25,000 \quad + \quad \$80,000$$

Exhibit 5-5 shows what happens when the company writes a $20,000 check to the bank to pay off a bank loan. The accounting equation stays in balance because assets and liabilities decrease by the same amount:

$$\text{Assets} = \text{Liabilities} + \text{Owner's Equity}$$
$$\$85,000 = \$5,000 + \$80,000$$

 Do you think that companies in other countries use this same structure? Why or why not?

ACCOUNTING FOR TRANSACTIONS TO START A BUSINESS

In Chapters 3 and 4 you saw how managers use accounting information to develop a business plan to show potential investors. Managers also use accounting information for internal decision making. C-V-P analysis and budgeting provide accounting information that helps managers answer questions such as the following: How much money does the company need to have on hand to start the business? How much inventory should the company have available? How much money can the company spend to advertise the grand opening of the business? These initial transactions are undertaken for one purpose—so that managers can pursue the goal of earning a satisfactory income (profit) for the owners. The profit goal is met by selling goods or services to customers at prices that are higher than the costs of providing the goods or services.

Once a company starts operating, it uses financial statements to report to external users about its operations. To prepare these financial statements, the company's accountant or owner identifies transactions and records them in the company's accounting system. This analysis uses the accounting equation and recognizes the dual effect of transactions.

 If you loaned money to a company, how often would you want the company to report on its operations? Why?

The transactions recorded in the accounting system are the basis for the internal and external reports that the company issues. Therefore, it is very important that the transactions are entered correctly, since they "live on" in the system, sometimes for many years.

With the accounting equation (and the dual effect of transactions) in mind, watch Anna Cox as she analyzes and records the December 2003 transactions involved in starting her business, Sweet Temptations. Note how Anna uses accounting concepts, the accounting equation, and the dual effect of transactions to build an effective accounting system.

Investing Cash (Transaction 1)

Anna Cox starts her business on December 15, 2003, by writing a $15,000 personal check and depositing it in a checking account she opened for Sweet Temptations. Anna wants to open the candy store to customers on January 2, 2004, so that she can build some customer traffic before people start buying their Valentine candy. This company checking account is separate, of course, from her personal account because of the entity concept.

Anna decides to establish a basic accounting system for Sweet Temptations on a sheet of paper by listing assets, liabilities, and owner's equity as headings of separate columns. Each column is called an **account**—that is, a place a company uses to record and retain information about the effect of its transactions on a specific asset, liability, or owner's equity item. Anna records each transaction by entering the amounts in the appropriate account columns. Anna uses the receipt issued by Sweet Temptations for her check and the sit slip she used to open the company's bank account as source documents for nsaction. Exhibit 5-6 shows how she records this first transaction.

Anna Cox's personal check serve as a source document for Sweet Temptations?
or why not?

EXHIBIT 5-6		ANNA INVESTS CASH IN SWEET TEMPTATIONS			

		Assets	=	Liabilities	+	Owner's Equity
Trans	Date	Cash				A. Cox, Capital
(1)	12/15/03	+ $15,000				+ $15,000
Balances		$15,000	=			$15,000

Note that Anna makes two $15,000 entries to record the dual effect of the transaction—one to an asset account (Cash) and one to an owner's equity account (A. Cox, Capital)—and that the accounting equation balances because she increases both sides of the equation by the same amount. At the end of each day Anna computes the **balance** of each account—the amount in the account at the beginning of the day plus the increases and minus the decreases recorded in the column that day. In this case, since Anna has recorded only one transaction, Sweet Temptations now has a balance in the Cash account of $15,000 and a balance in the A. Cox, Capital account of $15,000, as we indicate by the "double-lines" under the amounts.

Prepaying Rent (Transaction 2)

To open Sweet Temptations at the Westwood Mall, Anna signed, in the company's name, a rental agreement with the mall's manager on December 16, 2003. The monthly rent is $1,000, and the agreement requires that rent for six months be paid in advance. Since the space is empty, the mall manager agrees to let Anna begin setting up her business immediately but not to start charging her rent until January 1, 2004. This arrangement works well for Anna because now she can begin purchasing store equipment, supplies, and inventory.

So, on December 16, Anna writes a company check for $6,000 ($1,000 × 6 months) to the Westwood Mall. Sweet Temptations' check and the signed rental agreement are the source documents for the transaction. Exhibit 5-7 shows how Anna records this second transaction.

The benefit of using the mall space for six months to conduct business represents an economic resource, or asset, to Sweet Temptations. As a result, Anna records $6,000 as an increase in a new asset account—Prepaid Rent. Because cash is paid out, she decreases the asset Cash by the total amount paid, $6,000. At the end of the day, she subtracts this amount from the previous amount of Cash to show a new balance, $9,000. Then, she checks the accounting equation to see that it remains in balance. She does this by adding the assets ($9,000 + $6,000) and comparing this $15,000 amount with the to-

EXHIBIT 5-7			SWEET TEMPTATIONS PREPAYS RENT				

		Assets			=	Liabilities	+	Owner's Equity
Trans	Date	Cash	+	Prepaid Rent				A. Cox, Capital
(1)	12/15/03	+ $15,000						+ $15,000
(2)	12/16/03	− $ 6,000		+ $6,000				
Balances		$ 9,000	+	$6,000	=			$15,000

EXHIBIT 5-8			SWEET TEMPTATIONS PURCHASES SUPPLIES WITH CASH				
				Assets		= *Liabilities*	+ *Owner's Equity*
Trans	**Date**	**Cash**	+	**Prepaid Rent**	+ **Supplies**		**A. Cox, Capital**
(1)	12/15/03	+ $15,000					+ $15,000
(2)	12/16/03	− $ 6,000		+ $6,000			
(3)	12/17/03	− $ 700			+ $700		
Balances		$ 8,300 +		$6,000 +	$700	=	$15,000

tal of the liabilities ($0) plus owner's equity ($15,000). As you can see, the equation still balances.

 How many changes did Anna make in the accounting equation to record this transaction? Why?

Purchasing Supplies with Cash (Transaction 3)

On December 17, 2003, Sweet Temptations purchases $700 of office and store supplies from City Supply Company by writing a check for $700. Anna receives an *invoice* that lists the items purchased, the cost of each item, and the total cost. She uses the invoice as the source document to record this third transaction, as we show in Exhibit 5-8. Because the supplies will be used to conduct business, Anna records them as an asset, Supplies, of $700. Because the purchase is made with cash, she reduces the asset Cash by $700. Note that the changes in these two assets offset each other on the left side of the accounting equation, which thus remains in balance.

Purchasing Inventory on Credit (Transaction 4)

On December 20, 2003, Sweet Temptations purchases $1,620 of candy (360 boxes of candy for $4.50 each) on credit from Unlimited Decadence Corporation. Sweet Temptations agrees to pay for the candy within 30 days of purchase. An invoice from Unlimited Decadence is the source document for the transaction. Exhibit 5-9 shows how Anna records this fourth transaction. Sweet Temptations needs a way to keep track of the cost of the candy that it buys from manufacturers and has on hand to sell to retail candy customers.

EXHIBIT 5-9			SWEET TEMPTATIONS PURCHASES INVENTORY ON CREDIT					
				Assets			= *Liabilities*	+ *Owner's Equity*
Trans	**Date**	**Cash**	+	**Prepaid Rent**	+ **Supplies**	+ **Inventory**	= **Accounts Payable**	+ **A. Cox, Capital**
(1)	12/15/03	+ $15,000						+ $15,000
(2)	12/16/03	− $ 6,000		+ $6,000				
(3)	12/17/03	− $ 700			+ $700			
(4)	12/20/03					+ $1,620	+ $1,620	
Balances		$ 8,300 +		$6,000 +	$700 +	$1,620	= $1,620	+ $15,000

Anna thus adds an account column to assets to record Inventory. She increases Inventory by the cost of the candy, $1,620, but does not reduce Cash because none was paid out. Since Sweet Temptations agrees to pay for the inventory later, it incurs a debt, or a liability. Anna calls the liability Accounts Payable because it is an amount to be paid by the company, and she increases Accounts Payable by $1,620. Note that Unlimited Decadence Corporation, not Anna Cox, finances this increase in Sweet Temptations' assets (economic resources). Unlimited Decadence is now Sweet Temptations' creditor because it has a claim on $1,620 of the candy store's assets. The $1,620 increase in economic resources matches the $1,620 increase in the claims on those resources. So, the left side of the accounting equation and the liability component of the right side both increase by $1,620. The accounting equation balances after the transaction is recorded.

Purchasing Store Equipment with Cash and Credit (Transaction 5)

On December 29, 2003, Sweet Temptations purchases store equipment from Ace Equipment Company at a cost of $2,200. It pays $1,000 down and signs a note agreeing to pay the remaining $1,200 (plus interest of $24) at the end of three months. Anna uses the invoice, the check, and the note to record this fifth transaction, as we show in Exhibit 5-10. Because the store equipment is an economic resource to be used in the business, Anna increases the asset Store Equipment by the total cost of $2,200. She decreases the asset Cash by the amount paid, $1,000. Since Sweet Temptations incurs a $1,200 liability and issues a legal note, Anna increases the liability Notes Payable by this amount. She does not record any interest now because interest accumulates as time passes, and no time has passed since Sweet Temptations issued the note. This transaction affects two asset accounts and a liability account, but the accounting equation remains in balance.

Selling Extra Store Equipment on Credit (Transaction 6)

Sweet Temptations obtained a special price on the store equipment by buying a "package," which included an extra computer desk that the company did not need. So on December 30, 2003, Sweet Temptations sells the desk, which cost $400, for that same amount to The Hardware Store, another store in the mall. The Hardware Store agrees to pay for the desk on January 7. Exhibit 5-11 shows how Anna records this sixth transaction. Because Sweet Temptations sells one of its economic resources, Anna decreases the asset Store Equipment by $400, the cost of the desk. Because the amount to be received from The Hardware Store in January is an economic resource for Sweet Temptations, Anna also records an increase of $400 in the asset Accounts Receivable. Again, note the equality of the accounting equation.

| EXHIBIT 5-10 | | SWEET TEMPTATIONS PURCHASES STORE EQUIPMENT WITH CASH AND CREDIT | | | | | | | | | |

				Assets					=	Liabilities		+	Owner's Equity
Trans	Date	Cash	+	Prepaid Rent	+ Supplies	+ Inventory	+	Store Equipment =		Accounts Payable	+	Notes Payable +		A. Cox, Capital
(1)	12/15/03	+ $15,000												+ $15,000
(2)	12/16/03	− $ 6,000		+ $6,000										
(3)	12/17/03	− $ 700			+ $700									
(4)	12/20/03					+ $1,620				+ $1,620				
(5)	12/29/03	− $ 1,000						+ $2,200				+ $1,200		
Balances		$ 7,300	+	$6,000 +	$700 +	$1,620 +		$2,200 =		$1,620 +		$1,200 +		$15,000

								Assets							=	Liabilities			+	Owner's Equity
Trans	Date	Cash	+	Prepaid Rent	+	Supplies	+	Inventory	+	Store Equipment	+	Accounts Receivable	=		Accounts Payable	+	Notes Payable	+	A. Cox, Capital
(1)	12/15/03	+ $15,000																	+ $15,000
(2)	12/16/03	− $ 6,000		+ $6,000															
(3)	12/17/03	− $ 700				+ $700													
(4)	12/20/03							+ $1,620							+ $1,620				
(5)	12/29/03	− $ 1,000								+ $2,200							+ $1,200		
(6)	12/30/03									− $ 400		+ $ 400							
Balances		$ 7,300	+	$6,000	+	$700	+	$1,620	+	$1,800	+	$ 400	=		$1,620	+	$1,200	+	$15,000

EXHIBIT 5-11 — SWEET TEMPTATIONS SELLS EXTRA STORE EQUIPMENT ON CREDIT

EXPANDING THE ACCOUNTING EQUATION

Until now, we have focused on how a company records transactions that occur when it is preparing to open for business. You have learned how the accounting equation changes as the company uses its accounting system to record an owner's investment, the purchases of assets with cash and on credit, and the sale of equipment. After the company opens for business, internal and external users of financial statements need income information to evaluate how well the company has been operating. By recording the transactions of its day-to-day operations, a company develops this income information. As you continue reading, keep the accounting equation and the dual effect of transactions in mind.

5 What are revenues and expenses, and how is the accounting equation expanded to record these items?

Anna had no problem using the basic accounting equation to record the start-up transactions in the balance sheet columns. However, she needs to modify the accounting system to record the income-producing transactions, such as sales to customers; these transactions do not fit easily into the equation as it is currently stated.

 If you were Anna, how would you expand Sweet Temptations' accounting system so that you could record revenue and expense transactions? What column headings would you add to the accounting system?

To modify the accounting system, Anna separates the Owner's Equity part of the equation into two sections. The first, the Owner's Capital account, lets her record transactions relating to her investments and withdrawals of capital from the company. The second section lets her record net income (revenues and expenses). For recording both types of transactions, the equality of the accounting equation is maintained because of the dual effects of transactions. The expanded accounting equation is as follows:

$$\text{Assets} = \text{Liabilities} + \underbrace{\text{Owner's Capital} + \underbrace{\text{Net Income}}_{\text{Revenues} - \text{Expenses}}}_{\text{Owner's Equity}}$$

Recall from Chapter 1 that the income of a company is commonly referred to as net income. **Net income** is the excess of revenues over expenses *for a specific time period*. Net income is sometimes called *net profit*, *net earnings*, or simply *earnings*. **Revenues** are the prices a company charged to its customers for goods or services it provided during a specific time period. **Expenses** are the costs of providing the goods or services to customers during the time period.

Anna records revenue and expense transactions by expanding the columns in the simple accounting system she uses. To find out how much net income Sweet Temptations earned over a specific time period (e.g., the month of January), she subtracts the total in the expense column from the total in the revenue column.

We will demonstrate how to use the expanded accounting equation later in the chapter. But first we will discuss various principles and concepts related to net income, since the expanded equation deals with revenues and expenses—the two items that affect net income.

ACCOUNTING PRINCIPLES AND CONCEPTS RELATED TO NET INCOME

6 What are the accounting principles and concepts related to net income?

Earlier in the chapter we explained several basic concepts and terms used in accounting. Before you learn how a company records its daily transactions to determine net income, it is helpful to know several additional accounting principles and concepts that are part of GAAP.

Accounting Period

A company typically operates for many years. The company's owner (internal user) needs information about its net income on a regular basis to make operating decisions. External users of financial statement information also need to know about the company's net income on a regular basis to make timely business decisions. Suppliers need this information for granting credit, creditors need this information for renewing bank loans, and investors need this information for providing additional capital.

Given that both internal and external users benefit when a company routinely reports its net income, the question is: How often should a company do so? Earlier we said that net income is the excess of revenues over expenses for a specific time period. An **accounting period** is the time span for which a company reports its revenues and expenses. Most companies base their financial statements on a twelve-month accounting period called a *fiscal year* or a *fiscal period*. The fiscal year is often the same as the *calendar year;* however, a company whose operations are seasonal may use a year that corresponds more closely to its *operating cycle*. For instance, a retail company (such as **Wal-Mart**) may use January 31, 2004 through January 31, 2005 for its fiscal year so that its accounting period will include the purchase and sale of inventory that peaks for the December holiday season. Many companies also compute and report their net income on a quarterly basis. These accounting periods (and others shorter than a year) are referred to as *interim* periods. In this book we often present simplified examples that use one month as the accounting period.

http://www.wal-mart.com

Earning and Recording Revenues

Revenues result from a company's operating activities that contribute to its earning process. Broadly speaking, every activity of a company contributes to its earning process. More specifically, a company's **earning process** includes purchasing (or producing) inventory, selling the inventory (or services), delivering the inventory (or services), and collecting and paying cash. Although a company *earns* revenues continuously during this process, it generally *records* revenues near or at the end of the earning process.[2] This is because (1) the earning process is complete (the company has made the sale and delivered the product or performed the service) and (2) the prices charged to customers are collectible (a   ts receivable) or collected. So we can say that a company **records revenues** d
counting period in which they are earned and are collectible (or collected).

---

[2]Construction companies sometimes take several years to complete a project (e.g., an office building).
their yearly net income, these companies record a portion of their total revenues each year based on th
the year.

Matching Principle

Expenses are subtracted from revenues to calculate net income. Another way of saying this is that the costs used up are *matched* against the prices charged to customers to determine net income. The **matching principle** states that to determine its net income for an accounting period, a company computes the expenses involved in earning the revenues of the period and deducts the total expenses from the total revenues earned in that period. By matching expenses against revenues, a company has a good idea of how much better off it is at the end of an accounting period as a result of its operations during that period.

Accrual Accounting

Accrual accounting is related to both the recording of revenue and the matching principle. When a company uses **accrual accounting**, it records its revenue and related expense transactions in the same accounting period that it provides goods or services (in the period in which it *earns* the revenue), regardless of whether it receives or pays cash in that period. To accrue means to accumulate. Accrual accounting makes accounting information helpful to external users because it does not let cash receipts and cash payments distort a company's net income. Otherwise, the amount of revenues the company earned during an accounting period could be distorted because the company may have received cash earlier or later than it sold goods or provided services. The amount of expenses could be distorted because the company may have paid cash earlier or later than it incurred (or used up) the related costs.

 Do you think that, by requiring accrual accounting to be used in the preparation of a company's income statement, GAAP implies that the company's cash receipts and payments are not important? Why or why not?

Under accrual accounting, a company must be certain that it has recorded in each accounting period all revenues that it earned during that period, even if it received no cash during the period. Similarly, at the end of each accounting period, the company must be certain that it has matched all expenses it incurred during the period against the revenues it earned in that same period even if it paid no cash during the period.

Summary. How do these concepts relate to a company's accounting system? A company sets up and uses its accounting system based on the *accounting equation* and the *dual effect of transactions*. A company, which is a separate *entity* from its owners, analyzes *source documents* to record its *transactions*. It records the transactions in the accounting system in *monetary units* based on the *historical costs* involved in the company activity. In keeping with *accrual accounting*, a company records its revenue transactions when the revenues are *earned and collectible*, and it records expenses when it incurs the costs. The *matching* principle ensures that all expenses are matched with the revenues they helped earn so that the company can calculate its net income for each *accounting period*.

RECORDING DAILY OPERATIONS

Here's how Anna uses the expanded accounting equation to record the January 2004 day-to-day operations of the Sweet Temptations candy store at the Westwood Mall.

Cash Sale (Transaction 1)

On January 2, 2004, Anna Cox opens Sweet Temptations for business. Sweet Temptations sells a total of 30 boxes of candy at $10 a box for cash. For each sale, the cash register tape lists the date, the type and number of boxes of candy sold, and the total dollar amount of the sale. Anna uses the cash register tape as the source document for the 30 cash sales, which total $300. At the end of the day, Anna increases the Revenues column of Owner's Equity by $300. She also increases Cash by this amount. Of course, Sweet

Temptations had to give customers 30 boxes of candy in order to make the $300 in sales. By checking the purchasing records, Anna knows that the candy originally cost Sweet Temptations $135 ($4.50 × 30 boxes) when purchased from Unlimited Decadence Corporation. Because Sweet Temptations no longer owns the candy, Anna decreases Inventory by $135. In addition, because the cost of the candy is a cost of providing the goods that were sold to customers, she increases Expenses by $135.

Exhibit 5-12 shows Sweet Temptations' accounting equation at the close of its first business day, assuming no other transactions occur. The first line in Exhibit 5-12 shows the balances in each of Sweet Temptations' assets, liabilities, and owner's equity accounts at the *start* of its first business day. These balances came from the transactions that Anna recorded in December as she prepared for business (see the balances in Exhibit 5-11 on page 139). The next line shows how Anna records the cash sale on January 2. Note on this line that Anna shows that the increase in Expenses causes a decrease in Net Income (and therefore Owner's Equity) by putting a minus sign in the column before the increased amount [i.e., $|-|$ + $135] in the Expense column. Note also how the accounting equation remains in balance because of the dual effect of the cash sales transaction. Cash sales will take place every day that Sweet Temptations is open. Although Anna would record these cash sales transactions every day as they occur, to keep things simple in Exhibit 5-20 (later in the chapter), we include a transaction (#13) to represent *all* of the cash sales (770 boxes) that took place from January 3, 2004 through January 31, 2004. These sales total $7,700 ($10 × 770 boxes), so Anna increases both Revenues and Cash by that amount. She also increases Expenses by $3,465 ($4.50 × 770 boxes) and decreases Inventory by that same amount.

 Do you think it is typical to have customers purchase 30 boxes of candy from a small retail store on its first day of business? How could you verify your opinion?

Payment for Credit Purchase of Inventory and Additional Inventory Purchase (Transactions 2 and 3)

Recall that Sweet Temptations purchased its beginning inventory on credit from Unlimited Decadence Corporation on December 20, 2003. Anna recorded this transaction as a $1,620 increase in Inventory and a $1,620 increase in Accounts Payable. On January 3, 2004, Anna writes a Sweet Temptations check to Unlimited Decadence as payment for the December 20, 2003, purchase. An invoice from Unlimited Decadence is the source document for the transaction. To show the results of this transaction, Anna decreases the asset Cash by $1,620. Because the company no longer owes Unlimited Decadence for its purchase, she also decreases the liability Accounts Payable by $1,620.

On January 4, 2004, Sweet Temptations purchases 960 boxes of chocolates (at $4.50 per box) on credit from Unlimited Decadence for $4,320. As a result of the purchase, Anna increases Inventory by $4,320, and because the purchase is made on credit, she also increases Accounts Payable by the same amount. Exhibit 5-13 shows the changes in the accounting equation resulting from these two transactions.

 Does this purchase correspond to the expected purchase noted in Sweet Temptations' purchases budget for January, as we discussed in Chapter 4?

Credit Sale (Transaction 4)

On January 6, 2004, Sweet Temptations sells 10 boxes of candy for $100 on credit to Bud Salcedo, owner of Bud's Buds flower shop, next door to Sweet Temptations. The sales invoice lists the date, the type of candy sold, the flower shop's name and account number, and the total dollar amount of the sale. Anna assigns each of Sweet Temptations' credit customers a unique account number to help her identify transactions the company has with each of these customers. (This will be particularly useful as the number of credit

EXHIBIT 5-12 — SWEET TEMPTATIONS MAKES CASH SALES

| | | Assets | | | | | | = | Liabilities | | Owner's Equity | | |
| | | | | | | | | | | | Owner's Capital | Net Income | |
Trans	Date	Cash	Prepaid Rent	Supplies	Inventory	Store Equipment	Accounts Receivable	=	Accounts Payable	Notes Payable	A. Cox, Capital	Revenues	Expenses
Balances	1/1/04	$ 7,300	$6,000	$700	$1,620	$1,800	$400	=	$1,620	$1,200	$15,000		
(1)	1/2/04	+$ 300			–$ 135							+$ 300	+$ 135
Balances		$ 7,600	$6,000	$700	$1,485	$1,800	$400	=	$1,620	$1,200	$15,000	$ 300	$ 135

EXHIBIT 5-13 — SWEET TEMPTATIONS PAYS FOR CREDIT PURCHASES AND MAKES ADDITIONAL CREDIT PURCHASE OF INVENTORY

| | | Assets | | | | | | = | Liabilities | | Owner's Equity | | |
| | | | | | | | | | | | Owner's Capital | Net Income | |
Trans	Date	Cash	Prepaid Rent	Supplies	Inventory	Store Equipment	Accounts Receivable	=	Accounts Payable	Notes Payable	A. Cox, Capital	Revenues	Expenses
Balances	1/1/04	$ 7,300	$6,000	$700	$1,620	$1,800	$400	=	$1,620	$1,200	$15,000		
(1)	1/2/04	+$ 300			–$ 135							+$ 300	+$ 135
(2)	1/3/04	–$ 1,620							–$1,620				
(3)	1/4/04				+$4,320				+$4,320				
Balances		$ 5,980	$6,000	$700	$5,805	$1,800	$400	=	$4,320	$1,200	$15,000	$ 300	$ 135

EXHIBIT 5-14 — SWEET TEMPTATIONS SELLS CANDY ON CREDIT

| | | Assets | | | | | | = | Liabilities | | Owner's Equity | | |
| | | | | | | | | | | | Owner's Capital | Net Income | |
Trans	Date	Cash	Prepaid Rent	Supplies	Inventory	Store Equipment	Accounts Receivable	=	Accounts Payable	Notes Payable	A. Cox, Capital	Revenues	Expenses
Balances	1/1/04	$ 7,300	$6,000	$700	$1,620	$1,800	$400	=	$1,620	$1,200	$15,000		
(1)	1/2/04	+$ 300			–$ 135							+$ 300	+$ 135
(2)	1/3/04	–$ 1,620							–$1,620				
(3)	1/4/04				+$4,320				+$4,320				
(4)	1/6/04				–$ 45		+$100					+$ 100	+$ 45
Balances		$ 5,980	$6,000	$700	$5,760	$1,800	$500	=	$4,320	$1,200	$15,000	$ 400	$ 180

customers grows.) Having the account number on the sales invoice lets Sweet Temptations keep track of the money each customer owes.

Anna increases the Revenues column of Owner's Equity by $100. Because Sweet Temptations sold the candy on credit instead of receiving cash, Anna increases the asset Accounts Receivable by $100. Remember, Sweet Temptations has to dip into its candy inventory to make the sale. By checking the purchasing records, Anna knows that the boxes originally cost $45. Because the company no longer owns the candy, Anna decreases Inventory by $45. In addition, because the cost of the candy is a cost of providing the goods sold, she increases Expenses by $45. Exhibit 5-14 on page 143 shows the four changes in the accounting equation from this transaction. The accounting equation remains in balance.

Receipt of Payment for Credit Sale of Extra Store Equipment (Transaction 5)

Sweet Temptations receives a check for $400 from The Hardware Store on January 7, 2004. The check is to pay for the store equipment that Sweet Temptations sold on credit to The Hardware Store on December 30, 2003. As you can see in Exhibit 5-15, Anna reduces the asset Accounts Receivable by $400 because The Hardware Store has settled its account and no longer owes Sweet Temptations any money. She increases the asset Cash by $400 to show the receipt of the check.

Withdrawal of Cash by Owner (Transaction 6)

On January 20, 2004, Anna Cox withdraws $50 cash from the business for personal use, writing a $50 check to herself from the Sweet Temptations bank account. She then deposits the check in her personal bank account. The check is the source document for the transaction. A **withdrawal** is a payment from the company to the owner. Thus, it is a disinvestment of assets by the owner. Therefore, as we show in Exhibit 5-16, Anna records a decrease in Cash and a decrease in A. Cox, Capital by the amount of the withdrawal ($50).

 What do you think Anna would record if she took ten boxes of candy instead of cash?

We will discuss withdrawals again in Chapter 6.

 Can you think of any possible ethical issues involved in withdrawals?

Payments for Consulting and Advertising (Transactions 7 and 8)

To prepare for a Valentine's Day grand opening sale, Sweet Temptations hires the Dana Design Group to produce an advertisement. The design group charges $200 and presents the ad on January 25, 2004. Sweet Temptations writes a check for the full amount that day. The receipt received from Dana Design is the source document. As a result of this transaction, Anna increases Expenses by $200 and decreases Cash by $200.

Also on January 25, 2004, Sweet Temptations pays for the advertisement to be published in Westwood Mall's end-of-January promotional flyer. The quarter-page advertisement cost $300. The bill from Westwood Mall's management office is the source document for the transaction. As we show in Exhibit 5-17 on page 146, to record this transaction, Anna increases Expenses by $300 and decreases Cash by the same amount. Note that the accounting equation remains in balance after these transactions are recorded.

 If Cash was mistakenly decreased by only $100 when the last transaction was recorded, how would you find out that an error was made?

EXHIBIT 5-15 — SWEET TEMPTATIONS RECEIVES PAYMENT FROM CREDIT SALE OF EXTRA STORE EQUIPMENT

		Assets						=	Liabilities		+	Owner's Equity		
												Owner's Capital	Net Income	
Trans	Date	Cash	Prepaid Rent	Supplies	Inventory	Store Equipment	Accounts Receivable	=	Accounts Payable	Notes Payable	+	A. Cox, Capital	Revenues	Expenses
Balances	1/1/04	$ 7,300	$6,000	$700	$1,620	$1,800	$400		$1,620	$1,200		$15,000		
(1)	1/2/04	+$ 300			-$ 135								+$ 300	+$ 135
(2)	1/3/04	-$ 1,620							-$1,620					
(3)	1/4/04				+$4,320				+$4,320					
(4)	1/6/04				-$ 45		+$100						+$ 100	+$ 45
(5)	1/7/04	+$ 400					-$400							
Balances		$ 6,380	$6,000	$700	$5,760	$1,800	$100		$4,320	$1,200		$15,000	$ 400	$ 180

EXHIBIT 5-16 — ANNA WITHDRAWS CASH FROM SWEET TEMPTATIONS

		Assets						=	Liabilities		+	Owner's Equity		
												Owner's Capital	Net Income	
Trans	Date	Cash	Prepaid Rent	Supplies	Inventory	Store Equipment	Accounts Receivable	=	Accounts Payable	Notes Payable	+	A. Cox, Capital	Revenues	Expenses
Balances	1/1/04	$ 7,300	$6,000	$700	$1,620	$1,800	$400		$1,620	$1,200		$15,000		
(1)	1/2/04	+$ 300			-$ 135								+$ 300	+$ 135
(2)	1/3/04	-$ 1,620							-$1,620					
(3)	1/4/04				+$4,320				+$4,320					
(4)	1/6/04				-$ 45		+$100						+$ 100	+$ 45
(5)	1/7/04	+$ 400					-$400							
(6)	1/20/04	-$ 50										-$ 50		
Balances		$ 6,330	$6,000	$700	$5,760	$1,800	$100		$4,320	$1,200		$14,950	$ 400	$ 180

EXHIBIT 5-17   SWEET TEMPTATIONS PAYS FOR CONSULTING AND ADVERTISING

| | | Assets | | | | | | | | | | | = | Liabilities | | | | Owner's Equity | | | | |
| | | | | | | | | | | | | | | | | | | Owner's Capital | + | Net Income | | |
Trans	Date	Cash	+	Prepaid Rent	+	Supplies	+	Inventory	+	Store Equipment	+	Accounts Receivable	=	Accounts Payable	+	Notes Payable	+	A. Cox, Capital	+	Revenues	–	Expenses
Balances	1/1/04	$ 7,300	+	$6,000	+	$700	+	$1,620	+	$1,800	+	$400	=	$1,620	+	$1,200	+	$15,000	+		–	
(1)	1/2/04	+$ 300						–$ 135													+$ 300	–	+$ 135
(2)	1/3/04	–$ 1,620													–$1,620								
(3)	1/4/04							+$4,320							+$4,320								
(4)	1/6/04							–$ 45				+$100									+$ 100	–	+$ 45
(5)	1/7/04	+$ 400										–$400											
(6)	1/20/04	–$ 50																	–$ 50				
(7)	1/25/04	–$ 200																				–	+$ 200
(8)	1/25/04	–$ 300																				–	+$ 300
Balances		$ 5,830	+	$6,000	+	$700	+	$5,760	+	$1,800	+	$100	=	$4,320	+	$1,200	+	$14,950	+	$ 400	–	$ 680

EXHIBIT 5-18   SWEET TEMPTATIONS PURCHASES ANOTHER DISPLAY CASE

| | | Assets | | | | | | | | | | | = | Liabilities | | | | Owner's Equity | | | | |
| | | | | | | | | | | | | | | | | | | Owner's Capital | + | Net Income | | |
Trans	Date	Cash	+	Prepaid Rent	+	Supplies	+	Inventory	+	Store Equipment	+	Accounts Receivable	=	Accounts Payable	+	Notes Payable	+	A. Cox, Capital	+	Revenues	–	Expenses
Balances	1/1/04	$ 7,300	+	$6,000	+	$700	+	$1,620	+	$1,800	+	$400	=	$1,620	+	$1,200	+	$15,000	+		–	
(1)	1/2/04	+$ 300						–$ 135													+$ 300	–	+$ 135
(2)	1/3/04	–$ 1,620													–$1,620								
(3)	1/4/04							+$4,320							+$4,320								
(4)	1/6/04							–$ 45				+$100									+$ 100	–	+$ 45
(5)	1/7/04	+$ 400										–$400											
(6)	1/20/04	–$ 50																	–$ 50				
(7)	1/25/04	–$ 200																				–	+$ 200
(8)	1/25/04	–$ 300																				–	+$ 300
(9)	1/25/04	–$ 200								+$ 200													
Balances		$ 5,630	+	$6,000	+	$700	+	$5,760	+	$2,000	+	$100	=	$4,320	+	$1,200	+	$14,950	+	$ 400	–	$ 680

Do you think Sweet Temptations' location next to Bud's Buds will improve its sales of Valentine candy? Why or why not?

Acquisition of Store Equipment (Transaction 9)

On January 29, 2004, Sweet Temptations purchases an additional candy display case. Sweet Temptations pays $200 in cash by writing a check. As you can see in Exhibit 5-18, Anna increases Store Equipment by $200 and decreases Cash by the same amount.

Payment of Salaries (Transaction 10)

Sweet Temptations employs two people (you and Jaime Gonzales) to help stock and sell candy. On January 31, 2004, you both receive checks, totaling $2,050, as payment for your services during January. Your time cards, wage rate schedules, and paychecks are the source documents for the transactions. As you can see in Exhibit 5-19, Anna decreases the asset Cash by $2,050. Because paying an employee's salary is a cost of providing goods and services to customers, she also increases Expenses by the same amount.

Payment of Telephone and Utilities Bills (Transactions 11 and 12)

On January 31, 2004, Sweet Temptations pays its telephone bill and its utility bill (heat, light, and water) for January. The two checks are written for $60 and $190, respectively. Anna records each transaction separately, using the bills and checks as the source documents. As you can see in Exhibit 5-20, she decreases Cash and increases Expenses for both transactions.

Summary Cash Sales (Transaction 13)

Exhibit 5-20 also shows the summary transaction, which we discussed earlier, of all the cash sales from January 3 through January 31.

 How many additional boxes of candy did Sweet Temptations sell for cash from January 3 through January 31? Why were Expenses increased by $3,465?

EXHIBIT 5-19   SWEET TEMPTATIONS PAYS SALARIES

						Assets			=	Liabilities		+	Owner's Capital	+	Net Income		
Trans	Date	Cash	+	Prepaid Rent	+ Supplies	+ Inventory	+ Store Equipment	+ Accounts Receivable	=	Accounts Payable	+ Notes Payable	+	A. Cox, Capital	+	Revenues	−	Expenses
Balances	1/1/04	$ 7,300		$6,000	$700	$1,620	$1,800	$400		$1,620	$1,200		$15,000				
(1)	1/2/04	+$ 300				−$ 135									+$ 300		+$ 135
(2)	1/3/04	−$ 1,620								−$1,620							
(3)	1/4/04					+$4,320				+$4,320							
(4)	1/6/04					−$ 45		+$100							+$ 100		+$ 45
(5)	1/7/04	+$ 400						−$400									
(6)	1/20/04	−$ 50											−$ 50				
(7)	1/25/04	−$ 200															+$ 200
(8)	1/25/04	−$ 300															+$ 300
(9)	1/25/04	−$ 200					+$ 200										
(10)	1/31/04	−$ 2,050															+$2,050
Balances		$ 3,580	+	$6,000	+ $700	+ $5,760	+ $2,000	+ $100	=	$4,320	+ $1,200	+	$14,950	+	$ 400	−	$2,730

EXHIBIT 5-20   SWEET TEMPTATIONS PAYS TELEPHONE AND UTILITY BILLS AND RECORDS SALES FOR JANUARY 3–JANUARY 31

						Assets			=	Liabilities		+	Owner's Capital	+	Net Income		
Trans	Date	Cash	+	Prepaid Rent	+ Supplies	+ Inventory	+ Store Equipment	+ Accounts Receivable	=	Accounts Payable	+ Notes Payable	+	A. Cox, Capital	+	Revenues	−	Expenses
Balances	1/1/04	$ 7,300		$6,000	$700	$1,620	$1,800	$400		$1,620	$1,200		$15,000				
(1)	1/2/04	+$ 300				−$ 135									+$ 300		+$ 135
(2)	1/3/04	−$ 1,620								−$1,620							
(3)	1/4/04					+$4,320				+$4,320							
(4)	1/6/04					−$ 45		+$100							+$ 100		+$ 45
(5)	1/7/04	+$ 400						−$400									
(6)	1/20/04	−$ 50											−$ 50				
(7)	1/25/04	−$ 200															+$ 200
(8)	1/25/04	−$ 300															+$ 300
(9)	1/25/04	−$ 200					+$ 200										
(10)	1/31/04	−$ 2,050															+$2,050
(11)	1/31/04	−$ 60															+$ 60
(12)	1/31/04	−$ 190															+$ 190
(13)	1/3/04 thru 1/31/04	+$ 7,700				−$3,465									+$7,700		+$3,465
Balances		$11,030	+	$6,000	+ $700	+ $2,295	+ $2,000	+ $100	=	$4,320	+ $1,200	+	$14,950	+	$8,100	−	$6,445

END-OF-PERIOD ADJUSTMENTS

Remember, revenues are the prices charged to a company's customers for goods or services it provided during the accounting period, and expenses are the costs of providing those goods or services during the period. The net income is the excess of revenues over expenses for the period. To calculate net income for a month, for example, a company counts the dollar totals for all the revenue and expense transactions of that specific month and subtracts the expense total from the revenue total. That is, it matches the expenses against the revenues for the month.

To calculate a company's net income under accrual accounting, the company must make sure that all its revenues and expenses for the accounting period are included in the totals. For Sweet Temptations, Anna can easily verify that the revenue total is correct because every sale is listed on a source document (a sales invoice or a cash register tape), which she used to record each sales transaction. Anna can verify that *most* of Sweet Temptations' expenses are correct because they also have source documents (invoices, utilities bills, and time cards).

 Do you think it sometimes may be difficult to identify when a sale has occurred? Why?

It is more difficult, however, for a company to make sure that *all* of the expenses it incurred during the month are included in the net income calculation because some of the costs of providing goods or services occur without a source document. Since these expense transactions don't have source documents, there is no "automatic trigger" for recording the transactions. Before calculating its net income, then, a company must analyze its unique expenses (and a few unique revenues, which we will briefly discuss later) to see if it needs to adjust (increase) the total expenses (or revenues) to include those without source documents. These adjustments are called **end-of-period adjustments**.

 What types of expenses can you think of that occur without source documents?

In general, end-of-period adjustments involve assets that a company had at the beginning of the accounting period but that it used during the period to earn revenues. As assets lose their potential for providing future benefits, they are changed to expenses. Anna must analyze Sweet Temptations' assets to see what additional expenses to record. As you will see, the end-of-period adjustments may also include liabilities that a company owes because of expenses that must be recorded. Let's take a look at the four end-of-period adjustments that Anna makes before calculating Sweet Temptations' net income for January 2004, its first month of operations.

7 Why are end-of-period adjustments necessary?

Supplies Used (Transaction 14)

Recall that on December 17, 2003, Sweet Temptations purchased $700 of supplies from City Supply Company. At this time, Anna increased the asset Supplies by $700 to show the cost of this new asset. Sweet Temptations thus purchased the pens, paper, blank sales invoices, and other items it needed to operate the business.

Because Sweet Temptations operated during January, it used some of these supplies. Thus, at January 31, 2004, the $700 original amount of Supplies is not correct. Anna must adjust the amount to show that since some of the supplies were used, they now are an expense, and only part of the $700 of supplies is still an asset. Anna determines that the of-
lies used during January amount to $30. She makes an end-of-period adjustment
e Expenses by $30 and decrease the asset Supplies by $30, as we show in Ex-
When she subtracts the $30 from the $700 original amount, the $670 ending
the cost of supplies the company still owns at the end of January. Notice how

EXHIBIT 5-21

SWEET TEMPTATIONS MAKES END-OF-PERIOD ADJUSTMENTS

Trans	Date	Assets						=	Liabilities		+	Owner's Equity			
		Cash	+	Prepaid Rent	+	Supplies	+	Inventory	+	Store Equipment	+	Accounts Receivable		Accounts Payable	+	Notes Payable	+	Owner's Capital A. Cox, Capital	+	Net Income Revenues	−	Expenses

Trans	Date	Cash	+	Prepaid Rent	+	Supplies	+	Inventory	+	Store Equipment	+	Accounts Receivable	=	Accounts Payable	+	Notes Payable	+	A. Cox, Capital	+	Revenues	−	Expenses
Balances	1/31/04	$11,030		$6,000		$700		$2,295		$2,000		$100		$4,320		$1,200		$14,950		$8,100	−	$6,445
(14)	1/31/04					−$ 30																	+$ 30
(15)	1/31/04			−$1,000																			+$1,000
(16)	1/31/04									−$ 15													+$ 15
(17)	1/31/04																+$ 8						+$ 8
Balances	1/31/04	$11,030	+	$5,000	+	$670	+	$2,295	+	$1,985	+	$100	=	$4,320	+	$1,208	+	$14,950	+	$8,100	−	$7,498

this adjustment (and each of the following adjustments) maintains the equality of Sweet Temptations' accounting equation and has a dual effect on the equation.

 How do you think Anna determined the amount of supplies used?

Expired Rent (Transaction 15)

Recall that Sweet Temptations wrote a check for $6,000 to the Westwood Mall on December 16, 2003, to pay in advance for six months' rent starting on January 1, 2004. At that time Anna recorded a $6,000 asset, Prepaid Rent, to show that Sweet Temptations had purchased the right to use space in the Westwood Mall for six months (January through June) at a price of $1,000 per month. At the end of January, Sweet Temptations has used up one month of Prepaid Rent—for January—because the business occupied the mall space for that entire month. Therefore, Anna must include the cost of the mall space as an expense in the calculation of Sweet Temptations' net income.

Since Sweet Temptations made the $6,000 payment on December 16, 2003, no other source documents relating to the rental of the mall space exist. Although Sweet Temptations has used up one of its six months of rent, the amount listed for Prepaid Rent is still $6,000. Anna must adjust the Prepaid Rent amount to show that only five months of prepaid rent remain. To do so, she increases Expenses for January by $1,000 and reduces Prepaid Rent by the same amount, as we show in Exhibit 5-21. Now Prepaid Rent shows the correct balance of $5,000 ($1,000 $\times$ 5) for the remaining five months.

 What adjustment do you think Anna would make at the end of January if Sweet Temptations occupied the rental space for January but did not pay for any rent until February?

Depreciation of Store Equipment (Transaction 16)

At the beginning of January the amount for the asset Store Equipment was listed at the cost of $1,800. Sweet Temptations purchased the store equipment because it would help earn revenue. The equipment includes, for instance, display cases, a cash register, and a moving cart. Although Sweet Temptations doesn't expect any equipment to wear out completely after one month or even one year, it does not expect it all to last indefinitely. At some point in the future the display cases will become outdated, the cash register will quit working, and the moving cart will fall apart. At that time the company will decide to sell or dispose of the equipment.

The store equipment provides benefits to the company every period in which it is used. Because Sweet Temptations used the store equipment in January to help earn candy revenue and because the store equipment has a finite life, a portion of the cost of the store equipment is included as an expense in the January net income calculation. **Depreciation** is the part of the cost of a physical asset allocated as an expense to each time period in which the asset is used.

The simplest way to compute depreciation is to divide the cost by the estimated life of the asset. For now, assume that the depreciation for the store equipment is $15 a month. Anna makes an end-of-period adjustment for January's depreciation by increasing Expenses by $15 and decreasing the asset Store Equipment by $15, as we show in Exhibit 5-21.[3] Now store equipment shows the $1,985 remaining cost (called its "book value"). As Sweet Temptations uses the store equipment in each future month, it will record an additional $15 depreciation, which will reduce the book value of the equipment. Therefore, at any point in time the difference between the original cost and the book value is

---

[3]Sweet Temptations also purchased $200 of additional store equipment late in January. Sweet Temptations will include the depreciation on this store equipment as an expense in later months as it uses the equipment.

the "accumulated depreciation" to date. We will discuss the methods used to calculate depreciation in Chapter 21.

Accrual of Interest (Transaction 17)

At the end of December 2003, Sweet Temptations purchased store equipment by signing a $1,200 note payable to be paid at the end of three months. Generally, all notes payable also involve the payment of interest for the amount borrowed. This interest is an expense of doing business during the time between the signing of the note and the payment of the note. Sweet Temptations agreed to pay $24 total interest for the note, so that at the end of the three months Sweet Temptations will pay $1,224 ($1,200 + $24). Interest accumulates (*accrues*) over time until it is paid. Since Sweet Temptations owed the note during all of January, Anna must record one month of interest on the note as an expense of doing business during January. Because Sweet Temptations will not pay the interest until it pays the note, it records the January interest as an increase in a liability. For now, assume that the interest is $8 per month ($24 ÷ 3). Anna makes an end-of-period adjustment for the January interest by increasing Expenses by $8 and increasing the liability Notes Payable by $8, as we show in Exhibit 5-21 on page 150. We will discuss how to compute interest later in the book.

End-of-Period Revenue Adjustments

There are a few end-of-period adjustments that a company may need to make to ensure that its revenues are correct for the accounting period. Here, we briefly discuss two. First, a company may have a note receivable (asset) that earns interest that the company will collect when it collects the note. At the end of the accounting period, the company must record any interest that has accumulated *(accrued)* by increasing Revenues and increasing the asset Notes Receivable. Second, a company might collect cash in advance from a customer for sales of merchandise that it will deliver to the customer or for services that it will perform for the customer later in the current accounting period or in the next accounting period. In this case, the company has not earned the revenue at the time of the cash collecton. Therefore, it records the receipt by increasing Cash and increasing a liability (sometimes called *Unearned Revenue*). Then, at the end of the current accounting period the company must decrease the Unearned Revenue and increase Revenues for the amount of revenue it has earned during the period. We will discuss end-of-period revenue adjustments more completely in later chapters.

 At the end of January, 2004, what adjustment would Ace Equipment Company make for the $8 interest it has earned on the $1,200 note it received from Sweet Temptations at the end of December, 2003?

NET INCOME AND ITS EFFECT ON THE BALANCE SHEET

After recording the results of all the transactions and end-of-period adjustments (shown in Exhibit 5-21 on page 150), Anna calculates Sweet Temptations' net income for January:

Net Income = Revenues − Expenses
$602   = $8,100   − $7,498

A company will normally prepare an income statement that lists the various revenues and expenses included in net income. For simplicity, in this chapter simple accounting system, which does not help in the preparation of a deta' statement. In Chapter 6 we will expand the accounting system and show yo' pare an income statement. Anna can, however, compare the actual net income amount for January with the projected net income that she calculated when she planned Sweet Temptations' operations. In Chapter 3, she calculated a projected net income for Sweet Temptations of $110 for January 2004. Anna should be pleased; by achieving an actual net income of $602, Sweet Temptations has done better than she expected. Later in the book, we will discuss how internal and external users analyze the financial statements of a company to understand how well it did for a specific time period.

To prepare the January 31, 2004 balance sheet for Sweet Temptations, Anna uses the end-of-the-month balances for each asset, liability, and owner's equity account listed in Exhibit 5-21. Exhibit 5-22 shows Sweet Temptations' balance sheet at January 31, 2004.

You should be able to trace the asset and liability amounts directly to the ending balances listed on Exhibit 5-21 on page 150. The assets on the balance sheet are rearranged, however, to show them in the order of their *liquidity*, or how quickly the assets can be converted to cash or used up. We will discuss liquidity more in later chapters. Also notice that Anna must calculate the balance sheet amount for Owner's Equity (A. Cox, capital) at the end of January. It is the sum of all of the owner's equity items included in Exhibit 5-21:

A. Cox, capital	$14,950
Revenues	+8,100
Expenses	−7,498
Owner's Equity	$15,552

Expressed another way, it is the sum of A. Cox, capital and net income ($14,950 + $602). We will explain how to "update" the balance of the owner's capital account for net income in Chapter 6.

Since Anna is the owner, the net income (revenues minus expenses) is included in her capital amount on the January 31, 2004 balance sheet. Using the total amounts for the asset, liability, and owner's equity sections of Sweet Temptations' balance sheet, we can state the accounting equation on January 31, 2004 as follows:

$$\textbf{Assets} \ = \ \textbf{Liabilities} + \textbf{Owner's Equity}$$
$$\$21,080 \ = \ \$5,528 \ + \ \$15,552$$

Because Anna properly recorded Sweet Temptations' transactions, the company's accounting equation is in balance at January 31, 2004.

EXHIBIT 5-22	SWEET TEMPTATIONS' BALANCE SHEET

SWEET TEMPTATIONS
Balance Sheet
January 31, 2004

Assets		Liabilities	
Cash	$11,030	Accounts payable	$ 4,320
Accounts receivable	100	Notes payable	1,208
Inventory	2,295	Total Liabilities	$ 5,528
Supplies	670		
Prepaid rent	5,000	**Owner's Equity**	
Store equipment	1,985	A. Cox, capital	$15,552[a]
		Total Owner's Equity	$15,552
		Total Liabilities	
Total Assets	$21,080	and Owner's Equity	$21,080

[a]$14,950 + $8,100 − $7,498; from Exhibit 5-21.

Just as we expanded the accounting equation to record revenue and expense transactions, we will discuss other changes in the accounting system throughout the book. The changes make it easier to keep track of company activities and increase the usefulness of the accounting system. We will also introduce additional accounting concepts to help you understand why companies make changes to the accounting system. In the next three chapters, we will take a detailed look at three very important outputs of the accounting process—the income statement, the balance sheet, and the cash flow statement. We will also continue to answer questions concerning what accounting is, how accounting works, why accounting is performed, and how accounting information is used for problem-solving and decision-making. We will also discuss how to minimize errors that, among other things, can cause major embarrassments, as we discuss below.

BUSINESS ISSUES AND VALUES: A BILLION HERE, A BILLION THERE

http://www.fid-intl.com

In one year, Fidelity Investments estimated that it would make a year-end distribution of $4.32 per share to shareholders in its *Magellan Fund*. The company then admitted to an error. Included in a letter sent to shareholders was the following statement: " . . . The error occurred when the accountant omitted the minus sign on a net capital loss of $1.3 billion and incorrectly treated it as a net capital gain on (a) separate spreadsheet. This meant that the dividend estimate spreadsheet was off by $2.6 billion." The error had no effect on the fund's results or on the shareholders' taxes but was clearly an embarrassment to the company's management!

SUMMARY

At the beginning of the chapter we asked you several questions. During the chapter, we asked you to STOP and answer some additional questions to build your knowledge about specific issues. Be sure you answered these additional questions. Below are the questions from the beginning of the chapter, with a brief summary of the key points relating to the answers. Use your creative and critical thinking skills to expand on these key points to develop more complete answers to the questions and to determine what other questions you have that might lead you to learn more about the issues.

1 Why do managers, investors, creditors, and others need information about a company's operations?

Internal and external users need information about a company's operations to evaluate alternatives. For instance, a manager needs this information to decide which alternative best helps the company meet its goals of remaining solvent and earning a satisfactory profit. A banker also needs this information to decide the conditions for granting a loan.

2 What are the basic concepts and terms that help identify the activities that a company's accounting system records?

The basic concepts and terms that help identify the activities that a company's accounting system records are the entity concept (each company is separate from its owners), transactions (exchanges between a company and another entity), source documents (business records as evidence of transactions), the monetary unit concept (transactions are recorded in monetary terms), and the historical cost concept (transactions are recorded based on dollars exchanged).

3 What do users need to know about the accounting equation for a company?

Users need to know the accounting equation: Assets = Liabilities + Owner's Equity. They need to know that assets are a company's economic resources, liabilities are a company's debts, and owner's equity is the owner's current investment in the assets of the company.

4 Why are at least two effects of each transaction recorded in a company's accounting system?

A company's accounting system is designed so that two effects of each transaction are recorded in order to maintain the equality of the accounting equation. Under the dual effect of transactions, recording a transaction involves at least two changes in the assets, liabilities, and owner's equity of a company.

5 What are revenues and expenses, and how is the accounting equation expanded to record these items?

Revenues are the prices a company charged its customers for goods or services provided during an accounting period. Expenses are the costs of providing the goods or services during the period. Net income is the excess of revenues over expenses for the period. The accounting equation is expanded as follows to record revenues and expenses: Assets = Liabilities + [Owner's Capital + (Revenues − Expenses)].

6 What are the accounting principles and concepts related to net income?

The accounting principles and concepts related to net income are the accounting period, earning and recording revenues, the matching principle, and accrual accounting. The accounting period is the time span used by a company to report its net income. A company records revenues during the accounting period in which they are earned and collectible. The matching principle states that a company matches the total expenses of an accounting period against the total revenues of the period to determine its net income. Accrual accounting means that a company records its revenues and expenses in the accounting period in which it provides goods or services, regardless of whether it receives or pays cash.

7 Why are end-of-period adjustments necessary?

End-of-period adjustments are necessary to record any expenses that a company has incurred (or any revenues that the company has earned) during the accounting period but that it has not yet recorded. Adjustments ensure that these expenses (and revenues) are included in the company's net income calculation.

KEY TERMS

account (p. 135)
accounting equation (p. 131)
accounting period (p. 140)
accounting system (p. 130)
accounts payable (p. 131)
accounts receivable (p. 131)
accrual accounting (p. 141)
assets (p. 130)
balance (p. 136)
creditors (p. 131)
creditors' equity (p. 131)
depreciation (p. 151)
dual effect of transactions (p. 132)
earning process (p. 140)
end-of-period adjustments (p. 149)
entity (p. 127)
equity (p. 131)
expenses (p. 139)

historical cost concept (p. 129)
liabilities (p. 131)
matching principle (p. 141)
monetary unit concept (p. 128)
net assets (p. 132)
net income (p. 139)
owner's equity (p. 131)
partners' equity (p. 131)
prepaid insurance (p. 131)
records revenues (p. 140)
residual equity (p. 131)
revenues (p. 139)
source document (p. 128)
stockholders' equity (p. 131)
total equity (p. 131)
transaction (p. 128)
wages and salaries payable (p. 131)
withdrawal (p. 144)

dividends

INTEGRATED BUSINESS AND ACCOUNTING SITUATIONS

Answer the Following Questions in Your Own Words.

Testing Your Knowledge

5-1   Why do external users need financial accounting information about a company? How can financial statements help these external users?

5-2   Why is it important for external users to know that a company's financial statements are prepared according to GAAP?

5-3   Name and briefly define five concepts and terms that you need to understand to identify the activities that a company's accounting system records.

5-4   What is the entity concept? How does it affect the accounting for a specific company?

5-5   What is a transaction? Why is it important in accounting?

5-6   What is a source document? Why does a company need to prepare source documents?

5-7   What are the monetary unit and historical cost concepts? How do they affect the recording of transactions?

5-8   Define assets. Give four examples.

5-9   Define liabilities. Give two examples.

5-10   Define owner's equity. What items affect owner's equity?

5-11   Why is a company's statement of financial condition called a *balance sheet*?

5-12   What are a company's net assets? How do they relate to owner's equity?

5-13   What is meant by the dual effect of transactions? How does it relate to the accounting equation?

5-14   How is the accounting equation used to set up a basic accounting system for a company? What is an *account*? What is an *account balance*?

5-15   Define revenues, expenses, and net income. How is the accounting equation expanded to record income-related transactions?

5-16   Name and briefly define four principles and concepts relating to net income.

5-17   What is an accounting period? What is the usual length of an accounting period?

5-18   What is a company's earning process, and when does the company record revenues?

5-19   What is the matching principle? Why is it useful to a company?

5-20   What is accrual accounting, and why is it important?

5-21   How do the accounting concepts, principles, and terms discussed in the chapter relate to the accounting system of a company?

5-22   What are end-of-period adjustments? Why are they needed?

Applying Your Knowledge

5-23   Each of the following cases is independent of the others:

Case	Assets	Liabilities	Owner's Equity
1	A	$24,000	$54,000
2	$83,000	B	$42,000
3	$98,000	$32,000	C

Required: Determine the amounts for A, B, and C.

5-24 At the beginning of the year, the Thomas Lighting Company had total assets of $78,000 and total liabilities of $22,000. During the year, the total assets increased by $16,000. At the end of the year, owner's equity totaled $64,000.

Required: Determine (1) the owner's equity at the beginning of the year and (2) the total liabilities at the end of the year.

5-25 At the end of the year, a company's total assets are $75,000 and its total owner's equity is $48,000. During the year, the company's liabilities decreased by $11,000 while its assets increased by $7,000.

Required: Determine the company's (1) ending total liabilities, (2) beginning total assets, and (3) beginning owner's equity.

5-26 The following transactions are taken from the records of Phantom Security Company:

$$\text{Assets} = \text{Liabilities} + \text{Owner's Equity}$$

(a) Rex Simpson, the owner, invested $12,000 cash in the business.
(b) Phantom paid $6,000 cash to acquire security equipment.
(c) Phantom received a $7,000 cash loan from Story County Bank.

Required: Determine the overall effect of each transaction on the assets, liabilities, and owner's equity of Phantom Security Company. Use the symbols *I* for increase, *D* for decrease, and *N* for no change. Also show the related dollar amounts.

5-27 On August 31, 2004, the Hernandez Engineering Company's accounting records contained the following items (listed in alphabetical order):

Accounts payable	$3,700
Accounts receivable	4,000
Cash	5,200
L. Hernandez, capital	?
Notes payable	6,000
Office equipment	8,900
Office supplies	600
Prepaid insurance	800

Required: Prepare a balance sheet for the Hernandez Engineering Company at August 31, 2004. Insert the correct amount for L. Hernandez, capital.

5-28 Listed below, in random order, are all the items included in the Ridge Rental Company balance sheet at December 31, 2004:

Land	$ 2,200
Accounts receivable	3,500
Cash	?
Supplies	900
Accounts payable	4,600
Building	19,000
A. Ridge, capital	?
Rental equipment	6,800
Notes payable	5,700

Total assets on December 31, 2004 are $33,800.

Required: Prepare a balance sheet for the Ridge Rental Company on December 31, 2004. Insert the correct amounts for Cash and for A. Ridge, capital.

5-29 In the chapter, we stated that a transaction is an exchange of property or service by a company with another entity. We also explained that in the recording of a transaction, at least two changes must be made in the assets, liabilities, or owner's equity of a company.

Required: In each case on the following page, describe a transaction that will result in the following changes in the contents of a company's balance sheet:

(a) Increase in an asset and increase in a liability
(b) Decrease in an asset and decrease in a liability
(c) Increase in an asset and decrease in another asset
(d) Increase in an asset and increase in owner's equity
(e) Increase in an asset and increase in revenues
(f) Increase in expenses and decrease in an asset

5-30   Recall from the chapter that we defined a source document as a business record used by a company as evidence that a transaction has occurred.

Required: Name the source documents you think a company would use as evidence for each of the transactions listed below.
(a) Receipt of cash from the owner for additional investment in the company
(b) Payment by check to purchase office equipment
(c) Purchase of office supplies on credit
(d) Sale of office equipment at its original purchase price to a local CPA
(e) Purchase of fire and casualty insurance protection
(f) Sale of inventory on credit

5-31   During October, the Wilson Company incurred the following costs:
(a) At the beginning of the month, the company paid $1,200 to an insurance agency for a two-year comprehensive insurance policy on the company's building.
(b) The company purchased office supplies costing $970 on credit from Bailey's Office Supplies.
(c) The company paid the telephone company $110 for telephone service during October.
(d) The owner withdrew $1,200 for personal use.
(e) The company found that of the $970 of office supplies purchased in (b), only $890 remained at October 31.

Required: For each of the preceding items, identify whether it would be recorded as an asset or an expense by the Wilson Company for October. List the dollar amount and explain your reasoning.

5-32   Gertz Rent-A-Car is in the business of providing customers with quality rental cars at low rates. The company engaged in the following transactions during March:
(a) J. Gertz deposited an additional $1,900 of his personal cash into the company's checking account.
(b) The company collected $1,500 in car rental fees for March.
(c) The company borrowed $7,000 from the 1st National Bank to be repaid in one year.
(d) The company completed arrangements to provide fleet service to a local company for one year, starting in April, and collected $18,000 in advance.

Required: For each of the preceding transactions, identify which would be recorded as revenues by Gertz Rent-A-Car for March. List the dollar amount and explain your reasoning.

5-33   The Slidell Auto Supply Company entered into the following transactions during the month of July:

Date	Transaction
7/1	Joan Slidell, the owner, deposited $12,000 in the company's checking account.
7/11	Slidell Auto Supply purchased $800 of office supplies from Jips Paper Company, agreeing to pay for half of the supplies on July 31 and the rest of the supplies on August 15.
7/16	Slidell Auto Supply purchased a three-year fire insurance policy on a building owned by the company, paying $600 cash.
7/31	Slidell Auto Supply paid Jips Paper Company half the amount owed for the supplies purchased on July 11.

Required: Using the basic accounting equation that we presented in this chapter, record the preceding transactions. Use headings for the specific kinds of assets, liabilities, and owner's equity. Set up your answer in the following form:

Date	Assets	=	Liabilities	+	Owner's Equity

5-34 Amy Dixon opened the Dixon Travel Agency in January, and the company entered into the following transactions during January:

(a) On January 2, Amy deposited $23,000 in the company's checking account.

(b) To conduct its operations, the company purchased land for $3,000 and a small office building for $15,000 on January 3, paying $18,000 cash.

(c) On January 5, the company purchased $700 of office supplies from City Supply Company, agreeing to pay for half of the supplies on January 15 and the remainder on February 15.

(d) On January 12, the company purchased office equipment from Ace Equipment Company at a cost of $3,000. It paid $1,000 down and signed a note, agreeing to pay the remaining $2,000 at the end of one year.

(e) On January 15, the company paid City Supply Company half the amount owed for the supplies purchased on January 5.

(f) On January 28, Amy decided that the company did not need a desk it had purchased on January 12 for $400. The desk was sold for $400 cash to Chris Watson, an insurance agent, for use in his office.

(g) On January 30, the company collected $900 of commissions for travel arrangements made for customers during January.

(h) On January 31, the company paid Frank Jones $500 for secretarial work done during January.

(i) On January 31, the company received its utilities and phone bill, totaling $120 for January. It will pay for this bill in early February.

(j) On January 31, Amy withdrew $600 from the company for her personal use.

Required: (1) Using the accounting system we developed in the chapter, record the preceding transactions.

(2) Prove the equality of the accounting equation at the end of January.

(3) List the source documents that you would normally use in recording each of the transactions.

5-35 Parsons Fashion Designers was started on June 1. The following transactions of the company occurred during June:

(a) E. Parsons started the business by investing $18,000 cash.

(b) Land and an office building were acquired at a cost of $5,000 and $18,000, respectively. The company paid $6,000 down and signed a note for the remaining balance of $17,000. The note is due in two years.

(c) Design equipment was purchased. The cash price of $2,600 was paid by writing a check to the supplier.

(d) Office supplies totaling $250 were purchased on credit. The amount is due in 30 days.

(e) A one-year fire insurance policy was purchased for $800.

(f) Fashion design commissions (fees) of $1,200 were collected for June.

(g) An assistant's salary of $600 was paid for June.

(h) E. Parsons withdrew $500 from the company for personal use.

(i) Utility bills totaling $150 for June were received and will be paid in early July.

Required: (1) Using the accounting system shown in the chapter, record the preceding transactions.

(2) Prove the equality of the accounting equation at the end of June.

(3) List the source documents that you would normally use in recording each of the transactions.

5-36 L. Snider, a young CPA, started Snider Accounting Services on September 1. During September, the following transactions of the company took place:

(a) On September 1, Snider invested $7,000 to start the business.

(b) On September 1, the company paid $3,000 for one year's rent of office space in advance.

(c) On September 2, office equipment was purchased at a cost of $5,000. A down payment of $1,000 was made, and a $4,000, one-year note was signed for the balance owed.

(d) On September 5, office supplies were purchased for $600 cash.

(e) On September 18, $1,000 was collected from clients for accounting services performed.

(f) On September 28, a $500 salary was paid to an accounting assistant.

(g) On September 29, Snider withdrew $800 for personal use.

(h) On September 30, the company billed clients $1,200 for accounting services per-
formed during the second half of September.

(i) On September 30, the September utility bill of $100 was received; it will be paid in
early October.

(j) On September 30, Snider recorded the following adjustments:
1. Rent expense of $250
2. Depreciation of $60 on office equipment
3. Interest expense of $40 on note payable
4. Office supplies used of $50

Required: (1) Using the accounting system shown in the chapter, record the preceding
items.

(2) Prove the equality of the accounting equation at the end of September.

(3) Calculate the net income of the company for September.

(4) Prepare a balance sheet for the company on September 30.

5-37   The Johnson Drafting Company was started on March 1 to draw blueprints for building
contractors. The following transactions of the company occurred during March:

Date	Transactions
3/1	M. Johnson, the owner, started the business by investing $14,000 cash.
3/2	Land and a small office building were purchased at a cost of $4,000 and $20,000, respectively. A down payment of $8,000 was made, and a note for $16,000 was signed. The note is due in one year.
3/3	Cash of $4,800 was paid to purchase computer drafting equipment.
3/8	Drafting supplies totaling $850 were purchased on credit. The amount is due in early April.
3/15	The company collected $1,500 from contractors for drafting services performed.
3/28	M. Johnson withdrew $1,000 for personal use.
3/29	The company received a $110 utility bill for March, to be paid in April.
3/30	The company paid $600 in salary to a drafting employee.
3/30	The company billed contractors $2,000 for drafting services performed during the last half of March.
3/31	The company recorded the following adjustments:
	(a) Depreciation of $80 on the office building
	(b) Depreciation of $100 on computer drafting equipment
	(c) Interest of $160 on note payable
	(d) Drafting supplies used of $150

Required: (1) Using the accounting system shown in the chapter, record the preceding
transactions.

(2) Prove the equality of the accounting equation at the end of March.

(3) Calculate the net income of the company for March.

(4) Prepare a balance sheet for the company on March 31.

5-38   The five transactions that occurred during June, the first month of operations for
Brown's Gym, were recorded as follows:

				Assets						=	Liabilities		+	Owner's Equity
Trans	Date	Cash	+	Gym Supplies	+	Land	+	Building	+	Gym Equipment =	Accts. Payable	+	Notes Payable +	Tom Brown, Capital
(a)	6/01	+$24,000												+ $24,000
(b)	6/05	− 8,000				+ $5,000		+ $23,000					+ $20,000	
(c)	6/07	− 270		+ $270										
(d)	6/17	− 4,000								+ $10,000			+ 6,000	
(e)	6/26			+ 480							+ $480			
Balances	6/30	$11,730	+	$750	+	$5,000	+	$23,000	+	$10,000 =	$480	+	$26,000 +	$24,000

Required: (1) Describe the five transactions that took place during June.

(2) Prepare a balance sheet on June 30.

5-39 The following transactions were recorded by the Sutton Systems Design Company for May, its first month of operations:

Trans	Date	Cash	+	Office Supplies	+	Land	+	Building	+	Office Equipment	=	Accts. Payable	+	Notes Payable	+	Steve Sutton, Capital
						Assets					**=**	**Liabilities**			**+**	**Owner's Equity**
(a)	5/01	+ $55,000														+ $55,000
(b)	5/02	− 8,000				+ $6,000		+ $18,000						+ $16,000		
(c)	5/08	− 3,500								+ $7,500				+ 4,000		
(d)	5/10			+ $1,100								+ $1,100				
(e)	5/22	+ 300								− 300						
Balances	5/31	$43,800	+	$1,100	+	$6,000	+	$18,000	+	$7,200	=	$1,100	+	$20,000	+	$55,000

Required: (1) Describe the five transactions that took place during May.
(2) Prepare a balance sheet at May 31.

5-40 At the beginning of July, Patti Dwyer established PD Company by investing $20,000 cash in the business. On July 5, the company purchased land and a building, making a $6,000 down payment (which was 10% of the purchase price) and signing a 10-year mortgage for the balance owed. The land was 20% of the cost, and the building was 80% of the cost. On July 17, the company purchased $3,800 of office equipment on credit, agreeing to pay half the amount owed in 10 days and the remainder in 30 days. On July 27, the company paid the amount due on the office equipment. On July 31, the company sold $900 of the office equipment that it did not need to another company for $900. That company signed a note requiring payment of the $900 at the end of one year.

Required: Based on the preceding information, prepare a balance sheet for PD Company on July 31. Show supporting calculations.

Making Evaluations

5-41 Your friend Maxine plans to supplement her job salary by running her own company at night and on the weekends. When the company earns enough money so that she can pay for a vacation home in the Caribbean, she plans to pay the bills of the company, sell the company's remaining assets, withdraw all the company's cash, and shut down the company. Since she will be extremely busy with her regular job and with running her new company, she plans to wait until she is ready to shut down the company to prepare a balance sheet, income statement, and cash flow statement. You think this is a bad idea.

Required: Do your best to convince Maxine that she should prepare financial statements more often, giving her examples of how doing this can help her and her company.

5-42 Chris Schandling is a loan officer at the First National Bank in Rochester, Minnesota. One day Nathan Wooten, who owns KidzLand (an indoor playground for young children), comes to the bank to see Chris about getting a $50,000 loan.

Required: (1) What types of questions do you think Chris will ask Mr. Wooten? Come up with at least three types of questions.
(2) What types of financial information do you think Chris will ask Mr. Wooten to provide? If Mr. Wooten asks Chris why this financial information is needed, how should Chris respond?
(3) Is it important that KidzLand's financial statements follow GAAP? Why or why not?

5-43 Andrew Poist works for Nilakanta and Company, a public accounting firm in Florence, South Carolina. On October 4, 2004, Sydney Langston, who started selling decorative, carved-wood duck decoys out of a booth at Cypress Court Mall during the first week in September, comes to see Andrew for some accounting help.
    Mr. Langston walks into Andrew's office carrying a small cardboard box. He tells Andrew the following:

"After I retired, I decided I needed something to help keep me busy. I started this little business, 'The Woodshed,' a month ago. It is open only on Fridays when the mall has its Craft Day. I leased the booth for one year. So, every Friday until September 1, 2005, I will display my ducks in the booth and sell them.

I know I should have come to see you before I got started. I just kept putting it off. So, here's what I did. Throughout the month of September I tossed everything having to do with The Woodshed's finances into this box. It has all kinds of documents in it. I have all of my bank deposits for the month, checks I wrote that were paid by my bank, the receipts for the woodworking supplies I bought the day I started, etc. I sorted out some items, like checks I wrote to the grocery store and the electric company. Anyway, it's the first part of October, and I can't figure out how well The Woodshed did in September. Can you?"

"Of course I can," replies Andrew. "I'll have something for you in a couple of days."

Mr. Langston leaves, and Andrew opens the small box. Inside is a small pile of documents:

(a) Five deposit slips from Mr. Langston's checking account. They total $2,300. Andrew notices that on four of the deposit slips, Mr. Langston wrote "Craft Sales." Each one of the deposit dates corresponds to each of the four Fridays in September. On the other deposit slip, which is for $1,300, Mr. Langston wrote "Social Security."

(b) Six canceled checks from Mr. Langston's checking account. They total $3,350. Four checks written to Miranda's Woodworking Supplies Company total $600. One check for $350 was written to Circuit City, and one check for $2,400 was written to Cypress Court Mall Management.

(c) A handwritten schedule that reads as follows:

Mallard	$ 60	sold
Grey Goose	$100	sold
Baby Duck	$ 40	sold
Swan	$200	
Donald Duck	$ 70	sold
Large Mallard	$130	sold

Required: (1) Using the information Mr. Langston supplied to Andrew, calculate your best estimate of the revenues, expenses, and net income for The Woodshed for September 2004.

(2) How could your calculations of revenues, expenses, and net income be misstated? When Andrew meets with Mr. Langston to discuss The Woodshed's operating results for September, what questions should he ask concerning the information Mr. Langston supplied?

5-44 In this assignment, we are going to chronicle the changes in value and ownership of one asset—a one-acre plot of land on the corner of Cedar Springs Road and McKinney Avenue in Dallas, Texas—from January 2004 through December 2006. Here are the significant events that happened to that plot of land during this time period:

January 4, 2004:	The land is purchased for $450,000 by Dalton Realty Company.
April 25, 2005:	Dalton Realty receives a tax assessment notice from the city of Dallas stating that the city now values the land at $510,000 for local tax purposes.
December 12, 2005:	The land is sold by Dalton Realty Company to Park Cities Development Company for $515,000. Park Cities pays in cash.
May 22, 2006:	Using the land as collateral (meaning that if Park Cities fails to repay its loan, the bank may get ownership of the land), Park Cities borrows $550,000 from North Carolina National Bank.
June 14, 2006:	Park Cities rents the land to The Crescent Court office complex for six months. The Crescent Court will store construction equipment on the land while making renovations to its office space.
December 31, 2006:	Park Cities sells the land to The Crescent Court for $590,000.

Required: When business closes for each day listed below, state (1) which company shows this land in its accounting records as an asset and (2) at what dollar amount the land is shown in that company's accounting records.

Date	Company showing the land as its asset	Dollar amount shown
1-4-04		
4-25-05		
12-12-05		
5-22-06		
6-14-06		
12-31-06		

5-45   Five years ago, Linda Monroe became the sole owner of LM Electronics. LM Electronics sells home entertainment centers, car audio equipment, and computers. LM advertises that it sells only the best brands, purchasing its inventory from well-known manufacturers in Japan, Germany, Norway, and the United States. Before opening this company, Linda was the accountant for The Music Warehouse. She understands accounting extremely well and maintains LM Electronics' accounting records according to generally accepted accounting principles.

On Friday morning, September 12, one of Linda's best customers, Sandy Wheeler, purchased a German-made CD player for $600. Linda was excited about making the sale because LM had only recently started carrying this particular brand. Linda filled out the sales invoice, collected the money, and helped Sandy carry the CD player out to her car.

Later that same day, Linda's friend Chris Tucker came into the store, also wanting to purchase a CD player. After browsing through the store, Chris started to leave. Linda stopped him and asked, "Chris, didn't you find a CD player that you would like to own?" Chris responded: "Well, Linda, I saw several items I would love to own, but I hadn't realized how expensive the equipment was. I guess I really can't afford to buy a new CD player."

Except for the deposit of the day's cash sales in the bank, no other activity took place that day at LM Electronics. After the store closed, Linda began thinking about Chris's comment. Early that evening Linda telephoned Chris and said: "Chris, I know you were wanting a new CD player, but if you are interested in saving a bunch of money, I would like to sell you the CD player I use at home. It is about two years old, and it is in great shape. I would sell it to you for $100."

Chris was very excited about Linda's offer. He drove over to Linda's house that same night, gave Linda $100 in cash, and took the CD player home. Linda immediately deposited the $100 in the bank night depository.

Required: Given the facts presented above and the information you learned in the chapter, (1) indicate whether you agree or disagree with the following statements and (2) explain each answer (this is the most important part, so think through the following statements carefully).

(a) Linda Monroe sold two CD players on September 12.
(b) LM Electronics sold two CD players on September 12.
(c) LM Electronics should record CD player sales of $700 on September 12.
(d) Linda Monroe should deposit $700 in the bank on September 12.

5-46   Paul Jenkins is the sole owner of Friendly Pawn Shop. Friendly Pawn Shop buys and sells jewelry, musical instruments, televisions, telephones, and small kitchen appliances. Paul has owned the pawn shop for almost one year, and the shop has developed a reputation as an honest, reliable place for families to buy or sell their used items.

Up until now, Friendly Pawn Shop has bought and sold goods only from retail customers. Paul believes that Friendly Pawn Shop is overstocked with jewelry, and he thinks the shop does not have enough musical instruments to meet the demand that will occur after the new school year starts. Paul believes that the pawn shop needs to sell some jewelry, which cost about $1,500, and replace it with several trumpets, trombones, and flutes.

Friendly Pawn Shop advertises in the newspaper when it wants to buy particular types of used items. This way Paul has the opportunity to inspect the goods before they are purchased, and he has the opportunity to discuss the history of each item with its current owner. In the present situation, however, Paul is considering making a merchandise trade with a wholesale pawnbroker. Although Paul is almost convinced that the trade will be the best way for his company to obtain the musical instruments, he has two major concerns.

First, Paul is concerned about maintaining Friendly Pawn Shop's reputation for reliable merchandise. He knows almost all of his customers, and he has earned their trust. Because Paul does not know where the wholesaler's musical instruments were purchased, he worries that he will be trading good jewelry for inferior-quality musical instruments. He would not find out that the instruments are inferior until the customers told him of their dissatisfaction. Second, Paul does not know how to record the trade in Friendly Pawn Shop's accounting records. He knows that the jewelry he plans to trade cost $1,500 and that he was going to try to sell the jewelry for $4,000. Paul does not know how much the wholesaler paid for the musical instruments or what price to charge his customers for each item.

Required: (1) Using the four-step approach you learned earlier in this book, discuss how you think Paul should solve this business problem.

(2) Assuming Friendly Pawn Shop trades the jewelry for the musical instruments owned by the wholesale pawnbroker, discuss how you think this transaction should be recorded in the accounting records. Be sure to include references to the accounting concepts introduced in this chapter.

5-47   Your friend Jim Wilson is about to prepare the January 31 balance sheet for his new company, Cheap Fun Video Arcade. This is Cheap Fun's first month of operation, and Jim is also going to calculate the first month's net income. He needs to prepare the balance sheet and calculate net income so he can pass the information along to his parents. They loaned him $5,000 so that he could start Cheap Fun.

Although Jim thinks that business is booming, he has a big problem. He does not know enough about accounting to prepare the balance sheet or calculate January's net income. As a matter of fact, Jim had never heard the words "balance sheet" and "net income" until his parents asked him to promise to furnish these statements to them every month before they would agree to loan Jim the $5,000.

Luckily, Jim saves every piece of paper associated with Cheap Fun. He kept copies of all of the business agreements he signed. He deposited all of the money Cheap Fun earned in the company's bank account and retained copies of every deposit slip. Jim also paid every company bill with a check and saved all of the related documents.

Required: Assume Jim wants to prepare Cheap Fun's January 31 balance sheet and January's income statement according to generally accepted accounting principles. Describe to Jim, in your own words, how he should organize the information about Cheap Fun's January transactions so that he can prepare a balance sheet and an income statement and keep his promise to his parents.

5-48   Samson Construction Company is a small company that constructs buildings. Normally, the time for Samson to complete the construction of a building is about six months. At the beginning of this year, Samson signed a contract to build a three-story office building at a selling price of $2,000,000. Samson will collect this amount when it completes construction of the building. Because this is a larger building than it usually builds, Samson expected that it would take two years to complete the construction, at a total cost of $1,400,000. This is the only building that Samson worked on during the year. By the end of the year, construction was on schedule; the office building was half complete and Samson had paid $700,000 costs. At this time Samson's bookkeeper came to Bill Samson, the owner, and said, "Samson Construction Company didn't do very well this year; it had a net loss of $700,000 because its revenue was zero and its expenses were $700,000."

Bill comes to you for advice. He says, "Normally, my company records the revenue and related expenses for constructing a building when it is completed. However, this three-story building will take much longer than usual. My construction crews have already been working on the building for one year and will continue to work on it for another year. My company has paid for one year's worth of salaries, materials, and other costs and will continue to pay for all of these costs incurred next year, so a lot of money will be tied up in the contract and won't be recovered until my company collects the selling price when the building is completed. How and when should my company record the revenue and expenses on this building? Do I really have a $700,000 net loss for the current year?"

Required: Prepare a written answer to Bill Samson's questions.

5-49   Yesterday, you received the following letter for your advice column at the local paper:

DR. DECISIVE

Dear Dr. Decisive:

My girlfriend went with me and my family to Hawaii last month, and we had a great time. But we have a question that we hope you will answer. Suppose a company has accumulated frequent flier miles (which it hasn't used yet) from plane tickets that it purchased for business trips taken by its employees. Are the company's frequent flier miles an asset or an expense? My girlfriend says they're an asset, but I think they're an expense. We have a bet on your answer. If I lose, I have to take hula lessons. If she loses, she has to take sumo wrestling lessons.

Please help!! I don't look good in a grass skirt.

"Wrestling Fan"

Required: Meet with your Dr. Decisive team and write a response to "Wrestling Fan."

THE INCOME STATEMENT: ITS CONTENT AND USE

1 Why is a company's income statement important?

2 How are changes in a company's balance sheet and income statement accounts recorded in its accounting system?

3 What are the parts of a retail company's classified income statement, and what do they contain?

4 What is inventory and cost of goods sold, and what inventory systems may be used by a company?

5 What are the main concerns of external decision makers when they use a company's income statement to evaluate its performance?

6 What type of analysis is used by external decision makers to evaluate a company's profitability?

How much did you earn last year working during the summer or during the school year? How did you keep track of your earnings? Did you make enough to cover all of your expenses? Or did your parents have to help you out? If so, what percentage of your expenses did your earnings cover? Companies, like individuals, keep track of their earnings. For instance, in 2002, **Wal-Mart** reported $219,812 million of revenues on its income statement. It also reported $171,562 million as the cost of the merchandise that it sold to customers, as well as $41,579 million of various other expenses, so that its net income was $6,671 million. Wal-Mart obtained these numbers from its accounting system. Did Wal-Mart charge customers enough for the merchandise it sold to them compared to what it paid for the merchandise? Did Wal-Mart make enough net income as a percentage of its revenues?

http://www.walmartstores.com

 Overall, do these numbers show that Wal-Mart had a "good" or "bad" year? What other information would you like to have to answer this question?

In Chapter 5 we looked at the fundamentals of the financial accounting process. You saw how basic accounting concepts, such as the entity concept, the accounting equation, and accrual accounting, provide the framework for the accounting system that a company uses to record its day-to-day activities. The system provides internal users with valuable information that helps managers in their planning, operating, and evaluating activities. The revenue and expense transactions are also the basis of a company's income statement, which shows external users the company's profit (income) for the accounting period.

In this chapter, we discuss the importance of the income statement, expand the accounting system from Chapter 5, describe and present a classified income statement, and show how the income statement helps managers and external users make business decisions.

WHY THE INCOME STATEMENT IS IMPORTANT

A company's income statement plays a key role in the decision making of the users by communicating the company's revenues, expenses, and net income (or net loss) for a specific time period. A company earns income by selling inventory (goods) or by providing services to customers during an accounting period. Recall that revenues are the prices a company charges its customers for the goods or services. Expenses are the costs of providing the goods or services during the period. An income statement is based on the equation we showed in Chapter 5:

① Why is a company's income statement important?

Net Income = Revenues − Expenses

Companies may use different titles for their income statements, including *statement of income* (**AT&T, Avon Products**), *statement of earnings* (**Black & Decker, Eastman Kodak**), or *statement of operations* (**Apple Computer, Rocky Mountain Chocolate Factory, Inc.**). You may also hear the income statement referred to as a *profit and loss (P&L) statement*.

http://www.att.com
http://www.avon.com
http://www.blackanddecker.com
http://www.kodak.com
http://www.apple.com
http://www.rmcfusa.com

Recall from Chapter 4 that a company prepares a "projected" income statement for *internal* use as part of its master budget. Exhibit 6-1 shows how internal users (managers) use a company's *projected* income statement and actual income statement in their decision making, as well as how *external* users use a company's *actual* income statement to make economic decisions. We explain the impact of the income statement on users' decisions in the rest of this section.

The income statement summarizes the results of a company's operating activities for a specific accounting period. These operating activities stem from the planning and operating decisions that managers made during the period. Hence, a company's income statement shows the relationship between managers' decisions and the results of those decisions. This information helps both internal and external users evaluate how well the

| EXHIBIT 6-1 | USES OF A COMPANY'S INCOME STATEMENT |

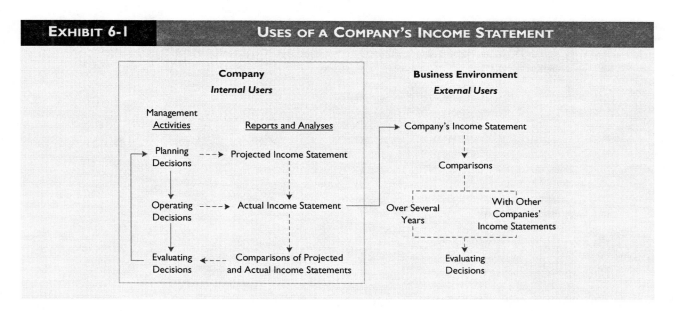

company's managers have "managed" during the period. By comparing a company's income statement information from period to period, users also can evaluate managers' ability over the longer run.

Let's first look at how managers use the income statement for making comparisons. Remember from Chapter 5 that a company keeps track of its activities by using an accounting system based on the accounting equation (Assets = Liabilities + Owner's Equity) and the dual effect of transactions. The accounting system provides the information that managers need to compare actual results with the expected (budgeted) results and to prepare external financial statements. At the end of an accounting period (e.g., one year), a company's income statement will show how well many of its managers' business decisions worked out.

For example, the revenue and expense information shows the results of managers' cost-volume-profit (C-V-P) analysis and budgeting decisions. In Chapter 3, you saw how Anna Cox used C-V-P analysis to develop her business plan. C-V-P analysis showed her how Sweet Temptations could break even and how it could earn a satisfactory profit. Anna calculated that Sweet Temptations needed to sell 700 boxes of candy for $10 a box for the company to break even. In addition to helping managers predict a company's break-even point, C-V-P analysis improves managers' operating decisions such as estimating how much inventory to purchase, what sales price to charge, and what effect on profit to expect from price changes. We determined that if Sweet Temptations was able to sell 720 boxes, it would earn a profit of $110.

Consider the decision that managers must make about what sales price to charge. If Anna sets the price too high, Sweet Temptations risks not selling enough boxes to break even. If she sets the price too low, Sweet Temptations may sell many boxes of candy but may not earn high enough revenues to cover the costs of selling the candy. Later, when the accounting system keeps track of every sale, it records those sales at the prices that the customers actually paid. (Remember that every sale generates a sales invoice to document the transaction and the amount of the sale.) If Anna did a good job of assessing the market and establishing an appropriate price, Sweet Temptations will make sales, will earn revenues high enough to cover its expenses, and will make a profit that it will report on its income statement.

In Chapter 4, we discussed how budgets help managers make plans, control company expenses, and evaluate company performance. If you were the manager of Sweet Temptations, budgeting would allow you to compare your expectations for revenue and expense amounts (reported in the projected income statement) with the actual amounts (reported in the actual income statement). If sales were higher or expenses lower than expected, you could find out what you did right and keep doing it. If, on the other hand, sales were lower or expenses higher than expected, you could analyze your mistakes and try to improve.

 How do you think a company's decision to decrease the price of its product will affect the revenues that it reports on its income statement? How will this decision affect its expenses?

As valuable as C-V-P analysis and budgets are for internal decision making, companies do not report to external decision makers much of the information they provide. For one thing, companies don't want to reveal specific cost or budget information to their competitors. For another, many companies prepare internal accounting reports daily, so external users may be more confused than helped by the sheer volume of information.

External users need accounting information that lets them compare a company's actual operating performance over several years or with that of other companies. For instance, if the company is a corporation, potential investors and current stockholders use its income statement information to help them decide whether to buy or hold capital stock of the corporation. By comparing the company's current operating performance with that of prior years, they can get a sense of the company's future operating performance. By comparing the company's current operating performance with that of other companies, they can get a sense of whether the company is doing "better" or "worse" than these other companies. Banks and other financial institutions also use a company's income statement in a similar way to evaluate whether or not to give the company a loan. Finally, suppliers also use a company's income statement information. Suppliers do not have the resources to grant credit to all customers. A supplier can compare its customers' income statements to determine which ones might be the best credit risks. Generally accepted accounting principles (GAAP) ensure that all companies calculate and publish financial statement information in a similar, and thus comparable, manner. Thus, understanding GAAP is important to the accountant who prepares financial statements and to the external decision maker who uses these statements to make business decisions.

In Chapter 5 we introduced a simple accounting system, as well as several concepts and terms that form the foundation of GAAP. In this chapter we expand that accounting system, extend our discussion of GAAP as it relates to the income statement, and begin to explain how external users evaluate income statement information for decision making.

EXPANDED ACCOUNTING SYSTEM

In Chapter 5 we kept track of Sweet Temptations' transactions using the accounting equation to set up columns for recording amounts for assets, liabilities, and/or owner's equity. We then expanded the accounting equation to include revenue and expense transactions. Adding revenue and expense columns let us keep track of these transactions separately from owner investments and withdrawals. However, a company needs to know more than its total revenues and total expenses. A company must know the total of each of its revenues and the total of each of its expenses for the accounting period so that it can report these items in a useful manner on its income statement for that period. In this chapter we expand the accounting system that we introduced in Chapter 5. We continue to use columns for each asset and liability account. However, we create a separate column under Owner's Equity for *each* revenue account and *each* expense account, while still retaining an owner's capital account column. A company uses these revenue and expense accounts *for only one accounting period* to record the effects of its transactions on its net income, so they are called **temporary accounts**. Asset, liability, and the owner's capital accounts are called **permanent accounts** because they are used *for the life of the company* to record the effects of its transactions on its balance sheet.

By using this expanded accounting system, we show how a company keeps track of the changes in (and balances of) each asset, liability, owner's capital, revenue, and

2 How are changes in a company's balance sheet and income statement accounts recorded in its accounting system?

expense account. After showing how a company records a transaction in this accounting system, we also include a marginal note to help you understand the effect of the transaction on the company's financial statements. (We will illustrate how this works in the next section of the chapter.) We use this columnar accounting system because it is easy to see the effects of a company's transactions on its various accounts, accounting equation, and financial statements. You should realize, however, that a real company has many (sometimes hundreds!) different types of assets, liabilities, revenues, and expenses. Imagine how wide the paper would need to be to record transactions involving hundreds of account columns! So a real company uses a computerized accounting system or a more complex manual accounting system involving items you may have heard of, such as "journals," "ledgers," "debits and credits," and different forms of accounts. We illustrate this manual system in Appendix A at the end of the book.

THE RETAIL COMPANY'S INCOME STATEMENT

3 What are the parts of a retail company's classified income statement, and what do they contain?

As we discussed earlier, the income statement is an important part of the decision-making process for both internal and external users. It is an expansion of the income equation that we presented earlier:

$$\text{Net Income} = \text{Revenues} - \text{Expenses}$$

Revenues may be thought of as the "accomplishments" of a company during an accounting period. Revenues are the prices charged to customers and *result in increases in assets (cash or accounts receivable) or decreases in liabilities (unearned revenues).* Expenses may be thought of as the "efforts" or "sacrifices" made by a company during an accounting period to earn revenue. Expenses are the costs of providing goods and services and *result in decreases in assets or increases in liabilities.*

Keep these definitions in mind while we discuss how a company provides revenue and expense information to external users in its "classified" income statement. Let's return to Sweet Temptations to see how Anna records and reports the results of its first month of operations. To reinforce your understanding of the columnar accounting system, we will show how to record a few revenue and expense transactions. We will also show Sweet Temptations' classified income statement. As you look at this income statement, focus on understanding the income statement sections but also think about how Anna recorded the individual revenue and expense transactions.

The classified income statement of a retail company like Sweet Temptations has two parts: an "operating income" section and an "other items" section. **Operating income** includes all the revenues earned and expenses incurred in the primary operating activities of the company. The operating income section has three subsections: (1) revenues, (2) cost of goods sold, and (3) operating expenses. **Other items** include any revenues and expenses that are not directly related to the primary operations of the company, items such as interest revenue and interest expense. Exhibit 6-2 shows Sweet Temptations' classified income statement for January 2004.

In the next sections, we will discuss various issues related to recording and reporting revenues and expenses. We refer to Exhibit 6-2 to show how Sweet Temptations reports certain items.

REVENUES

A retail company sells goods to customers either for cash or on credit. When goods are sold on credit, some retail companies offer an incentive for prompt payment. Whether the sales are for cash or on credit, customers sometimes return the goods they purchased. Let's see how companies record these aspects of sales.

EXHIBIT 6-2	SWEET TEMPTATIONS' CLASSIFIED INCOME STATEMENT

SWEET TEMPTATIONS
Income Statement
For the Month Ended January 31, 2004

Sales revenues (net)..		$ 8,100
Cost of goods sold...		(3,645)
Gross profit..		$ 4,455
Operating expenses (see Exhibit 6-4):		
Selling expenses	$2,961	
General and administrative expenses	884	
Total operating expenses		(3,845)
Operating income..		$ 610
Other item:		
Interest expense...		(8)
Net Income		$ 602

Sales Revenue

Whether a customer buys goods for cash or on credit, retail companies use a Sales Revenue, or simply Sales, account to record the transaction. Recall from Chapter 5 that the source document for a sale is a sales invoice, or simply an invoice. Some companies that sell only a few products or have a computerized accounting system may use a cash register tape or a credit card receipt as the source document. Exhibit 6-3 shows the sales invoice that Sweet Temptations used for one of its sales. This invoice shows you that on January 6, 2004, Sweet Temptations sold 10 boxes of milk chocolate candy for $10 per box, totaling to a $100 sale. Notice that the invoice also tells you that the invoice number was 0001, that the boxes of milk chocolate had an inventory identification number (ID #) of 0036, that it was a credit sale, and that the credit sale was made to Bud's Buds. It is important that the invoice includes all of the sales information needed to record this transaction.

Anna records the January 6 credit sale by first increasing Accounts Receivable by $100 to show that Bud's Buds owes Sweet Temptations that amount, as we show on the next page. Notice that the beginning balance of accounts receivable was $400; this is the amount owed to Sweet Temptations by The Hardware Store (see Exhibit 5-11 on page 139). So, Accounts Receivable now has a balance of $500. Anna also increases the Sales Revenue account column under the *Net Income* heading of Owner's Equity by $100 to show that Sweet Temptations earned that amount from the sale. Notice that the previous

EXHIBIT 6-3	SWEET TEMPTATIONS' SALES INVOICE

Invoice #0001		SWEET TEMPTATIONS				
		Sales Invoice			Cash____	Credit_X_
Date	Description	ID #	# of Boxes		Unit Price	Total
1/6/04	milk chocolate	0036	10		$10	$100
Sold To: ___Bud's Buds___		Acct # ___0103___				

balance of sales revenue was $300 due to a sale on January 2, so that Sales Revenue now has a balance of $400.

Assets	=	Liabilities	+	Owner's Equity		
					Net Income	
				Revenues	−	Expenses
Accounts Receivable				Sales Revenue		
Bal $400				$300		
1/6/04 +100				+100		
Bal $500				$400		

To illustrate how an account balance changes, we showed the beginning and ending balances of both the Accounts Receivable account and the Sales Revenue account. **For simplicity, later in this chapter and future chapters we will include the balance of an account only when it is critical to the discussion.** At first glance, in the previous example it does not appear that Sweet Temptations' accounting equation is in balance. Remember, however, that Sweet Temptations' has many accounts in its accounting system and that we are showing only two of its accounts in this example. Sweet Temptations' accounting equation was in balance before Anna recorded this transaction, as we showed in Exhibit 5-13 on page 143. What is important to notice is that in the above example, assets (Accounts Receivable) increased by $100 and owner's equity (Sales Revenue) increased by $100. So Sweet Temptations' accounting equation remains in balance after Anna recorded the transaction, as we showed in Exhibit 5-14 on page 143.

Remember that Sweet Temptations had to dip into its candy inventory to make its sale. Anna uses the inventory identification number from the boxes of candy that Sweet Temptations sold to determine the $45 cost of the candy (10 boxes at $4.50 per box). So Anna records the cost of the sale by first decreasing Inventory by $45, as we show below. Note that the balance of the Inventory account prior to the sale was $5,805 (see Exhibit 5-13 on page 143) and the balance is $5,760 after the sale. Anna also increases the Cost of Goods Sold account column (an expense) under the *Net Income* heading of Owner's Equity by $45 to show the cost of the boxes of candy that Sweet Temptations sold on January 6. Remember that as cost of goods sold (an expense) *increases,* both net income and owner's equity *decrease.* That is why we include a minus (−) sign in the column in front of the Cost of Goods Sold column. Notice that the previous balance of cost of goods sold was $135 due to the sale on January 2, so that Cost of Goods Sold now has a balance of $180.

Assets	=	Liabilities	+	Owner's Equity		
					Net Income	
				Revenues	−	Expenses
						Cost of Goods Sold
Inventory						
Bal $5,805						$135
1/6/04 − 45					−	+ 45
Bal $5,760					−	$180

Although we do not show all the account balances in Sweet Temptations' accounting system, notice that the accounting equation continues to remain in balance after Anna records these two transactions because the $55 total increase in assets ($100 increase in Accounts Receivable less $45 decrease in Inventory) is equal to the $55 total increase in owner's equity [$100 increase in Sales Revenue (a revenue) less $45 increase in Cost of Goods Sold (an expense)]. Anna records each sales transaction for January in the same way (except she records cash sales in the Cash account column rather than the Accounts Receivable account column). At the end of the accounting period, she calculates the $8,100 balance in the Sales Revenue account column and Sweet Temptations reports it as revenue on its income statement for January, as we show in Exhibit 6-2. She calculates the $3,645 balance in the Cost of Goods Sold account column and Sweet Temptations reports this as an expense on its income statement.

How Sales Policies Affect Income Statement Reporting

Companies may have several policies related to the sales of their goods or services. There are three types: discount policies, sales return policies, and sales allowance policies. Companies want to encourage customers to buy their merchandise or services, and sales policies help them do this. A retail company's specific policies will also have an impact on its net sales—the net dollar amount of sales reported on its income statement—because its revenues for an accounting period should include only the prices actually charged to customers for goods sold during the period. In the following sections, we will discuss each of these sales-related policies.

Discounts

Have you ever taken advantage of a two-for-one special, paid a lower price because you bought a larger quantity of the same item, or used a coupon to get three dollars off the price of your pizza? If so, the company you bought from offered you a discount. A **quantity** (or *trade*) **discount** is a reduction in the sales price of a good or service because of the number of items purchased or because of a sales promotion.

Companies use discounts to attract customers and increase sales. Suppose that in early February, Sweet Temptations puts in its front window a sign that reads, "Valentines Day Special—Buy four or more boxes of chocolates and receive a 10% discount." By using this sales promotion, Sweet Temptations hopes that people walking by will notice the sign, come into the store, and buy candy. In addition, the company hopes that customers who had planned to buy only one or two boxes will instead buy four so that they can get the discount. Anna also hopes the policy will encourage repeat customers.

Before deciding to start a specific quantity discount policy, Sweet Temptations uses C-V-P analysis to determine the discount that will most likely improve company profits. Once a quantity discount policy is set, the company keeps track of the impact the policy has on sales, costs, and profits. However, the company does *not* record quantity discounts in its accounting system.

A company also may decide to offer a discount for early payment on credit sales. A **sales discount** is a percentage reduction of the invoice price if the customer pays the invoice within a specified period. A sales discount is frequently called a **cash discount** because when taken by a customer, the discount reduces the cash received. The sales invoice shows the terms of payment. These terms vary from company to company, although most competing companies have similar credit terms.

Sales (cash) discount terms might read 2/10, n/30 ("two ten, net thirty"). The first number is the percentage discount (2%), and the second number (10) is the number of days in the discount period. The discount period is the time, starting from the date of the invoice, within which the customer must pay the invoice to get the sales discount. The term n/30 means that the total invoice price is due within 30 days of the invoice date. Thus 2/10, n/30 is read as "a 2% discount is allowed if the invoice is paid within 10 days; otherwise, the total amount of the invoice is due within 30 days." If Sweet Temptations makes a $50 sale on credit with terms 2/10, n/30 and the customer pays the invoice within 10 days, the customer would pay $49 [$50 − (0.02 × $50)], and $1 would be the sales discount taken. Sometimes companies offer cash discounts by charging a lower price for cash ͏ es (rather than credit sales). Some gas stations have this policy. A company's ac- system keeps track of sales (cash) discounts by reducing sales revenue by the sales discounts taken when customers pay for their credit purchases.

do you think a sales discount taken by a credit customer when paying the account able is recorded in the customer's accounting system?

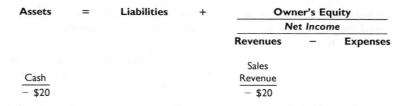

FINANCIAL STATEMENT EFFECTS

Decreases assets on **balance sheet**. Decreases revenues, which decreases net income on **income statement** (and therefore decreases owner's equity on **balance sheet**).

Sales Returns and Allowances

When a customer buys merchandise, both the company and the customer assume that it is not damaged and is acceptable to the customer. Occasionally, on checking the merchandise after the purchase, the customer may find that it is damaged, is of inferior quality, or simply is the wrong size or color. Most retail companies have a policy allowing customers to return merchandise. For example, The Limited, Inc.'s Express store prints its return policy on all its price tags. A **sales return** occurs when a customer returns previously purchased merchandise. The effect of a sales return is to cancel the sale (and the related cost of goods sold).

 Have you ever returned merchandise to a store? Did the customer service representative ask to see your sales receipt? Did you or the customer service representative fill out additional source documents? Why?

If a customer discovers that merchandise is damaged, a company may offer the customer a sales allowance. A **sales allowance** occurs when a customer agrees to keep the merchandise, and the company refunds a portion of the original sales price.

Although this transaction is not part of our ongoing analysis of Sweet Temptations, assume that one of its customers, Roger Leslie, purchased four boxes of chocolates for $10 per box and paid cash. Remember, each box of candy costs Sweet Temptations $4.50. Anna would have recorded this transaction by increasing both Sweet Temptations' Cash and Sales Revenue by $40 ($4 \times 10), decreasing Inventory by $18 ($4 \times 4.50), and increasing Cost of Goods Sold (an expense) by $18.

What would have happened if later that day when Roger opened the candy, he noticed that half of it was melted? If he returned to the store, Anna might have asked him if he wanted to exchange the candy for new boxes, return the candy for a refund, or accept a $20 sales allowance and keep the candy. If the candy still tasted fine, Roger might have decided to accept the sales allowance.

Because Roger paid for the candy with cash, Sweet Temptations would have granted the sales allowance by refunding him $20 cash. Anna would have recorded this sales allowance transaction in Sweet Temptations' accounts as follows:

Assets	=	Liabilities	+	Owner's Equity		
					Net Income	
				Revenues	−	Expenses
Cash				Sales Revenue		
− $20				− $20		

If Roger originally purchased the candy on credit, Sweet Temptations would have granted the sales allowance by decreasing Roger's Account Receivable balance, instead of the Cash balance, by $20.

 If Roger returned the candy for a refund, how would you record the transaction?

Whether a company grants a sales return or a sales allowance, it prepares a source document called a credit memo. (Remember that a source document serves as evidence that a transaction has occurred.) A **credit memo** is a business document that lists the information for a sales return or allowance. It includes the customer's name and address, how the original sale was made (cash or credit), the reason for the sales return or allowance, the items that were returned or on which the allowance was given, and the amount of the return or allowance. The credit memo is the source document used to record the return or allowance. As we will discuss in Chapter 18, it is also the document used to keep track of "external failure costs," a measure of customer dissatisfaction. The effect of recording sales discounts, sales returns, and sales allowances is to reduce sales revenue (as we will discuss in the next section).

How can a company's sales return policy help increase profits? Do you think a sales return policy ever can hurt more than it helps? How?

Net Sales

At the end of the accounting period, the balance of a company's Sales Revenue account column includes the initial sales revenue, less the sales returns and allowances, and the sales (cash) discounts taken. The balance of the Sales Revenue account is called Sales Revenue (net), or Net Sales, and is reported on the company's income statement.

In January 2004 Sweet Temptations did not allow any cash discounts and did not have any customers return their purchases or ask for an allowance. The company thus reports total sales revenue of $8,100 on its income statement, as we show in Exhibit 6-2. It seems that Sweet Temptations' customers were satisfied with the quality of the candy they bought. In general, the amounts that a company records as sales returns and allowances (and sales discounts) provide useful information about the quality of the company's products (and the effect of its cash discount policy).

Do you think a company should report to its managers a single net sales amount or both the total sales and the sales returns, allowances, and discounts? To external users? Why?

EXPENSES

An old business phrase says, "You have to spend money to make money." But a company should understand that planning and controlling its expenses is an important part of running a business. In the previous section you saw how Sweet Temptations, a retail company, recorded and reported its revenues. In this section we focus on expenses.

Cost of Goods Sold

One of the major expenses of a retail company is the cost of the goods (merchandise) that it sells during the accounting period. A classified income statement shows this expense as the **cost of goods sold**.

Although all retail companies report their costs of goods sold, *how* a retail company calculates the amount depends on the type of inventory system it uses. Remember, **inventory** is the merchandise a retail company is holding for resale. A company uses an inventory system to keep track of the inventory it purchases and sells during an accounting period and, thus, the inventory it still owns at the end of the period. Companies use either a perpetual inventory system or a periodic inventory system. Because the type of inventory system that a company uses affects its managers' decisions and the income statement calculations, we briefly discuss the cost of goods sold under each type of system.

4 What is inventory and cost of goods sold, and what inventory systems may be used by a company?

Perpetual Inventory System

A **perpetual inventory system** keeps a continuous record of the cost of inventory on hand and the cost of inventory sold. Under the perpetual inventory system, when a company purchases an item of inventory, it increases the asset Inventory by the invoice cost of the merchandise plus any freight charges (sometimes called *transportation-in*) it paid to have the inventory delivered. When the company sells merchandise, it records the sale in the usual way. It also reduces Inventory and increases Cost of Goods Sold by the cost of the inventory that it sold. (We illustrated this earlier for one of Sweet Temptations' sales.) So, the company has Inventory and Cost of Goods Sold accounts that are perpetually up-to-date, and the company always knows the physical quantity of inventory it should have on hand.

Do you think perpetual inventory records could be wrong? What could cause the records to show either too much or too little inventory?

Because of computer technology, many retail stores use a perpetual inventory system. When you buy something in a store, if the salesperson uses a scanner to record your purchase, the company is using a perpetual inventory system. Computers help stores record sales transactions and keep their perpetual inventory records. For instance, a grocery store uses an optical scanner to read a bar code and record the price of the item into the cash register. The store's computer simultaneously increases Cash (or Accounts Receivable) and Sales Revenue for the item's sales price, reduces Inventory and increases Cost of Goods Sold by the amount of that item's cost, and updates the count of the quantity of inventory on hand. Most department stores use a perpetual inventory system, as do most retail stores that sell a relatively small number of very expensive items, such as automobiles and jewelry.[1]

Whether a company sells expensive jewelry or generic grocery items, the company's perpetual inventory system keeps up-to-date amounts for both Inventory and Cost of Goods Sold. This information helps managers with decisions about day-to-day operations. By monitoring the daily changes in inventory amounts, managers can decide when to make inventory purchases, thus making sure that inventory items are always in stock. Because the cost of goods sold information is current, managers can also compare the revenues and costs of recent sales and estimate the company's profitability. However, managers should evaluate the costs of computerized equipment, employee training, and the other support needed to operate a perpetual inventory system before deciding to use this type of system. In some cases, the benefits may not justify the added costs of keeping perpetual records. As computer technology becomes more affordable and as competition increases, companies typically find that perpetual systems are worth the costs.

At the end of an accounting period, a company includes the balance of its Inventory account on its balance sheet. The company includes the balance of its Cost of Goods Sold account on its income statement. As we illustrated earlier, Sweet Temptations uses a perpetual inventory system. Sweet Temptations also uses the *specific identification method* for determining its cost of goods sold because it identifies the cost of each box of candy sold based on the identification number of the box. At the end of January 2004, its Inventory account has a balance of $2,295, and its Cost of Goods Sold account has a balance of $3,645 from all the purchases and sales transactions recorded in January. We show these account columns below (the amounts are the same as those listed in Exhibit 5-20):

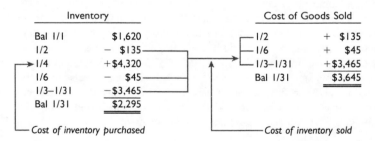

The $3,645 cost of goods sold is reported on the income statement as we show in Exhibit 6-2. The $2,295 ending inventory is reported on the balance sheet shown in Exhibit 5-22. We will discuss perpetual inventory systems in more detail in Chapter 19.

Periodic Inventory System

A **periodic inventory system** does not keep a continuous record of the inventory on hand and sold, but determines the inventory at the end of each accounting period by physically

[1] In Chapter 19 we will discuss accounting for inventory in more detail. Accounting systems may keep track only of inventory quantities, instead of inventory costs, on a perpetual basis.

Assuming this person is "taking inventory," what do you suppose is in his notebook?

counting it. Because a periodic inventory system does *not* reduce the Inventory account each time a sale occurs, the *only* time the company knows the cost of its inventory on hand is when it counts the inventory.

Why would a company choose not to keep perpetual inventory records? There are two common reasons. First, many companies that use a periodic inventory system are small enough that they can manage their inventory without perpetual records. Second, many companies sell a high volume of similar, inexpensive goods. Because the items are not expensive, perpetual records are not as important for keeping day-to-day physical control over the inventory. For these reasons, a company may decide that the costs of a perpetual system (i.e., recordkeeping costs, computer hardware and software costs) are not worth the benefits.

Because a company using a periodic system does not keep perpetual records, it must physically count its inventory at year-end. Physically counting the inventory is the only way for the company to determine an accurate inventory amount to be reported in the company's ending balance sheet. Therefore, a company usually counts its inventory immediately after the last working day of its fiscal year.[2] This is a difficult and time-consuming task. Thus, most companies end their fiscal year when inventory levels are likely to be low and business is slow. For example, most department stores take their inventory following the "after Christmas" sales, whereas a ski shop might count its inventory in June.

Perhaps you have noticed a company's advertisement that says something like the following:

> **YEAR-END INVENTORY
> CLEARANCE SALE!!!!
> WE'D RATHER SELL IT
> THAN COUNT IT!**

This company is reducing its prices to sell more goods so that it will not have to spend as much time counting inventory. Near the end of the year it is not unusual for a company

---

[2]A company using a perpetual inventory system also physically counts its inventory at year-end. Even though its accounting records show what *should* be in the inventory, the company takes a physical count to determine its *actual* inventory so it can test the accuracy of its accounting records and estimate the amount of lost or stolen inventory.

to close temporarily, so that it can count its inventory. If you saw the company's sign (like the one in the picture on the previous page) and peeked in the window, you would see people moving from one aisle to the next, counting the merchandise on each shelf.

How does a company using a periodic system know its cost of goods sold? Since the company does not record the cost of the goods sold when each sales transaction takes place, it must *calculate* its cost of goods sold for an accounting period as follows:

$$\text{Cost of Goods Sold} = \overbrace{\underset{\text{Inventory}}{\text{Beginning}} + \underset{\text{Purchases}}{\text{Net}}}^{\substack{\text{Cost of goods}\\\text{available for sale}}} - \overbrace{\underset{\text{Inventory}}{\text{Ending}}}^{\substack{\text{Cost of goods}\\\textbf{not}\text{ sold}}}$$

A company knows the cost of its beginning inventory because the beginning inventory for a new accounting period is the same as the ending inventory for the previous accounting period. A company's cost of net purchases is the dollar amount it recorded during an accounting period for the merchandise it bought for resale. The term **net purchases** is used because the amount of merchandise purchases (invoice cost and transportation-in) is adjusted (reduced) for purchases returns, allowances, and discounts. (These adjustments are similar to the net sales adjustments that we discussed earlier.) A company's **cost of ending inventory** is the dollar amount of merchandise on hand, based on the physical count, at the end of the accounting period.

Cost of Goods Sold and Gross Profit

Because cost of goods sold is usually a retail company's largest expense, many companies subtract cost of goods sold from net sales to determine **gross profit**. Gross profit is the amount of revenue that a company has "left over" (after recovering the cost of the products it sold) to cover its operating expenses. Sweet Temptations subtracts its $3,645 cost of goods sold from its $8,100 sales revenue (net) to get its $4,455 gross profit, as we showed in Exhibit 6-2 on page 171.

 Do you think Anna Cox is pleased with Sweet Temptations' gross profit? Why or why not?

Operating Expenses

Of course, the cost of goods sold is not the only expense that a retail company incurs. Activities such as having a sales staff, occupying building space, or running advertisements in the newspaper also cost money. These types of expenses are called *operating expenses*. **Operating expenses** are the expenses (other than cost of goods sold) that a company incurs in its day-to-day operations.

A company records its operating expenses in account columns as we discussed earlier. For instance, when Sweet Temptations paid $300 for advertising on January 25, 2004, Anna recorded the transaction as follows:

FINANCIAL STATEMENT EFFECTS

Decreases assets on *balance sheet*. Increases operating expenses, which decreases net income on *income statement* (and therefore decreases owner's equity on *balance sheet*).

Assets	=	Liabilities	+	Owner's Equity		
					Net Income	
				Revenues	−	Expenses
Cash						Advertising Expense
−$300					−	+$300

Notice again that an increase in an expense causes a decrease in owner's equity, as indicated by the minus (−) sign in the column in front of the $300 advertising expense.

Likewise, on January 31, 2004, when Anna prepared the end-of-period adjustment of $15 for depreciation, she recorded the expense as follows:

Assets	=	Liabilities	+	Owner's Equity		
					Net Income	
				Revenues	−	Expenses
Store Equipment						Depreciation Expense
Bal $2,000					−	+ $15
− $15						
Bal $1,985					−	$15

FINANCIAL STATEMENT EFFECTS

Decreases assets on **balance sheet**. Increases operating expenses, which decreases net income on **income statement** (and therefore decreases owner's equity on **balance sheet**).

Companies refer to the recording of end-of-period adjustments as making **adjusting entries**.

A company may divide its operating expenses section of the income statement into two parts, one for selling expenses and the other for general and administrative expenses. **Selling expenses** are the operating expenses related to the sales activities of a company. Sales activities are activities involved in the actual sale and delivery of merchandise to customers. Selling expenses include such items as sales salaries expense, advertising expense, and delivery expense (sometimes called *transportation-out*) for merchandise sold. **General and administrative expenses** are the operating expenses related to the general management of a company. They include such items as office salaries expense, insurance expense, and office supplies expense.

Why do you think that companies may report selling expenses separately from general and administrative expenses?

Some operating expenses involve both sales activities and the general management of a company. Consider utilities, for example. A company keeps telephones at sales desks and office desks, and both sales areas and office spaces are provided with electricity. In these cases, a company allocates part of the total expense to selling expenses and the remainder to general and administrative expenses based on an estimate of how much is used for each activity, as we discussed in Chapter 4.

Exhibit 6-4 shows a detailed schedule of Sweet Temptations' operating expenses for January 2004. Anna developed this schedule from the balances in Sweet Temptations'

EXHIBIT 6-4   SWEET TEMPTATIONS' OPERATING EXPENSES

SWEET TEMPTATIONS
Schedule 1: Operating Expenses
For Month Ended January 31, 2004

	Amount	Chapter 5 Transaction #
Selling expenses		
Consulting expense	$ 150	(7)
Advertising expense	300	(8)
Sales salaries expense	1,537	(10)
Sales telephone expense	45	(11)
Sales utilities expense	142	(12)
Sales supplies expense	22	(14)
Rent expense	750	(15)
Depreciation expense	15	(16)
Total selling expenses	$2,961	
General and administrative expenses		
Consulting expense	$ 50	(7)
Office salaries expense	513	(10)
Office telephone expense	15	(11)
Office utilities expense	48	(12)
Office supplies expense	8	(14)
Rent expense	250	(15)
Total general and administrative expenses	$ 884	

expense accounts at the end of January. These balances are based on all the expense transactions Anna recorded in the accounts during January. Although we do not show these expense accounts here, the amounts she recorded are the same as those in the Expenses column in Exhibits 5-20 and 5-21 on pages 148 and 150. For clarity, we have identified the number of the 2004 transaction listed in Chapter 5 that caused each expense. Most of Sweet Temptations' operating expenses are selling expenses. However, Anna estimates that one-quarter of each total for consulting ($200), salaries ($2,050), telephone ($60), utilities ($190), supplies ($30), and rent ($1,000) expenses are general and administrative expenses.[3] Sweet Temptations includes the $2,961 total selling expenses and the $884 total general and administrative expenses on its income statement in Exhibit 6-2. The detailed schedule of expenses is included with its income statement so that users interested in specific types of expenses can get the information they need.

OPERATING INCOME, OTHER ITEMS, AND NET INCOME

On a company's income statement, the total operating expenses are deducted from gross profit to determine **operating income**. In Exhibit 6-2, Anna adds the total selling expenses to the total general and administrative expenses to determine the $3,845 total operating expenses. She deducts the total operating expenses from the $4,455 gross profit to determine Sweet Temptations' operating income of $610.

The **other items** (sometimes called the *nonoperating income*) section of a company's income statement includes items that are not related to the primary operations of the company. Reported in this section are revenues and expenses related to investing activities or to financing the company's operations (e.g., interest revenue and interest expense), revenues and expenses (called *gains* and *losses*) related to selling property and equipment assets, and incidental revenues and expenses (e.g., miscellaneous rent revenue, losses due to theft or fire). We will discuss these items more fully later in the book. Sweet Temptations includes interest expense of $8 in its *other items* section. This interest expense relates to the note payable that it owed for the entire month of January; the interest was recorded in an adjusting entry (#17) that we showed in Exhibit 5-21.

The total amount of the *other items* (nonoperating income) section is added to (or subtracted from) the operating income to determine a company's net income. The net income of Sweet Temptations (shown in Exhibit 6-2) for the month ended January 31, 2004 is $602, determined by subtracting the $8 other item (interest expense) from the $610 operating income.

USES OF THE INCOME STATEMENT FOR EVALUATION

To help you understand *how* internal and external decision makers use income statements, we will first briefly review *why* companies prepare financial statements and *what* the statements show. Recall that accounting information helps managers plan, operate, and evaluate company activities. Managers use accounting information on a day-to-day basis to help make decisions (e.g., what type of sales return policy to use or how much inventory to order) to achieve their objective of earning a profit and thereby increasing the company's value. At the end of a specific time period, managers prepare financial statements to report to external users the cumulative results of their day-to-day decisions. By analyzing a company's financial statements, external users can evaluate how well managers' decisions worked and decide whether to do business with the company.

If you are a creditor, a company's financial statements help you decide whether to loan money to the company and, if so, under what loan arrangements (e.g., the interest

---

[3]For simplicity, the salaries, supplies, and utilities expense allocations are rounded to the nearest dollar.

rate to charge, the amount of time to allow before the loan must be repaid, the restrictions to place on the company's ability to borrow additional money). If you are an investor, a company's financial statements help you estimate the return you may expect on your investment and whether you want to become or continue to be an owner.

 Who else do you think is interested in a company's financial statements? Why?

Investors use the income statement to help judge their return on investment, and creditors use it to help make loan decisions. On what do these users base their evaluations? To make their business decisions, financial statement users evaluate a company's risk, operating capability, and financial flexibility. Although these may sound like complicated terms, once we have explained them you will see that they describe the main concerns of most investors and creditors.

When investors or creditors use the income statement to evaluate a company's risk, they are estimating the chances that the company will *not* earn a satisfactory profit or that it will earn a higher-than-expected profit in the future. So **risk** is the uncertainty about the future earnings potential of a company. The greater the chances are that a company will earn a satisfactory profit or will earn a higher-than-expected profit, the less risk there is in investing in that company. As the chances decrease that a company will earn a satisfactory profit or the chances increase that it will earn a lower-than-expected profit, the risk of investing in that company increases. A company's "risk factor" affects the expected investment return that is needed to attract investors and affects the interest rate that creditors charge on that company's loans. The greater the risk, the higher will be the required rate of return and the interest rate.

External users evaluate a company's operating capability and financial flexibility because these factors help determine a company's level of risk. **Operating capability** refers to a company's ability to continue a given level of operations in the future. For example, by comparing a company's current set of financial statements with those of prior years, external users can learn about the company's ability to earn a stable stream of operating income. If the statements show that the company can do this, chances are good that the company will be able to maintain its current level of operations in the future.

Financial flexibility refers to a company's ability to adapt to change in the future. External users want to see evidence of financial flexibility because this means that a company will be able to take advantage of business opportunities, such as introducing a new product or building a new warehouse. As you would expect, investors want the company to grow, so they prefer companies that have financial flexibility.

5 What are the main concerns of external decision makers when they use a company's income statement to evaluate its performance?

 Do you think your personal financial flexibility is high or low? Why?

Business Issues and Values

Have you ever had a part-time job? If so, then you know that many companies depend on part-time employees in their operations. For some companies, part-time employees make up a large percentage of their employee group. By using part-time employees, these companies may significantly enhance their financial flexibility because they can hire and lay off employees quickly. They also avoid having to pay for items such as health insurance and retirement benefits, which companies normally pay for full-time employees. Other companies, though still using some part-time employees to help the companies improve their financial flexibility, have a different view regarding their commitment to their employees. They believe that it is part of their social responsibility to hire, train, and retain full-time employees. Although these companies may have less financial flexibility than those that depend more on part-time employees, some investors and creditors feel that a commitment to full-time employees offsets this limitation.

NEEDED: Part-time receptionist and technician. Office hours 12 noon–8 p.m. Apply in person. NO telephone calls. Horton Animal Hospital—Forum Blvd.

For example, consider Malden Mills, a Lawrence, Massachusetts manufacturer of Polartec and Polar fleece (fabrics in demand by such retailers as **Eddie Bauer** and **L.L. Bean**). When it burned to the ground just before Christmas in 1995, the owner gave every employee a $275 Christmas bonus. Then he announced that all employees would continue to receive full pay and benefits for at least 90 days. His decision was based on the philosophy that "Loyalty and profit go hand in hand! Superior employees produce a superior product, and loyal customers and loyal employees are cut from the same fabric."[4]

Ratios

6 What type of analysis is used by external decision makers to evaluate a company's profitability?

To evaluate a company's operating performance, managers and external users may perform ratio analysis. **Ratio analysis** consists of computations in which an item on the company's financial statements is divided by another, related item. Although individual users may compute ratios themselves, groups that specialize in financial analysis compute and publish ratios for many companies and industries. The ratios are "benchmarks" used to compare a company's performance with that of previous periods and with that of other companies. There are many commonly computed ratios, which we will discuss in later chapters. As an introduction, here we discuss two that relate to profitability, since profitability affects risk, operating capability, and financial flexibility.

Profit Margin
One ratio is the *profit margin* (sometimes called the *return on sales*), which is usually expressed as a percentage. A company's **profit margin** is calculated as follows:

$$\text{Profit Margin} = \frac{\text{Net Income}}{\text{Net Sales}}$$

If a company's profit margin is higher than that of previous years or higher than that of other companies, it usually means that the company is doing a better job of controlling its expenses in relation to its sales.

The profit margin of Sweet Temptations for January 2004 is calculated as follows, based on the information in Exhibit 6-2:

$$7.43\% = \frac{\$602}{\$8,100}$$

This means that, on average, 7.43 cents of every sales dollar is profit (net income) for Sweet Temptations. Since this is Sweet Temptations' first month of operations, we cannot compare the 7.43% profit margin for January with the profit margin of previous months. However, this profit margin during initial operations is a positive sign.

Gross Profit Percentage
A second ratio is the *gross profit percentage* (sometimes called the *gross profit margin*), which relates a company's gross profit to its net sales. A company's **gross profit percentage** is calculated as follows:

$$\text{Gross Profit Percentage} = \frac{\text{Gross Profit}}{\text{Net Sales}}$$

A retail company's gross profit generally ranges from 20% to 60% of net sales depending on the types of products it sells or its "pricing" strategy. For example, some companies use a pricing strategy of offering lower selling prices to increase their sales volume,

---

[4]"Mill Owner Keeps Faith with Workers," *Columbia Daily Tribune,* December 22, 1996, sec. A, 1.

thereby increasing their total gross profit. The gross profit percentage of Sweet Temptations for January 2004 is calculated as follows:

$$55\% = \frac{\$4,455}{\$8,100}$$

This means that, on average, 55 cents of every sales dollar (after the cost of goods sold is subtracted) is left to cover operating expenses and other expenses, and to increase Sweet Temptations' net income. Again, we cannot make comparisons with previous months, but this 55% gross profit margin for January is within the range of a retail company's usual gross profit. This is another positive sign of Sweet Temptations' successful initial operating capability. The managers of a retail company keep a close watch on the company's gross profit because changes in gross profit typically result in large changes in net income.

Profitability Ratios of Actual Companies

To illustrate ratio analysis, we will use information from the financial statements of two retail companies, **JCPenney Company Inc.** (JCPenney) and **Sears Roebuck and Co.** (Sears). Their profit margins[5] for 2001 were as follows:

http://www.jcpenney.com
http://www.sears.com

	JCPenney	Sears
Profit Margin	0.3%	2.1%

Neither Sears nor JCPenney had a very good year in 2001 because, in part, of the poor economy. However, when we compare these two ratios, it is clear that Sears was more successful at generating net income from its sales than was JCPenney, since Sears made 7 times more profit from each dollar of sales.

Would you expect Sears' gross profit percentage also to be 7 times higher than JCPenney's? Why or why not?

Now let's compare the gross profit percentages for the two companies.

	JCPenney	Sears
Gross Profit Percentage	28.8%	26.6%

Notice that Sears had a *lower* gross profit percentage than JCPenney, even though it had a *higher* profit margin. Based on a comparison of these ratios, we can say that JCPenney was more efficient than Sears in controlling the costs of merchandise but that Sears was more efficient than JCPenney in controlling operating expenses.

How else might you explain the differences in the ratios of the two companies?

We will expand the discussion of operating capability and financial flexibility in Chapter 7, adding new ratios for analysis. We also will continue our comparison of JCPenney and Sears.

STATEMENT OF CHANGES IN OWNER'S EQUITY

A company's owner's equity is affected by the owner's investments and withdrawals, as well as by the company's revenue and expense transactions. Although an income statement and its supporting schedules help external users understand the results of revenue

---

[5]We use well-known corporations in this illustration because the financial statements of most small entrepreneurial companies are not publicly available. For simplicity, we do not show the calculations of the ratios, although the numbers were taken from ompany's financial statements. For instance, we calculated JCPenney's profit margin by dividing its net income of $98 by its net sales of $32,004 million. We calculated Sears' profit margin by dividing its net income of $735 million by ales of $35,843 million.

and expense activities, the statement and schedules do not include all the activities that affect owner's equity. A company prepares a supplementary schedule, called a *statement of changes in owner's equity,* for this purpose. The **statement of changes in owner's equity** summarizes the transactions that affected owner's equity during the accounting period. A company presents this statement to "bridge the gap" between its income statement and the amount of owner's capital it reports on its balance sheet.

The schedule begins with the balance in the owner's capital account at the beginning of the accounting period. Then, the total amount of the owner's investments for the accounting period is added because this amount increases the owner's claim on the company's assets. Next, the amount of the company's net income is added because this amount also increases the owner's claim on the company's assets as a result of its operating activities for the accounting period. Finally, the amount of withdrawals that the owner made during the accounting period is subtracted.

Note that the owner's withdrawals are recorded directly in the owner's capital account, as we illustrated in Chapter 5 for Anna's $50 withdrawal in transaction #6 and in Exhibit 5-18. It is important to understand that *withdrawals are not expenses* because they are not the costs of providing goods or services to customers. Withdrawals are recorded as *reductions* of the owner's capital account because they are *dis*investments of assets by the owner. The final amount on a company's statement of changes in owner's equity is the owner's capital balance at the end of the accounting period. The company reports this amount on its ending balance sheet.

By summarizing all the transactions affecting the owner's equity of a company, the statement of changes in owner's equity helps to complete the picture of the company's financial activities for the accounting period. External users find this information helpful in evaluating the changes in the claims on the company's assets, changes that have an impact on its risk, operating capability, and financial flexibility.

 If you saw a large amount of withdrawals reported in a company's statement of changes in owner's equity, how would that affect your evaluation of its risk? Why?

Exhibit 6-5 shows Sweet Temptations' statement of changes in owner's equity for the month ended January 31, 2004. The $15,000 beginning amount of owner's capital comes from the A. Cox, Capital account (shown in Exhibit 5-6 on page 136). Anna made no additional investments during the accounting period, so the next item is net income. The $602 net income comes from the income statement in Exhibit 6-2. The $50 of withdrawals comes from the A. Cox, Capital account in Sweet Temptations' accounting records, as we just discussed. The $15,552 ending amount for A. Cox, Capital is the amount reported as owner's equity in Sweet Temptations' January 31, 2004 balance sheet (which is prepared next, and which we showed in Exhibit 5-22 on page 153).

EXHIBIT 6-5	SWEET TEMPTATIONS' STATEMENT OF CHANGES IN OWNER'S EQUITY

SWEET TEMPTATIONS
Statement of Changes in Owner's Equity
For Month Ended January 31, 2004

A. Cox, capital, January 1, 2004	$15,000
Add: Net income	602
	$15,602
Less: Withdrawals	(50)
A. Cox, capital, January 31, 2004	$15,552

Managers and external users know that a company's income statement and statement of changes in owner's equity do not provide all of the financial information needed for business decisions. Information that is not reported on the company's income statement can have a big impact on its ability to earn profits in the future. In the next chapter, we will discuss how managers, investors, and creditors use the balance sheet in conjunction with the income statement to make business decisions.

CLOSING THE TEMPORARY ACCOUNTS

Earlier in the chapter we explained that revenue and expense accounts are *temporary accounts* used to accumulate a company's net income amounts for the accounting period. After a company prepares its income statement, statement of changes in owner's equity, balance sheet, and cash flow statement for the accounting period, it prepares "closing entries." **Closing entries** are entries made by a company to transfer the ending balances from its temporary revenue and expense accounts into its permanent account for owner's capital. A company uses closing entries so that when a new accounting period starts, (1) the revenue and expense accounts (temporary accounts) have zero balances because they no longer contain the amounts of any transactions from previous periods, (2) the accounting system keeps the revenue and expense transactions of the current period separate from the revenue and expense transactions of other periods, and (3) the permanent (balance sheet) accounts are up-to-date (net income has been added to the previous balance of the owner's capital account).

We do not illustrate the closing entries for each revenue and expense account because it is too time-consuming and not necessary for your understanding of how a company's accounting system works. To give you an idea of closing entries, however, we will show you a "summary closing entry" for Sweet Temptations. Recall from Exhibit 5-21 on page 150 that the amounts of the A. Cox, Capital, Revenue, and Expense columns at the end of January 2004 (prior to closing) were $14,950, $8,100, and $7,498, respectively, as we show in the following schedule.

| | Owner's Equity | | | |
| | Owner's Capital | + | Net Income | |
	A. Cox, Capital		Revenues	−	Expenses
Balances prior to closing	$14,950		$8,100	−	$7,498
Summary closing entry	+ $602		−$8,100	−	−$7,498
Balan ces after closing	$15,552		$ 0		$ 0

Anna prepares the summary closing entry as follows. To decrease the Revenue account column to zero, she subtracts $8,100 from this column. Similarly, to decrease the Expenses column to zero, she subtracts $7,498 from this column. Anna then adds the $602 difference (which is the net income shown in Exhibit 6-2) to the A. Cox, Capital account column. After this summary closing entry, the Revenue and Expense columns both have zero balances and are ready to accumulate the revenue and expense information for February 2004. The A. Cox, Capital account column has a balance of $15,552, which is the amount shown in Exhibit 6-5. We emphasize that this is a "summary closing entry" because in actual closing entries a company would close *each* revenue account and *each* expense account to the owner's capital account. Furthermore, a company's accounting period usually is one year so it would normally prepare its closing entries at the end of the year, rather than at the end of each month.

Which accounts in Sweet Temptations' accounting system have nonzero balances on February 1, 2004?

SUMMARY

At the beginning of the chapter we asked you several questions. During the chapter, we asked you to STOP and answer several additional questions to build your knowledge about specific issues. Be sure you answered these additional questions. Below are the questions from the beginning of the chapter, with a brief summary of the key points relating to the answers. Use your creative and critical thinking skills to expand on these key points to develop more complete answers to the questions and to determine what other questions you have that might lead you to learn more about the issues.

1 Why is a company's income statement important?

A company's income statement is important because it summarizes the results (revenues, expenses, and net income) of the company's operating activities for an accounting period. This information is useful in the decision making of both internal and external users because it helps to show how well the company's management has performed during the period and from period to period.

2 How are changes in a company's balance sheet and income statement accounts recorded in its accounting system?

Changes in a company's balance sheet accounts are recorded in its accounting system by creating a separate column for *each* asset, liability, and owner's capital account. These accounts are called *permanent accounts* because they are used for the life of the company to record its balance sheet transactions. Changes in a company's income statement accounts are recorded in its accounting system by creating a separate column under Owner's Equity for *each* revenue account and *each* expense account, while still retaining the owner's capital account column. A company uses these revenue and expense accounts to record its net income transactions for only one accounting period, so they are called *temporary accounts*.

3 What are the parts of a retail company's classified income statement, and what do they contain?

The classified income statement of a retail company includes two parts, an operating income section and an other items section. The operating income section includes revenues, cost of goods sold, and operating expenses subsections related to a company's primary operating activities. The other items section includes any revenues or expenses that are not directly related to the company's primary operations.

4 What is inventory and cost of goods sold, and what inventory systems may be used by a company?

Inventory is the merchandise a retail company is holding for resale. Cost of goods sold is the cost to the company of the merchandise that it sells during the accounting period. A company may use either a perpetual inventory system or a periodic inventory system. A perpetual inventory system keeps a continuous record of the cost of inventory on hand and the cost of inventory sold. A periodic inventory system does not keep a continuous record of the inventory on hand and sold, but uses a physical count to determine the inventory on hand at the end of the accounting period.

5 What are the main concerns of external decision makers when they use a company's income statement to evaluate its performance?

When external decision makers use a company's income statement to evaluate its performance, they are concerned about the company's risk, operating capability, and financial flexibility. Risk is uncertainty about the future earnings potential of the company. Operating capability refers to the company's ability to continue a given level of operations. Financial flexibility refers to the company's ability to adapt to change.

6 **What type of analysis is used by external decision makers to evaluate a company's profitability?**

Ratio analysis is used by external users to evaluate a company's profitability. Ratio analysis involves computations in which an item on the company's financial statements is divided by another, related item. The ratios are compared with the company's ratios in previous periods or with other companies' ratios. The ratios used to evaluate a company's profitability include the profit margin (net income divided by net sales) and the gross profit percentage (gross profit divided by net sales).

KEY TERMS

adjusting entries (p. 179)
cash discount (p. 173)
closing entries (p. 185)
cost of ending inventory (p. 178)
cost of goods sold (p. 175)
credit memo (p. 174)
financial flexibility (p. 181)
general and administrative expenses
  (p. 179)
gross profit (p. 178)
gross profit percentage (p. 182)
inventory (p. 175)
net purchases (p. 178)
operating capability (p. 181)
operating expenses (p. 178)
operating income (pp. 170, 180)

other items (pp. 170, 180)
periodic inventory system (p. 176)
permanent accounts (p. 169)
perpetual inventory system (p. 175)
profit margin (p. 182)
quantity discount (p. 173)
ratio analysis (p. 182)
risk (p. 181)
sales allowance (p. 174)
sales discount (p. 173)
sales return (p. 174)
selling expenses (p. 179)
statement of changes in owner's equity
  (p. 184)
temporary accounts (p. 169)

SUMMARY SURFING

Here is an opportunity to gather information on the Internet about real-world issues related to the topics in this chapter. Go to http://www.cunningham.swlearning.com and click on the Interactive Study Center. Click on this chapter number then click on Summary Surfing and answer the following questions.

- Click on **JCPenney**. Find the company's income statements. Compute the profit margin and the gross profit percentage for the most current year. How do these results compare with the 2001 ratios we discussed in this chapter?

- Click on **Sears**. Find the company's income statements. Compute the profit margin and the gross profit percentage for the most current year. How do these results compare with the 2001 ratios we discussed in this chapter?

INTEGRATED BUSINESS AND ACCOUNTING SITUATIONS

Answer the Following Questions in Your Own Words.

Testing Your Knowledge

6-1 Write out the income statement equation, and explain its components.

6-2 Explain how managers use a company's income statement for decision making.

6-3 How are changes in a company's income statement accounts recorded in its accounting system?

6-4 What is the difference between temporary and permanent accounts?

6-5 Identify the parts and subsections of a retail company's classified income statement. What is included in each part?

6-6 Explain the difference between a quantity discount and a sales (cash) discount.

6-7 Explain the difference between a sales return and a sales allowance.

6-8 What is a perpetual inventory system? How is a company's cost of goods sold determined under this system?

6-9 What is a periodic inventory system? How is a company's cost of goods sold determined under this system?

6-10 What are operating expenses? Explain the difference between selling expenses and general and administrative expenses.

6-11 Explain the meaning of the terms *risk, operating capability,* and *financial flexibility.*

6-12 What is ratio analysis, and what is it used for?

6-13 Explain how to compute a company's profit margin. What is this ratio used for?

6-14 Explain how to compute a company's gross profit percentage. What is this ratio used for?

6-15 Explain what is included in a company's statement of changes in owner's equity and how the statement is used.

6-16 What are closing entries and why are they used?

Applying Your Knowledge

6-17 On July 1, Drexel's Appliance purchased $5,000 of goods for resale. On July 15, it sold $2,600 of these goods to customers at a selling price of $4,000. The company uses a perpetual inventory system, and all transactions were for cash.

Required: Prepare account column entries to record this information.

6-18 On April 6, Piper Model Shop made a cash sale of $127 in merchandise to a customer. The company uses a perpetual inventory system; the merchandise had cost Piper $88. On April 8, the customer was given a sales allowance of $25 cash for a defective model that the customer chose to keep.

Required: (1) Prepare account column entries to record this information.
(2) What source documents would be used to record each transaction?

6-19 The Jardine Tax Services Company was established on January 1 of the current year to help clients with their tax planning and with the preparation of their tax returns. During January, the company entered into the following transactions:

Date	Transactions
Jan. 2	D. Jardine set up the company by investing $5,000 in the company's checking account.
3	The company paid $2,400 in advance for one year's rent of office space.
4	Office equipment was purchased at a cost of $6,000. A down payment of $1,000 was made, and a $5,000 note payable was signed for the balance owed. The note is due in one year.
7	Office supplies were purchased for $800 cash.
16	Fees of $1,700 were collected from clients for tax services provided during the first half of January.
29	A salary of $700 was paid to the office secretary.
30	Jardine withdrew $900 for personal use.
31	The January utility bill of $120 was received; it will be paid in early February.

31 Clients were billed $2,300 for tax services performed during the second half of January.

31 Jardine recorded the following adjustments:
  a. Rent expense for the month
  b. Depreciation of $45 on office equipment
  c. Interest expense of $50 on the note payable
  d. Office supplies used (the office supplies on hand at the end of the month cost $720)

Required: (1) Record the preceding transactions in appropriate account columns.
  (2) Prepare a classified income statement for the company for January.
  (3) Prepare a balance sheet for the company on January 31.
  (4) Briefly comment on how well the company did during January.

6-20 The Salanar Answering Service Company was started on April 1 of the current year to answer the phones of doctors, lawyers, and accountants when they are away from their offices. The following transactions of the company occurred and adjustments were made during April:

Date	Transaction
Apr. 1	P. Salanar started the business by investing $3,000 cash.
2	The company paid cash of $900 in advance for six months' rent of office space.
3	The company purchased telephone equipment costing $5,500, paying $1,500 down and signing a $4,000 note payable for the balance owed.
6	Office supplies totaling $450 were purchased on credit. The amount is due in early May.
15	The company collected $800 from clients for answering services performed during the first half of April.
28	P. Salanar withdrew $600 for personal use.
29	The April $110 utility bill was received; it is to be paid in May.
30	The company paid $300 salary to a part-time employee.
30	Clients were billed $700 for answering services performed during the last half of April.
30	Salanar recorded the following adjustments:

  a. Rent expense for April
  b. Depreciation of $42 on telephone equipment
  c. Interest of $40 on the note payable
  d. Office supplies used of $58

Required: (1) Record the preceding transactions in appropriate account columns.
  (2) Prepare a simple income statement for the company for April.
  (3) Prepare a balance sheet for the company on April 30.
  (4) Briefly comment on how well the company did during April.

6-21 The Steed Art Supplies Company sells various art supplies to local artists. The company uses a perpetual inventory system, and the balance of its inventory of art supplies at the beginning of August was $2,500. Its cash balance was $800 and the J. Steed, capital balance was $3,300 at the beginning of August. Steed entered into the following transactions during August:

Date	Transactions
Aug. 1	J. Steed invested another $1,000 cash into the company.
2	Purchased $400 of art supplies for cash.
4	Made a $900 sale of art supplies on credit to P. Tarlet, with terms of n/15; the cost of the inventory sold was $550.
6	Purchased $700 of art supplies on credit from the Rony Company, with terms of n/20.
10	Returned, for credit to its account, $100 of defective art supplies purchased on August 6 from the Rony Company.
12	Made cash sales of $330 to customers; the cost of the inventory sold was $200.
13	Granted a $25 allowance to a customer for damaged inventory sold on August 12.
15	Received payment from P. Tarlet of the amount due for inventory sold on credit on August 4.
25	Paid balance due to the Rony Company for purchase on August 6.

Required: (1) Record the preceding transactions in appropriate account columns.

(2) Determine the balances in all the accounts at the end of August.

(3) Compute the gross profit and the gross profit percentage for August.

6-22   The Kerem Heater Company sells portable heaters and related equipment. The company uses a perpetual inventory system, and its inventory balance at the beginning of November was $2,600. Its cash balance was $1,500, and the B. Kerem, capital balance was $4,100 at the beginning of November. Kerem entered into the following transactions during November:

Date	Transactions
Nov. 1	B. Kerem invested another $900 cash into the company.
2	Made $480 cash sales to customers; the cost of the inventory sold was $280.
3	Purchased $1,700 of heaters for cash from Jokem Supply Company.
5	Received $250 cash allowance from Jokem Supply Company for defective inventory purchased on November 3.
6	Paid $210 for parts and repaired defective heaters purchased from Jokem Supply Company on November 3.
8	Made a $1,500 sale of heaters on credit to Arvin Nursing Home, with terms of 2/10, n/20; the cost of the inventory sold was $850.
15	Purchased $1,100 of heaters on credit from Duwell Supplies, with terms of n/15.
18	Received amount owed by Arvin Nursing Home for heaters purchased on November 8, less the cash discount.
30	Paid for the inventory purchased from Duwell Supplies on November 15.

handwritten note: +250 cash / −250 inv

handwritten note: −210 cash + 210 inv.

Required: (1) Record the preceding transactions in appropriate account columns.

(2) Determine the balances in all the accounts at the end of November.

(3) Compute the gross profit and the gross profit percentage for November.

6-23   The following information is available for the Arnhold Horn Company for the year:

Beginning inventory	$ 45,000
Ending inventory	50,000
Purchases	102,000
Purchases returns and allowances	4,000

Required: Prepare a schedule that computes the cost of goods sold for the year.

6-24   The income statement information of the Weeden Furniture Company for 2004 and 2005 is as follows:

	2004	2005
Cost of goods sold	$ (a)	$59,300
Interest expense	600	0
Selling expenses	(b)	10,800
Operating income	21,800	(d)
Sales (net)	96,000	(e)
General expenses	7,900	(f)
Net income	(c)	21,600
Interest revenue	0	600
Gross profit	39,000	40,200

Required: Fill in the blanks lettered (a) through (f). All the necessary information is listed. (*Hint:* It is not necessary to find the answers in alphabetical order.)

6-25   The following information is taken from the accounts of the Harburn Hobby Shop for the month of October of the current year.

Cost of goods sold	$54,000
Sales revenue (net)	88,000
Selling expenses	5,000
Interest expense	1,000
General and administrative expenses	12,000

Required: (1) Prepare a classified income statement for Harburn.

(2) Compute Harburn's profit margin.

6-26 The following information is taken from the accounts of Foile's Music Store for the current year ended December 31.

Depreciation expense: office equipment	$ 1,600
Interest revenue	725
Sales salaries expense	8,200
Rent expense	1,800
Depreciation expense: store equipment	2,400
Sales revenue (net)	94,200
Office salaries expense	4,000
Interest expense	250
Office supplies expense	600
Cost of goods sold	59,400
Advertising expense	360

Of the rent expense, 5/6 is applicable to the store and 1/6 is applicable to the office.

Required: (1) Prepare a classified income statement for Foile's Music Store for the current year.

(2) Compute the profit margin.

(3) Compute the gross profit percentage. Does this percentage fall near the high or the low end of the range of typical retail companies' gross profit percentages?

6-27 The December 31, 2004 income statement accounts and other information of Lyon's Hardware are shown below:

Advertising expense	$ 4,300
Depreciation expense: store equipment	1,600
Depreciation expense: building (store)	3,700
Depreciation expense: office equipment	2,300
Depreciation expense: building (office)	1,100
Interest revenue	1,700
Interest expense	900
Cost of goods sold	63,900
Insurance expense	350
Sales (net)	102,000
Office supplies expense	480
Store supplies expense	800
Sales salaries expense	5% of net sales
Office salaries expense	2,600
Utilities expense (store)	1,500
Utilities expense (office)	400

Required: (1) Prepare a classified 2004 income statement for Lyon's Hardware.

(2) Compute the profit margin for 2004. If the profit margin for 2003 was 12.5%, what can be said about the 2004 results?

6-28 Four independent cases related to the owner's equity account of the Cox Company follow:

Case	L. Cox, Capital May 1	Net Income for May	Withdrawals in May	L. Cox, Capital May 31
1	$ A	$2,700	$1,000	$25,700
2	37,000	B	1,720	40,250
3	28,200	900	C	24,800
4	34,000	3,820	1,500	D

Required: Determine the amounts of A, B, C, and D.

6-29 The beginning balance in the R. Barnum, Capital account on October 1 of the current year, was $23,000. For October, the Barnum Company reported total revenues of $8,000 and total expenses of $4,250. In addition, R. Barnum withdrew $1,400 for his personal use on October 25.

Required: Prepare a statement of changes in owner's equity for October for the Barnum Company.

6-30 Rodgers Company shows the following amounts in its owner's equity accounts at the end of December: B. Rodgers, Capital, $32,200; Revenues, $57,300; Expenses, $42,800.

Required: Set up the account balances in account columns and prepare summary closing entries at the end of December.

Making Evaluations

6-31 A company engages in many types of activities.

Required: For each of the following sets of changes in a company's accounts, give an example of an activity that the company could engage in that would cause these changes and explain why you think the activity would cause these particular changes:

(1) Increase in an asset and decrease in another asset
(2) Increase in an asset and increase in a liability
(3) Increase in an asset and increase in owner's equity
(4) Increase in an asset and increase in a revenue
(5) Decrease in an asset and increase in an expense
(6) Decrease in an asset and decrease in a liability

6-32 During the current accounting period, the bookkeeper for the Nallen Company made the following errors in the year-end adjustments:

| | | | Effect of Error on: | | | |
| | | | Net | | | Owner's |
Error	Revenues	Expenses	Income	Assets	Liabilities	Equity
Example: Failed to record $200 of salaries owed at the end of the period	N	U $200	O $200	N	U $200	O $200
1. Failed to adjust prepaid insurance for $400 of expired insurance						
2. Failed to record $500 of interest expense that had accrued during the period						
3. Inadvertently recorded $300 of annual depreciation twice for the same equipment						
4. Failed to record $100 of interest revenue that had accrued during the period						
5. Failed to reduce unearned revenues for $600 of revenues that were earned during the period						

Required: Assuming that the errors are not discovered, indicate the effect of each error on revenues, expenses, net income, assets, liabilities, and owner's equity at the end of the accounting period. Use the following code: O = Overstated, U = Understated, and N = No effect. Include dollar amounts. Be prepared to explain your answers.

6-33 Suppose you own a retail company and are considering whether to allow your customers to have quantity discounts, sales discounts, and sales allowances.

Required: How do you think quantity discounts, sales discounts, and sales allowances would affect the results of a company's C-V-P analysis and its budgets? Explain what the effects would be. If a company gives quantity discounts, sales discounts, and sales allowances, what information would it need in order to conduct C-V-P analysis and develop budgets? (What questions would you have to ask?)

6-34 Your friend Allison is planning to open an automobile parts store and has come to you for advice about whether to use a perpetual or a periodic inventory system.

Required: Before you advise her, list the questions you would like to ask her. How would the answer to each question help you advise her? Explain to her the advantages and disadvantages of each system.

6-35 Cara Agee owns a hairstyling shop, Air Hair Company. It is now November 2004, and Cara thinks she might need a bank loan. Her bank has asked Cara to prepare a "projected" income statement and to compute the "projected" profit margin for next year. Although she has never developed this information before, she understands that to do so, she must make a "best guess" of her revenues and expenses for 2005 based on past activities and future estimates. She asks for your help and provides you with the following information.

(a) Styling revenues for 2004 were $70,000. Cara expects these to increase by 10% in 2005.
(b) Air Hair employees are paid a total "base" salary of $30,000 plus 20% of all styling revenues.
(c) Styling supplies used have generally averaged 15% of styling revenues; Cara expects this relationship to be the same in 2005.
(d) Air Hair recently signed a two-year rental agreement on its shop, requiring payments of $400 per month, payable in advance.
(e) The cost of utilities (heat, light, phone) is expected to be 25% of the yearly rent.
(f) Air Hair owns styling equipment that cost $12,000. Depreciation expense for 2005 is estimated to be 1/6 of the cost of this equipment.

Required: Prepare a projected income statement for Air Hair Company for 2005 and compute its projected profit margin. Show supporting calculations.

6-36 The Gray Service Company had a fire and lost some of the accounting records it needed to prepare its 2004 income statement. Stan Gray, the owner, has been able to determine that his capital in the business was $32,000 at the beginning of 2004 and was $33,000 at the end of 2004. During 2004 he withdrew $14,000 from the business. Stan has also been able to remember or determine the following information for 2004.

(a) Cash service revenues were three times the amount of net income; credit service revenues were 40% of cash service revenues.
(b) Rent expense was $500 per month.
(c) The company has one employee, who was paid a salary of $20,000 plus 20% of the service revenues.
(d) The supplies expense was 15% of the total expenses.
(e) The utilities expense was $100 per month for the first nine months of the year and $200 per month during the remaining months of the year due to the cold winter.

Stan also knows that the company owns some service equipment, but he cannot remember the cost or the amount of depreciation expense.

Required: Using the preceding information, prepare Gray Service Company's 2004 income statement and compute its profit margin. Show supporting calculations.

6-37 Your boss has given you last year's income statements of two companies and asked you to recommend one in which your company should invest. The income statements include the following information (in thousands):

	Amalgamated Snacks	Gourmet Goodies
Net sales	$1,360,000	$2,000
Cost of goods sold	884,000	1,360
Selling expenses	205,294	312
General and administrative expenses	114,541	110
Net income	156,165	218

Required: Based on this information alone, which company would be the better investment choice? Explain your answer. What other information would you like to have in order to make a more informed decision? How would this information help you recommend the one in which you think your company should invest?

6-38   A paragraph accompanying recent financial statements of **Dillard's Department Stores Inc.** begins: "Advertising, selling, administrative and general expenses increased as a percentage of sales in [the current year] compared to [last year]. This occurred because of the slower growth rate of sales during the year as compared to prior years."

Required: How does the second sentence explain the first? Explain in more detail how this could happen.

6-39   On January 3, 2005, Ken Harmot agreed to buy the Ace Cleaning Service from Janice Steward. They agreed that the purchase price would be five times the 2004 net income of the company. To determine the price, Janice prepared the following condensed income statement for 2004.

Revenues	$ 48,000
Expenses	(36,000)
Net Income	$ 12,000

Janice said to Ken, "Based on this net income, the purchase price of the company should be $60,000 ($12,000 × 5). Of course, you may look at whatever accounting records you would like." Ken examined the accounting records and found them to be correct, except for several balance sheet accounts. These accounts and their December 31, 2004 balances are as follows: two asset accounts—Prepaid Rent, $3,600; and Equipment, $4,800; and one liability account—Unearned Cleaning Service Revenues, $0.

Ken gathered the following company information related to these accounts. The company was started on January 2, 2002. At that time, the company rented space in a building for its operations and purchased $6,400 of equipment. At that time, the equipment had an estimated life of 8 years, after which it would be worthless. On July 1, 2004, the company paid one year of rent in advance at $300 per month. On September 1, 2004, customers paid $600 in advance for cleaning services to be performed by the company for the next 12 months. Ken asks for your help. He says, "I don't know how these items affect net income, if at all. I want to pay a fair price for the company."

Required: (1) Discuss how the 2004 net income of the Ace Cleaning Service was affected, if at all, by each of the items.
(2) Prepare a corrected condensed 2004 income statement.
(3) Compute a fair purchase price for the company.

6-40   The bookkeeper for Powell Import Service Agency was confused when he prepared the following financial statements.

POWELL IMPORT SERVICE AGENCY
Profit and Expense Statement
December 31, 2004

Expenses:		
Salaries expense		$ 21,000
Utilities expense		3,400
Accounts receivable		1,600
C. Powell, withdrawals		20,000
Office supplies		1,500
Total expenses		$(47,500)
Revenues:		
Service revenues		$ 47,000
Accounts payable		1,100
Accumulated depreciation: office equipment		1,800
Total revenues		$ 49,900
Net Revenues		$ 2,400

POWELL IMPORT SERVICE AGENCY
Balancing Statement
For Year Ended December 31, 2004

Liabilities		Assets	
Mortgage payable................................	$27,000	Building...	$44,000
Accumulated depreciation:		Depreciation expense:	
building...	6,400	building..	1,600
Total Liabilities.....................................	$33,400	Office equipment..........................	9,700
		Depreciation expense:	
C. Powell, capital[a]...............................	27,000	office equipment.......................	900
Total Liabilities and		Cash..	4,200
Owner's Equity..................................	$60,400	Total Assets.................................	$60,400

[a]$24,600 beginning capital + $2,400 net revenues

C. Powell asks for your help. He says, "Something is not right! My company had a fantastic year in 2004; I'm sure it made more than $2,400. I don't remember much about accounting, but I do recall that 'accumulated depreciation' should be subtracted from the cost of an asset to determine its book value." You agree based on your understanding of the depreciation discussion on pages 151 and 152 of this book. After examining the financial statements and related accounting records, you find that, with the exception of office supplies, the *amount* of each item is correct even though the item might be incorrectly listed in the financial statements. You determine that the office supplies used during the year amount to $800 and that the office supplies on hand at the end of the year amount to $700.

Required: (1) Review each financial statement and indicate any errors you find.
(2) Prepare a corrected 2004 income statement, statement of changes in owner's equity, and ending balance sheet.
(3) Compute the profit margin for 2004 to verify or refute C. Powell's claim that his company had a fantastic year.

6-41 Yesterday, the letter shown on the following page arrived for your advice column in the local paper.

DR. DECISIVE

Dear Dr. Decisive:

I can't believe I am writing to you. In the past, I have always tried to solve my own problems, but now I have one I can't solve on my own. My roommate and I are taking an accounting course together. One night, or maybe I should say very early one morning, we were debating where a bank's interest expense goes on its income statement. I think it should go in the "other items" section, but my roommate thinks it should go in "operating expenses." We could argue about this forever, but neither one of us is willing to give in. My roommate agrees that we will accept your answer. If I win, I don't have to pay my roommate interest on the money I owe him.

"Interested"

Required: Meet with your Dr. Decisive team and write a response to "Interested."

CHAPTER 7

THE BALANCE SHEET: ITS CONTENT AND USE

"THERE ARE BUT
TWO WAYS OF
PAYING A DEBT:
INCREASE INCOME,
OR INCREASE
THRIFT."

—THOMAS CARLYLE

1. Why is a company's balance sheet important?

2. What do users need to know about a company's classified balance sheet?

3. What is a company's liquidity, and how do users evaluate it?

4. What is a company's financial flexibility, and how do users evaluate it?

5. Why and how do users evaluate a company's profitability?

6. What is a company's operating capability, and how do users evaluate it?

What assets do you own? How do you keep track of them? How much is in your checking account? Savings account? Do you currently have some bills that you need to pay next month? Do you own a car? A computer? A house? What did these assets cost? Did you take out a loan to pay for any of them? Do you still owe money on this loan? If so, what percentage of your total assets is the amount that you owe? Do your assets exceed your debts? Companies also keep track of their assets and debts. For instance, **Wal-Mart** reported $28,246 million of "current" assets, $42,556 million of property and equipment, and $83,451 million of total assets on its January 31, 2002 balance sheet. On this balance sheet, among other amounts, Wal-Mart also reported $27,282 million of "current" liabilities, $15,687 million of long-term debt, and $35,102 of owners' equity. Wal-Mart obtained these numbers from its accounting system. Does Wal-Mart have enough "current" assets on hand to pay its "current" liabilities? Has Wal-Mart taken out too much of a "loan" as long-term debt to pay for its property and equipment?

http://www.walmartstores.com

 Overall, do these numbers show that Wal-Mart was in a "good" or "bad" financial position on this date? What other information would you like to have to answer this question?

In Chapter 6 you saw how a company's income statement provides managers and external users with important information about its activities. By describing the revenues, expenses, and net income (or net loss) for an accounting period, the income statement helps show whether a company is earning a satisfactory profit. A company's net income (net loss) for an accounting period is the increase (decrease) in owner's equity that resulted from the operating activities of that period. An income statement prepared according to generally accepted accounting principles (GAAP) also enables users to compare financial results from period to period or across companies.

Although the income statement provides useful information for business decision making, managers and external users don't use it alone. They also study the balance sheet. In this chapter, we discuss the importance of the balance sheet. First, we look at the principles, concepts, and accounting methods related to the balance sheet. Second, we describe and present a classified balance sheet. Finally, we explore how managers and external users use a balance sheet to help them make business decisions.

WHY THE BALANCE SHEET IS IMPORTANT

A balance sheet provides information that helps internal and external users evaluate a company's ability to achieve its primary goals of earning a satisfactory profit and remaining solvent. You may recall that the income statement provides information that is used for similar purposes. The income statement and the balance sheet provide different yet related types of information.

1 Why is a company's balance sheet important?

An income statement presents a summary of a company's operating activities for an accounting period: revenues earned, expenses incurred, and the net income that resulted. So, the income statement reports on a company's actions *over a period of time* or, as some say, the "flow of a company's operating activities." The income statement answers questions such as the following: "How much sales revenue did the company earn last year?" "What was the cost of advertising for the year?"

In contrast, a balance sheet presents a company's financial position *on a specific date,* allowing users to "take stock" of a company's assets, liabilities, and owner's equity on that date. Managers and external users need this "financial position" information in order to make business decisions. By examining the balance sheet, users can answer questions such as the following: "What types of resources does the company have available for its operations?" "What are the company's obligations?" They can find out how much money customers owe the company (accounts receivable), see the total dollar amount of the inventory on hand at year-end, and discover how much money the company owes its creditors (accounts payable).

Some people say that the balance sheet is a "snapshot" of a company's assets, liabilities, and owner's equity on a given date. What do they mean? Do you agree? Why or why not?

Why Users Need Both the Balance Sheet and the Income Statement

Remember the creative thinking strategies we discussed in Chapter 2? Let's try a couple of analogies to understand why internal and external users need both the income statement and the balance sheet. You will get the most out of these analogies if you read them actively. In other words, every time you see a question, don't just read ahead, but try to come up with your own answers first. Making analogies really will help you understand accounting—we promise!

Let's say that you want to predict whether your friend Chuck can bake a delicious loaf of bread. What do you need to know to increase your chances of making an accurate prediction? We think you need to know three related pieces of information. First, before baking delicious bread, Chuck must have all of the cooking equipment and the ingredients for the bread on hand: flour, butter, yeast, salt, sugar, bread pans, an oven, etc. So, your first question should be, "Does Chuck have everything he needs to bake the bread?" However, even if he has all of the necessary equipment and ingredients, does that mean he can bake delicious bread? Certainly not! The second question you would ask is, "Has he baked delicious bread before?" If the answers to both these questions are yes, it is likely that he can bake a delicious loaf of bread. If the answer to either of these questions is no, then you are much less sure about his ability to bake. Do you agree? The third question (and probably the most important question if you plan on eating his bread) is not as easy to answer. That question is, "Does he still know how to bake?" You won't know the answer to this question until you taste the next loaf out of his oven.

You would follow a similar strategy if you were trying to determine whether a company can earn a satisfactory profit. You want to know if the company has the assets, liabilities, and owner's equity (does it have the "ingredients"?) needed to earn a satisfactory profit (to bake a delicious loaf of bread). You also need to know if the company has been able to use its resources in the past to earn such a profit (has it baked delicious bread before?). Because a company's balance sheet and income statement provide this financial information, analyzing both statements helps you make an informed decision about the

How do you think these legs would be listed on a sprinter's balance sheet?

© KARL WEATHERLY/GETTY IMAGES

company's ability to earn a satisfactory profit (can it still "bake"?). If either financial statement is missing, it is much more difficult to predict how well the company will perform.

For example, to estimate a company's sales revenue for 2004, it is important to know the amount of cash and the amount of inventory available for sale at the beginning of 2004, and the amount of sales revenues in 2003. You would look at the beginning balance sheet to see if the company has sufficient cash to pay for expenses such as advertising, rent, and salaries and whether it has enough inventory to meet customers' demands for its product. Last year's sales revenue gives an indication of how well a company will perform in the current period. You look at the income statement to find that amount.

Now let's try a sports analogy. Suppose you want to estimate how long it will take a friend, Barb, to run a marathon. What questions would you want answered? You would probably want information about her physical characteristics—things like her age, her height, or her muscle development. These physical characteristics will help you understand her potential running ability. Equally important is information about how well she has used her physical attributes—how long it took her to run her last marathon, when she last ran, and how often she exercises.

When you ask about Barb's physical characteristics, you are "taking stock" of her physical resources, much as an investor analyzes a balance sheet to take stock of a company's financial resources. By learning how well she has used her physical attributes, you are getting information about her past activities that can help you predict her ability to run a fast race. Running a fast race is to sports what earning a satisfactory profit is to business.

 Can you think of any other analogies? What are the goals of the activities in your analogies? What information can help you predict whether the goal will be attained?

The lesson to be learned from our analogies is this: whenever you try to estimate if a goal can be reached—whether the goal is baking a delicious loaf of bread, running a marathon in a certain amount of time, or earning a satisfactory profit—many different types of related information are helpful. A company's balance sheet is one important source of unique and valuable information for predicting the company's financial performance.

Cost-Volume-Profit Analysis, Budgeting, and the Balance Sheet

In Chapter 6, we discussed the relationship between cost-volume-profit (C-V-P) analysis, budgeting, and the income statement. Remember that a company's income statement reports on the results of many of its managers' operating decisions. At least in part, the revenue and expense information shows the results of past C-V-P analysis and budgeting decisions.

Balance sheet information also summarizes the results of managers' decisions. For instance, in Chapter 4 we saw how Anna Cox used a sales budget to help her decide how much inventory Sweet Temptations should keep on hand. A company also uses the sales budget to decide how often to purchase inventory and how many units to order. The accounting system keeps track of inventory balances to help the company evaluate these budgeting decisions and report the inventory as an asset on the balance sheet. If the amount of inventory on hand grows at a faster rate than sales from year to year, the company probably overestimated both sales and its need to make inventory purchases. If the amount of inventory on hand is decreasing as a proportion of sales from year to year, the company may have underestimated sales and the need to make additional inventory purchases.[1]

 What other balance sheet information helps you evaluate the budgeting decisions of a company's managers?

---

[1]Some companies intentionally minimize the amount of inventory they keep on hand. These companies buy their inventory "just-in-time" to meet their sales. We will discuss this just-in-time philosophy more in Chapter 18.

Remember that financial statements help external users decide if they want to do business with a company. External users are interested in a company's assets, liabilities, and owner's equity because, as we explained in the previous section, these items describe a company's financial characteristics. Because external users may be trying to decide *which* company to do business with, companies prepare balance sheets according to GAAP. Since U.S. companies follow the same set of accounting rules when preparing their financial statements, external users can reliably compare the financial positions of any of these companies as part of their decision-making processes. In the sections that follow, we will briefly review the basic accounting principles that underlie the balance sheet and we will discuss the components of a classified balance sheet.

THE ACCOUNTING EQUATION AND THE BALANCE SHEET

Recall from Chapter 5 that the financial accounting process is based on a simple equation:

Economic Resources = Claims on Economic Resources

Using accounting terminology, we restated the equation as follows:

Assets = Liabilities + Owner's Equity

Remember that this mathematical expression is known as the basic **accounting equation** and that the equality of the assets to the liabilities plus owner's equity is the reason a company's statement of financial position is often called a *balance* sheet. The monetary total for the economic resources (assets) of a company must always be *in balance* with the monetary total for the claims to the economic resources (liabilities + owner's equity).

 Do you remember the "transaction scales" we showed in Chapter 5? How did they work?

The **balance sheet** is a financial statement that reports the types and the monetary amounts of a company's assets, liabilities, and owner's equity on a specific date. A company prepares a balance sheet at the end of each accounting period, although it can prepare a balance sheet at any other time to give a current "snapshot" of the company's financial position. Before preparing a balance sheet, though, the company must be certain that the monetary totals for each of its assets and liabilities and the monetary total for owner's equity are correct. By "correct," we mean that since the date of the last balance sheet, all of the company's transactions and events have been recorded in its accounting system according to GAAP. As we discussed in Chapter 6, we also mean that the balances of the revenue, expense, and owner's withdrawals accounts at the end of the accounting period have been transferred to the owner's capital account.

Exhibit 7-1 shows Sweet Temptations' balance sheet accounts in its accounting system on January 31, 2004 (to save space, we have listed the accounts vertically). The accounts include the correct balance for each of Sweet Temptations' assets, liabilities, and owner's capital (after including the revenue, expense, and withdrawals amounts for January) accounts.

Exhibit 7-2 shows Sweet Temptations' January 31, 2004 balance sheet. By comparing Exhibit 7-2 with Exhibit 7-1, you can see how the balances in the accounts provide the information needed to prepare a balance sheet. Notice that the balance sheet shows the balance for each account. Also notice that the liability and owner's equity items are listed below the assets. This format is called a *report form* of balance sheet (as compared with the *account form,* shown in Exhibit 5-22 on page 153, which lists the components in an accounting equation format—assets on the left and liabilities and owner's equity on the right). The report form of balance sheet is common because it is easier to show the accounts vertically on a standard sheet of paper.

EXHIBIT 7-1	SWEET TEMPTATIONS' BALANCE SHEET ACCOUNTS

Assets	=	Liabilities	+	Owner's Equity
				[Owner's Capital + (Revenues − Expenses)]

Cash	Accounts Payable	A. Cox, Capital
1/31/04 Bal $11,030	1/31/04 Bal $4,320	1/31/04 Bal $15,552

Accounts Receivable	Notes Payable
1/31/04 Bal $100	1/31/04 Bal $1,208

Inventory
1/31/04 Bal $2,295

Supplies
1/31/04 Bal $670

Prepaid Rent
1/31/04 Bal $5,000

Store Equipment
1/31/04 Bal $1,985

EXHIBIT 7-2	CLASSIFIED BALANCE SHEET

SWEET TEMPTATIONS
Balance Sheet
January 31, 2004

Assets

Current Assets
Cash	$11,030	
Accounts receivable	100	
Inventory	2,295	
Supplies	670	
Prepaid rent	5,000	
Total current assets		$19,095

Property and Equipment
Store equipment (net)	$ 1,985	
Total property and equipment		1,985
Total Assets		$21,080

Liabilities

Current Liabilities
Accounts payable	$ 4,320	
Note payable	1,208	
Total current liabilities		$ 5,528
Total Liabilities		$ 5,528

Owner's Equity

A. Cox, capital	$15,552
Total Liabilities and Owner's Equity	$21,080

2 What do users need to know about a company's classified balance sheet?

Sweet Temptations' balance sheet is called a **classified balance sheet.** By "classified," we mean that the balance sheet shows subtotals for assets, liabilities, and owner's equity in related groupings. A company decides on the classifications based on the type of business it is in. Later in the chapter, you will see that these groupings make it easier for financial statement users to evaluate a company's performance and to compare its performance with that of other companies.

Notice the way that Sweet Temptations organizes its classified balance sheet. The company adds together the asset groupings to report Total Assets of $21,080 and adds together the liability groupings to report Total Liabilities of $5,528. Finally, it adds the total liabilities to the total owner's equity to report Total Liabilities and Owner's Equity of $21,080. Because Sweet Temptations kept the accounting equation in balance as it recorded its transactions, Total Assets equals Total Liabilities plus Owner's Equity. Let's look at the balance sheet classifications in more detail.

Assets

Assets are a company's economic resources that it expects will provide future benefits to the company. A large company may own hundreds of different types of assets. Some are physical in nature—such as land, buildings, supplies to be used, and inventory that the company expects to sell to its customers. Others do not have physical characteristics but are economic resources because of the legal rights they give to the company. For instance, accounts receivable give the company the right to collect cash in the future.

Think about the different types of assets owned by the grocery store where you shop. We would need several pages to present a complete list, but here are a few examples: cash, accounts receivable, cleaning supplies, grocery carts, storage shelves, forklifts, refrigerators, freezers, bakery equipment, cash registers, the building, the parking lot . . . whew! We'll stop there, but notice that we did *not* even mention any inventory items (groceries) which, of course, are also assets.

 If you work, list the types of assets owned by the company for which you work. If you don't work, talk with someone who does, and find out what types of assets are owned by the company for which he or she works.

Because Sweet Temptations started with a very small amount of capital, it has only a few assets. This small-company example makes it easier for you to see the basic framework that a company uses in accounting for and reporting its resources. When presenting a classified balance sheet, a small retail candy store, a large grocery store, or almost any other type of retail company uses the same classifications for its assets. Look at the balance sheet for a large retail company, and you will likely find subtotals for current assets, long-term investments, and property and equipment.

Current Assets
Current assets are cash and other assets that the company expects to convert into cash, sell, or use up within one year.[2] Current assets include (1) cash, (2) marketable securities, (3) receivables, (4) inventory, and (5) prepaid items. The current assets section presents these items in the order of their liquidity—that is, according to how quickly they can be converted into cash, sold, or used up. Because companies need cash to pay currently due liabilities, grouping current assets together helps financial statement users evaluate a company's ability to pay its current debts.

Cash includes cash on hand (i.e., cash kept in cash registers or the company's safe) and in checking and savings accounts. *Marketable securities,* sometimes called *temporary investments* or *short-term investments,* are items such as government bonds and capital stock of corporations in which the company has temporarily invested (and which the com-

---
[2]As we discussed in Chapter 4, some companies, such as lumber, distillery, and tobacco companies, have operating cycles of longer than one year, so they use their operating cycle to define current assets.

Accumulated
Depreciation

↓

Contra
Asset

Book
Value

pany expects to sell within a year). A company usually makes these short-term investments because it has cash it does not need immediately for purchasing inventory or paying liabilities. Instead of just keeping the cash in the bank, the company purchases the marketable securities to earn additional revenue through interest or dividends. Investment companies such as **UBS PaineWebber**, **Merrill Lynch**, and **Charles Schwab** help other companies (and individual investors) buy and sell marketable securities.

http://www.ubspainewebber.com
http://www.ml.com
http://www.schwab.com

Receivables include accounts receivable (amounts owed by customers) and notes receivable (and related interest). *Inventory* is goods held for resale. *Prepaid items* such as insurance, rent, office supplies, and store supplies will not be converted into cash but will be used up within one year. Note in Exhibit 7-2 that the current assets for Sweet Temptations on January 31, 2004 total $19,095, consisting of cash ($11,030), accounts receivable ($100), inventory ($2,295), supplies ($670), and prepaid rent ($5,000). Remember, the amount for each current asset listed on Sweet Temptations' balance sheet comes from the related account balance in its accounting system.

Assets that are not classified as current assets are called *noncurrent assets*. The balance sheet shows noncurrent assets, such as long-term investments and property and equipment, in separate categories.

Long-Term Investments

Long-term investments include items such as notes receivable, government bonds, bonds and capital stock of corporations, and other securities. Sometimes these are called *noncurrent marketable securities*. A company must *intend* to hold the investment for more than one year to classify it in the long-term investments section of the balance sheet.

Why would a company purchase a corporation's stock? A company makes investments for many reasons, which we will discuss in more detail later. The most basic reason, however, is because the company believes the investment will increase the company's profit. For example, a company may invest in the stock of a corporation because it expects the price of that stock to increase. If that happens, when the company sells the stock later, it will have a gain, which will increase its net income.

Sweet Temptations has not made any long-term investments. At this early stage of the company's life, Sweet Temptations uses its cash to replenish inventory and meet other basic business needs such as paying salaries, rent, and advertising. Therefore Sweet Temptations shows no long-term investments on its January 31, 2004 balance sheet.

Property and Equipment

Property and equipment includes all the physical, long-term assets used in the operations of a company. Often these assets are referred to as *fixed assets* or *operating assets* because of their relative permanence in the company's operations. Assets that have a physical existence, such as land, buildings, equipment, and furniture, are listed in this category. Land is listed on the balance sheet at its original cost. The remaining fixed assets are listed at their book values. The **book value** of an asset is its original cost minus the related accumulated depreciation. **Accumulated depreciation** is the total amount of depreciation expense recorded over the life of an asset to date; thus, it is the portion of the asset's cost that has been "used up" to earn revenues to date. The book values for fixed assets change from period to period as the company sells and/or buys these assets and as accumulated depreciation increases. The balance sheet thus helps report on related budgeting and operating decisions made by the company's managers. We will discuss accumulated depreciation in detail in Chapter 21.

You may be wondering how to reconcile the historical cost concept with the reported book value of a company's property and equipment. As a company uses up assets other than property or equipment, the assets that have *not* been used remain on its balance sheet at their historical cost. The assets that *have* been used no longer exist in the company. But when the company uses property or equipment, the asset still exists in the company until the company has finished using it. Every year, the company uses a portion of the asset, but the entire asset still continues to physically exist in the company. What the company reports (the book value) on its balance sheet represents the portion of the property or equipment that the company has not yet used.

Notice in Exhibit 7-2 that Sweet Temptations' store equipment is classified as property and equipment on its balance sheet. Also notice that the store equipment is listed at $1,985 (net). The "(net)" tells the reader that accumulated depreciation has been deducted from the cost (the $1,985 book value consists of the $2,000 cost less $15 accumulated depreciation, as we discussed in Chapter 5). Sweet Temptations does not include any amounts for land or buildings because it rents space in the Westwood Mall and thus does not own such items.

 Sometimes a company has difficulty deciding how to classify its assets. For example, do you think the cars owned by a rental car company are classified as inventory or equipment? Why? Can you think of more examples that present a dilemma?

Liabilities

Liabilities are the economic obligations (debts) of a company. The external parties to whom the company owes the economic obligations are the company's *creditors*. Legal documents often serve as evidence of liabilities. These documents establish a claim *(equity)* by the creditors (the *creditors' equity*) against the assets of the company.

Companies have many different types of liabilities. For instance, consider the claims that creditors may have on the grocery store's assets we listed earlier. The grocery store probably borrowed money from a bank (by signing a mortgage) to finance its purchases of land and a building. The company also could have obtained the funds used to purchase refrigerators, freezers, baking equipment, and other types of equipment from a bank by signing a long-term (e.g., ten-year) note payable. Most likely, it purchased grocery items from suppliers on credit, resulting in accounts payable. Generally, a company has two types of liabilities—current and noncurrent.

Current Liabilities

Current liabilities are obligations that the company expects to pay within one year by using current assets. Current liabilities include (1) accounts payable and salaries payable, (2) unearned revenues, and (3) short-term notes (and interest) payable. Like current assets, current liabilities are usually listed in the order of their liquidity—that is, how quickly they will be paid.

Accounts payable (amounts owed to suppliers) and *salaries payable* (amounts owed to employees) are common examples of obligations to pay for goods and services.

Unearned revenues are advance collections from customers for the future delivery of goods or the future performance of services. For instance, if a customer pays a company in advance for rent or for some service, the company owes the customer the future use of the rental space or the service, and it records the liabilities as unearned rent or unearned fees.

Short-term *notes payable* (and related interest owed) are obligations that arise because a company signs a note (legal document) that it will pay within one year. The portion of noncurrent liabilities (discussed next) that the company will pay during the next year is also included in current liabilities.

In Exhibit 7-2, Sweet Temptations' current liabilities total $5,528. Sweet Temptations has two kinds of current liabilities—accounts payable ($4,320) and a short-term note payable ($1,208, which includes the $1,200 borrowed plus $8 accrued interest, as we discussed in Chapter 5). Remember that the amount for each current liability comes from the related account balance in the company's accounting system.

Noncurrent Liabilities

Noncurrent liabilities are obligations that a company does not expect to pay within the next year. Noncurrent liabilities are also called *long-term* liabilities because a company usually won't pay them for several years. This category includes such items as long-term notes payable, mortgages payable, and bonds payable (we will look at these in Chapter 22 when we discuss corporations). In most cases, a company incurs a long-term liability when it purchases property or equipment because it "finances" the purchase by borrowing the money to buy the item, and then pays back the amount borrowed over a period longer than a year.

The noncurrent liabilities section shows the past financing decisions of the company's managers. The balance sheet for Sweet Temptations in Exhibit 7-2 does not include a noncurrent liabilities section because the company has not yet incurred any long-term debt.

 Is the fact that Sweet Temptations has no long-term liabilities good or bad? Why?

Owner's Equity

Owner's equity is the owner's current investment in the assets of the company. It is the company's assets less its liabilities. For a sole proprietorship, such as Sweet Temptations, the balance sheet lists the total ending owner's equity in a single *capital* account. The balance sheet shows the owner's equity by listing the owner's name, the word *capital,* and the amount of the current investment. The balance sheets of a partnership and corporation show the owner's equity slightly differently, as we will discuss later in the book. *Residual equity* is a term sometimes used for owner's equity because creditors have first legal claim to a company's assets. Once the creditors' claims have been satisfied, the owner is entitled to the remainder (residual) of the assets.

The ending balance in the account for the owner's capital is affected by the owner's additional investments or withdrawals and by net income. As we discussed in Chapter 6, the company prepares a separate schedule, the statement of changes in owner's equity, to report these items. It also makes closing entries to update the owner's capital account. We show the statement of changes in owner's equity for Sweet Temptations for January 2004 in Exhibit 7-3. Note that we show the $15,552 ending amount of A. Cox, Capital on the balance sheet in Exhibit 7-2. The $21,080 total liabilities and owner's equity ($5,528 total liabilities + $15,552 owner's equity) is equal to the $21,080 total assets.

EXHIBIT 7-3	STATEMENT OF CHANGES IN OWNER'S EQUITY

SCHEDULE A

Sweet Temptations
Statement of Changes in Owner's Equity
For Month Ended January 31, 2004

A. Cox, capital, January 1, 2004	$15,000
Add: Net income	602
	$15,602
Less: Withdrawals	(50)
A. Cox, capital, January 31, 2004	$15,552

USING THE BALANCE SHEET FOR EVALUATION

Remember our bread-baking analogy? We said that without the proper equipment and ingredients, your friend Chuck will have difficulty making delicious bread, no matter how skilled he is. We also noted that the balance sheet informs users of a company's "financial ingredients." Company managers, investors, and creditors are very interested in this information. Without enough resources ("ingredients"), a company will have difficulty remaining solvent and earning a satisfactory profit, no matter how skilled its managers.

However, just having the necessary baking ingredients for the baker is not sufficient. The ingredients must be mixed in the proper proportions at the proper times to improve the chances of baking good bread. Likewise, a company can manage its mix (i.e., types and amounts) of assets, liabilities, and owner's equity to improve its chances of remaining solvent and earning a profit.

A manager is concerned with the company's balance sheet because it is used to evaluate his or her own performance. Also, since external users make investment and credit decisions based in part on balance sheet information, a manager knows that the company's balance sheet affects its ability to get a bank loan or attract new investors.

External users analyze a company's balance sheet to determine whether the company has the right amount and mix of assets, liabilities, and owner's equity to justify making an investment in the company. What they look at in a balance sheet depends on the type of investment they are considering. Short-term creditors are mostly interested in a company's short-term liquidity—whether it can pay current obligations as they are due. Long-term creditors are concerned about whether their interest income is safe and whether the company can continue to earn income and generate cash flows to meet its financial commitments. Investors are concerned about whether they will receive a return on their investment, and how much of a return they will receive. Some potential investors are interested in "solid" companies, that is, companies whose financial statements indicate stable earnings (and, therefore, a steady return). Others want to invest in newer companies that may earn higher income (and, therefore, a higher return) but have more risk.

Notice that in describing the information that external users need from financial statements, we use the words *short-term, long-term, liquidity, stability,* and *risk.* Balance sheet items are classified in a way that help address these needs.

Remember from Chapter 2 that decision making consists of four stages—recognizing the problem, identifying the alternatives, evaluating the alternatives, and making the decision itself. Financial accounting information becomes especially useful when managers and external users want to evaluate the alternatives they have identified. In the next sections, we will discuss a few of the main financial characteristics ("financial ingredients") that managers and external users study when making business decisions. In addition, we will discuss the types of analyses that they use to evaluate a company's performance. Some of these analyses include calculating and evaluating financial statement ratios, which we introduced in Chapter 6. The ratios are "benchmarks" against which decision makers compare a company's performance with its performance in prior periods and with the performance of other companies.

Evaluating Liquidity

3 What is a company's liquidity, and how do users evaluate it?

Liquidity is a measure of how quickly a company can convert its assets into cash to pay its bills. It is an important financial characteristic because to remain solvent, a company must have cash, for instance, to run its operations and pay its liabilities as they become due. The need for adequate liquidity is a major reason a company prepares a cash budget.

External users assess how well a company manages its liquidity by studying its working capital. **Working capital** is a company's current assets minus its current liabilities. The term "working capital" is used because this excess of current assets is the dollar amount of liquid resources a company has to "work with" after it pays all of its short-term debts. Often, users make slightly different computations for the same purpose. The current ratio and the quick (acid-test) ratio are two common indicators of a company's liquidity.

 Given the definition of liquidity, do you think working capital is a good measure of a company's ability to pay its liabilities? Why or why not?

© PHOTODISC/GETTY IMAGES

When we refer to water as "liquid" and a company as "liquid," do we mean the same thing? How are the meanings similar?

Current Ratio

The **current ratio** shows the relationship between current assets and current liabilities and is probably the most commonly used indicator of a company's short-run liquidity. It is calculated as follows:

$$\text{Current Ratio} = \frac{\text{Current Assets}}{\text{Current Liabilities}}$$

current assets − current liabilities

The current ratio is more useful than working capital for measuring a company's liquidity because the current ratio allows comparisons of different-sized companies.

In the past, as a "rule of thumb," users thought a current ratio of 2.0, or 2 to 1 (signifying that a company has twice the amount of current assets as current liabilities) was satisfactory. If a company's current assets were twice its current debt, creditors generally believed that even if an emergency arose requiring an unexpected use of cash, the company could still pay its short-term debts.

Today, however, users pay more attention to (1) industry structure, (2) the length of a company's operating cycle, and (3) the "mix" of current assets. The mix is the proportion of different items that make up the total current assets. This mix has an effect on how quickly the current assets can be converted into cash. For instance, if a company has a high proportion of prepaid items within its current assets, it may be in a weak liquidity position because prepaid assets are used up rather than being converted into cash. Also, if a company has too *high* a current ratio compared with the ratios of similar companies in the same industry, this may indicate poor management of current assets. For example, maybe the company keeps too much cash on hand rather than investing its excess cash. Finally, the shorter a company's operating cycle, the less likely it is to need a large amount of working capital or as high a current ratio to operate efficiently.

 How do you think the length of a company's operating cycle would affect the amount of working capital it needs or the size of its current ratio?

Sweet Temptations has working capital of $13,567, calculated by subtracting the $5,528 total current liabilities from the $19,095 total current assets, shown in Exhibit 7-2. The company's current ratio is 3.45 ($19,095 total current assets divided by $5,528 total current liabilities), which would be high for an older company. Because Sweet Temptations is a new company, a high current ratio is good because it indicates a strong ability to pay current debts. Anna Cox will want to keep track of this ratio as she makes decisions about future credit purchases.

Quick Ratio

The **quick ratio** is a more convincing indicator of a company's short-term debt-paying ability. Short-term lenders often use this ratio when deciding whether to extend credit. The quick ratio uses only the current assets that may be easily converted into cash—referred to as quick assets. *Quick assets* consist of cash, short-term marketable securities, accounts receivable, and short-term notes receivable. The quick ratio excludes inventory because it may not be sold soon and it may be sold on credit; in both cases, inventory cannot be turned into cash as quickly. The quick ratio also excludes prepaid items because they are not convertible into cash. This is why the ratio is sometimes called the *acid-test* ratio. Thus, the quick ratio is calculated as follows:

$$\text{Quick Ratio} = \frac{\text{Quick Assets}}{\text{Current Liabilities}}$$

The quick ratio shows potential liquidity problems when a company has a poor mix of current assets. For instance, the quick ratio will show that a company with a lot of inventory has less liquidity than indicated by its current ratio. The current ratio won't show this because it includes inventory in the numerator. A quick ratio of 1.0, or 1:1 (showing that a company's quick assets and current liabilities are equal) has generally been considered satisfactory, but users also consider the industry structure as well as the length of the company's operating cycle.

 Suppose a company has a quick ratio of 0.5. What do you think of its liquidity? Does your opinion change if you learn that unearned rent revenue amounts to 60 percent of the company's current liabilities? Why or why not?

Sweet Temptations' quick ratio is 2.01, calculated by dividing Sweet Temptations' $11,130 total quick assets ($11,030 cash + $100 accounts receivable) by its $5,528 total current liabilities. Again, it seems that Sweet Temptations is in a good short-term financial position. Note that Sweet Temptations' quick ratio is about two-thirds of its current ratio because the quick ratio calculation excludes the inventory, supplies, and prepaid rent.

 Why do you think words like "liquid" and "quick" are used in reference to a company's current assets and current liabilities?

Liquidity Ratios of Actual Companies

http://www.jcpenney.com
http://www.sears.com

To illustrate ratio analysis, we continue our evaluation of **JCPenney Company Inc.** and **Sears Roebuck and Co.**, which we began in Chapter 6. The companies' current ratios and quick ratios at the end of 2001 were as follows:

	JCPenney	Sears
Current Ratio	1.93	2.32
Quick Ratio	0.79	1.87

Both the current ratio and the quick ratio of Sears were higher than those of JCPenney. Furthermore, Sears' current ratio was slightly higher than the 2.0 "rule of thumb," and its quick ratio was considerably higher than the 1.0 "rule of thumb." On the other hand, both the current ratio and the quick ratio of JCPenney were below the respective norms. This means that JCPenney is more likely to have short-term liquidity concerns than Sears.

Evaluating Financial Flexibility

4 What is a company's financial flexibility, and how do users evaluate it?

Recall from Chapter 6 that **financial flexibility** is the ability of a company to adapt to change. It is an important financial characteristic because it enables a company to increase or reduce its operating activities as needed. For example, a company with financial flexibility can revise its purchasing plan to take advantage of temporary reductions in wholesale inventory prices. The current ratio and the acid-test ratio can be used to assess short-term financial flexibility.

Managers, owners, and creditors are also interested in a company's ability to take advantage of major, long-term business opportunities. For a company to be able to purchase additional retail stores, build another manufacturing plant, or adopt new information technologies, it must have enough available resources or must be able to raise additional resources. To assess long-term financial flexibility, financial statement users evaluate a company's debt levels. To do this, they calculate a company's debt ratio.

Debt Ratio

The **debt ratio** shows the percentage of total assets provided by creditors and is calculated as follows:

$$\text{Debt Ratio} = \frac{\text{Total Liabilities}}{\text{Total Assets}}$$

The higher a company's debt ratio, the lower its financial flexibility. This is because a higher debt ratio indicates that a company may not be able to borrow money (or may need to pay a higher interest rate to borrow money) to adapt to business opportunities. *Creditors* also prefer that a company have a lower debt ratio because if business declines, a lower debt ratio indicates that the company is more likely to be able to pay the interest it owes as well as its other fixed costs. Up to a point, *owners* prefer a higher debt ratio, particularly when the return earned on assets purchased by the company with the borrowed money is higher than the interest the company has to pay to its creditors. We will discuss this in more detail later.

The debt ratio is subtracted from 100% to show the percentage of total assets contributed by the owner. The desired mix between debt and owner's equity depends on the type of business and the country in which the company is located. For example, in Japan, historically investors have preferred a higher debt ratio than is typical for U.S. companies. This is true primarily because Japanese creditors and investors have worked together more closely. (However, in Japan's current economic environment, this relationship may change.)

 Debt ratios vary from industry to industry. What economic factors do you think would account for these differences?

Sweet Temptations has a debt ratio of 0.26 ($5,528 total liabilities divided by $21,080 total assets). Its debt ratio indicates that most of its assets (74%) are financed by owner's equity. Because Sweet Temptations is new and has no long-term debts, the debt ratio and the current ratio show that it has no immediate problems with solvency or liquidity—that is, it has financial flexibility. If Sweet Temptations decides to expand, creditors will like the fact that, so far, it has relied on Anna's investments and short-term liabilities to finance its operations.

Debt Ratios of Actual Companies

The debt ratios of JCPenney and Sears at the end of 2001 were as follows:

	JCPenney	Sears
Debt Ratio	66.0	86.1

Because of its lower proportion of debt, we can conclude that JCPenney relied less on creditors to finance its assets. Therefore, JCPenney had higher financial flexibility because borrowing more money would be easier for the company. JCPenney may also be able to borrow money at a lower interest rate because the lenders may think it has a lower level of risk.

RELATIONSHIP BETWEEN THE INCOME STATEMENT AND THE BALANCE SHEET

Although decision makers find the ratios we just presented—the current ratio, the quick ratio, and the debt ratio—to be very helpful, these ratios do have one limitation: they use only balance sheet information. *It is very important for you to know that many significant business questions can be answered only by analyzing a company's income statement and balance sheet together.* This is the only way to determine whether a company has made a "satisfactory" profit and to calculate other measures of its "operating capability" (which we will discuss later).

 Do you think creditors always need to evaluate a company's balance sheet and income statement before granting a loan? Why or why not?

Say, for instance, that on its income statement, a company reports that net income for the accounting period is $5 million. Five million dollars may sound like a lot, but did the company earn a satisfactory profit? You can't tell without comparing the $5 million with the dollar amount of resources the company used to earn the income (and with the income it earned in each of the last few years). The $5 million may or may not be satisfactory depending on the size of the company (and how well it has done in prior periods).

Let's say the company reports $50 million of total assets on its balance sheet at the end of the accounting period. We can divide the company's net income for the period by its total assets (a net income to total assets ratio) and calculate the company's rate of return on assets. Using the dollar amounts given, this company earned a 10% return on assets ($5 million ÷ $50 million). Just how satisfied a company's managers, investors, and creditors are with a 10% return on assets depends on how well similar types of companies performed, and on whether this return met or exceeded their expectations. Financial statements work as a *set* of information because, as we noted in the example above, external users need both the income statement and the balance sheet to evaluate a company's performance and financial position.

Exhibit 7-4 illustrates the relationships among the financial statements.[3] Here we show a balance sheet for January 1, 2004, on the far left of the exhibit. This balance sheet reports the resources and claims on resources of a company on the first day of January. It shows the mix of "financial ingredients" that the company had available to work with when starting the accounting period. During the accounting period (2004), the company had many transactions and events affecting assets, liabilities, and owner's equity (owner's investments and withdrawals, revenues and expenses), and all of these transactions and events were recorded in its accounting system. At the end of the accounting period, the company prepares its financial statements. Take some time to study Exhibit 7-4 carefully before you move on; it contains some very important concepts.

The 2004 income statement summarizes how the company used its financial "ingredients" to earn net income and remain solvent. It shows the results of the operating decisions the company's managers made to improve the company's financial position (e.g., how much advertising it used, how much salary expense it incurred, how many sales it made). The year's operating activities, owner's investments, and owner's withdrawals all affect the company's mix of financial resources and the claims to its resources. The 2004 ending balance sheet shows the effects of the net changes.

With the financial information contained in the beginning and ending balance sheets, along with the income statement, you can see what resources the company started with, how it used those resources, and what resources it owns at the end of the accounting period. With this information, you can calculate liquidity, solvency, and performance ratios, as we will discuss in the following section.

---

[3]The cash flow statement is also a useful financial statement, along with the balance sheet and the income statement. We will discuss this statement in Chapter 8, so we have not included it in our present discussion.

EXHIBIT 7-4	RELATIONSHIPS AMONG FINANCIAL STATEMENTS

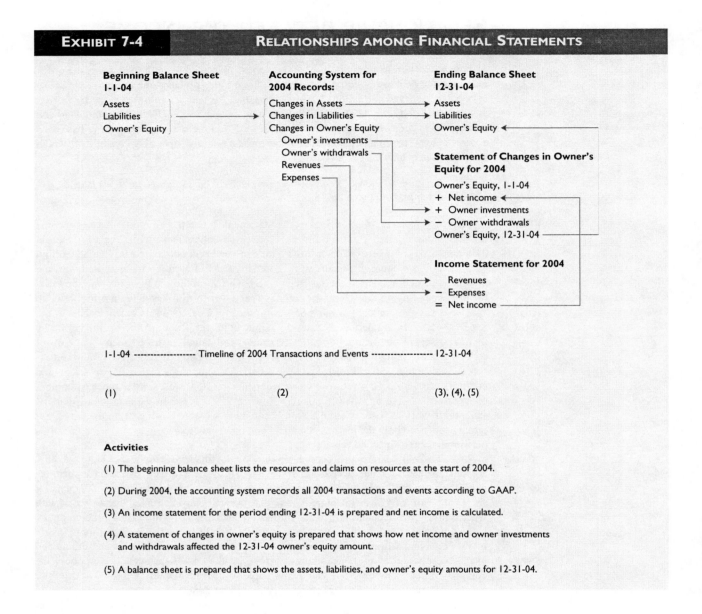

Activities

(1) The beginning balance sheet lists the resources and claims on resources at the start of 2004.

(2) During 2004, the accounting system records all 2004 transactions and events according to GAAP.

(3) An income statement for the period ending 12-31-04 is prepared and net income is calculated.

(4) A statement of changes in owner's equity is prepared that shows how net income and owner investments and withdrawals affected the 12-31-04 owner's equity amount.

(5) A balance sheet is prepared that shows the assets, liabilities, and owner's equity amounts for 12-31-04.

EVALUATIONS USING THE BALANCE SHEET AND THE INCOME STATEMENT

A company's managers, investors, and creditors use information from its income statement and balance sheet to calculate ratios for measuring the company's financial success. The numerator of each ratio is an income statement amount showing the "flow" into or out of the company (e.g., net income, net credit sales, cost of goods sold) *during* the accounting period. The denominator of each ratio is a balance sheet amount showing the "resources" used to obtain the "flow." Some of these ratios use an *average* figure for the denominator. This is because balance sheet amounts are measured at a *point in time* (the beginning and the end of the accounting period). By using an average amount for the accounting period (average total assets, average owner's equity, average inventory) in the denominator, the balance sheet amount "covers" the same time period as the income statement amount. To determine the average amount, add the beginning and the ending amounts together and then divide by 2. We will discuss some common ratios and their calculations in the following sections.

Evaluating Profitability

5 Why and how do users evaluate a company's profitability?

Decision makers use profitability ratios to evaluate how well a company has met its profit objectives in relation to the resources invested. Two of these profitability ratios are the return on total assets ratio and the return on owner's equity ratio.

Return on Total Assets

A company's managers have the responsibility to use the company's assets to earn a satisfactory profit. The amount of net income earned compared with total assets shows whether a company used its economic resources efficiently. A company's **return on total assets** is calculated as follows:

$$\text{Return on Total Assets} = \frac{\text{Net Income} + \text{Interest Expense}}{\text{Average Total Assets}}$$

The net income is obtained from the company's income statement. If a company reports any interest expense on its income statement, decision makers using this ratio add the amount back to net income in the numerator. They make this adjustment because the interest expense is a financing cost paid to creditors and not an operating expense of earning revenue. (The company could have earned the same income before interest expense if the owner had contributed assets rather than financing them through creditors.) Because the company uses its assets to earn net income over the entire accounting period, decision makers use the *average* total assets for the period as the denominator.

When comparing one company's return on total assets with that of another company, you should consider the age of the assets of each company. With increasing prices today, a company using recently purchased assets (at higher costs) will show a lower return on these assets. Also, older assets have higher amounts of accumulated depreciation and therefore lower book values.

 Do you think companies' liabilities affect how their return on total assets ratios are interpreted? Why or why not?

Let's calculate Sweet Temptations' return on total assets for January 2004. Listed below is the information needed to make the calculation (we took the information from its financial statements):

Net income for January 2004:	$ 602
Interest expense for January 2004	8
Total assets, January 1, 2004	17,820
Total assets, January 31, 2004	21,080

The return on total assets ratio of Sweet Temptations for January 2004 is calculated as follows:

$$3.14\% = \frac{\$602 + \$8}{[(\$17,820 + \$21,080)/2]}$$

Although this 3.14% return on assets for Sweet Temptations seems low, remember that it is for one month and not for an entire year, as is typical. (Multiplying the ratio by 12 [months] gives an estimate of Sweet Temptations' ratio for the year.) If Sweet Temptations can keep up January's level of income for the rest of the year, it is likely to earn a satisfactory rate of return on its assets for 2004.

Return on Owner's Equity

A company's managers also have the responsibility to earn a satisfactory return on the owner's investment in the company. Dividing net income by the *average* owner's equity shows the company's return (in percentage terms) to the owner—resulting from all of the company's activities during the accounting period. A company's **return on owner's equity** is calculated as follows:

$$\text{Return on Owner's Equity} = \frac{\text{Net Income}}{\text{Average Owner's Equity}}$$

Note that in contrast to the return on total assets ratio, the return on owner's equity ratio does not add interest expense back to net income. This is because net income is a measure of a company's profits available to owners *after* incurring the financial cost related to creditors.

We can calculate Sweet Temptations' return on owner's equity for January 2004 by taking the information listed below from its financial statements:

Net income for January 2004	$ 602
A. Cox, capital, January 1, 2004	15,000
A. Cox, capital, January 31, 2004	15,552

The return on owner's equity ratio of Sweet Temptations for January 2004 is calculated as follows:

$$3.94\% = \frac{\$602}{[(\$15,000 + \$15,552)/2]}$$

Sweet Temptations' 3.94% return on owner's equity for January 2004 is low, but again it is for only one month. (Multiplying the ratio by 12 gives an estimate of Sweet Temptations' ratio for the year.) Also, Sweet Temptations' return on owner's equity is higher than its return on total assets. This shows users that the company has benefited from using debt to help finance its assets.

Profitability Ratios of Actual Companies

The ratios used to evaluate the profitability of JCPenney and Sears for 2001 were as follows:

	JCPenney	Sears
Return on Total Assets	2.6%	5.3%
Return on Owners' Equity	1.5%	11.4%

How do you think Ernie's plan would boost his company's return-on-assets ratio? Do you think it is a good idea? Why or why not?

DILBERT® REPRINTED BY PERMISSION OF UNITED FEATURES SYNDICATE, INC.

Based on Sears' higher return on total assets and its higher return on owners' equity, we can say that it used its assets more efficiently and earned a more satisfactory profit for its owners than did JCPenney.

Evaluating Operating Capability

6 What is a company's operating capability, and how do users evaluate it?

Recall from Chapter 6 that **operating capability** refers to a company's ability to sustain a given level of operations. Information about a company's operating capability is important in evaluating how well it is maintaining its operating level, and in predicting future changes in its operating activity. The current ratio helps predict a company's ability to continue to purchase inventory. If a company's current ratio is less than 2.0, investors may worry that the company's operations won't generate enough cash to replenish inventory. The debt ratio helps evaluate whether a company has the resources to replace property and equipment.

In this section we discuss how evaluating the level of a company's activities can provide insights into its operating capability. This is done through activity ratios, used to show the length of the parts of the company's operating cycle. This knowledge lets users evaluate the liquidity of selected current assets. Recall that a retail company's operating cycle is the length of time it takes to invest cash in inventory, make credit sales, and convert the receivables into cash. Two common activity ratios are the (1) inventory turnover and (2) accounts receivable turnover.

Inventory Turnover

A company purchases, sells, and replaces inventory throughout its accounting period. Dividing the company's cost of goods sold (from its income statement) for the period by the average inventory (from its beginning and ending balance sheets) shows the number of times the company *turns over* (or sells) the inventory during that period. A company's **inventory turnover** is calculated as follows:

$$\text{Inventory Turnover} = \frac{\text{Cost of Goods Sold}}{\text{Average Inventory}}$$

As a general rule, the higher the inventory turnover, the more efficient the company is in its purchasing and sales activities and the less cash it needs to invest in inventory. A company with a higher turnover generally purchases its inventory more often and in smaller amounts than it would if it had a lower inventory turnover. It is also less likely to have obsolete inventory (because it holds on to its inventory for only a short time before selling it). These efficiencies "free up" a company's cash—it needs less cash and can invest excess cash in other earnings activities. However, a company's inventory turnover can be too high. If a company's inventory turnover is too high, the company may not be keeping enough inventory on hand to meet customer demand, and it may be missing out on additional sales.

Let's calculate Sweet Temptations' inventory turnover for January 2004. Listed below is the information needed to make the calculation (we took the information from its financial statements):

Cost of goods sold for January 2004	$3,645
Inventory, January 1, 2004	1,620
Inventory, January 31, 2004	2,295

We can calculate Sweet Temptations' inventory turnover ratio for January 2004 as follows:

$$1.86 \text{ times} = \frac{\$3,645}{[(\$1,620 + \$2,295)/2]}$$

This ratio shows that Sweet Temptations turned over its inventory almost two times in January. (To estimate Sweet Temptations' inventory turnover for the year, assuming every month has the same rate of turnover, multiply the January turnover by 12.) Since Sweet Temptations is a candy store, this turnover is a good sign that Sweet Temptations is operating efficiently (who wants to buy old candy?). Over the next few months, Anna should continue to monitor Sweet Temptations' inventory turnover to see if she needs to make any changes in its purchasing budget.

Users sometimes want a different measure of how efficient a company is in its inventory activities—how long it takes a company to sell its inventory. This measure is called the **number of days in the selling period.** Dividing the number of operating days in a company's business year (a company that does business seven days a week has 365 days in its business year) by its inventory turnover shows the number of days in its selling period, as follows:

$$\text{Number of Days in Selling Period} = \frac{\text{Number of Days in Business Year}}{\text{Inventory Turnover}}$$

This ratio estimates the average time (in days) it takes the company to sell its inventory. Because we are calculating these ratios for Sweet Temptations for only one month, we use 30 days (the number of days Sweet Temptations was open in January, excluding New Year's Day) as the numerator. With that in mind, the number of days in its selling period is calculated as follows:

$$16.13 \text{ days} = \frac{30}{1.86}$$

This ratio tells us that in January, Sweet Temptations sold its inventory about every 16 days. To evaluate how well it is managing its inventory, we should compare these results with how long a box of candy stays "fresh," with Sweet Temptations' ratio in previous years (if it were not a new company), and with other companies' performances.

 What do you think Anna should do if a box of candy stays "fresh" about two weeks?

Accounts Receivable Turnover

If a company sells inventory on credit, it must collect the accounts receivable from the sales to complete its operating cycle. Dividing a company's net credit sales for the period (from its income statement) by its average accounts receivable (from its beginning and ending balance sheets) shows how many times the average receivable turns over (is collected) each period. A company's **accounts receivable turnover** is calculated as follows:

$$\text{Accounts Receivable Turnover} = \frac{\text{Net Credit Sales}}{\text{Average Accounts Receivable}}$$

The accounts receivable turnover measures how efficiently a company collects cash from its credit customers. Users prefer to see a higher turnover, which shows that the company has less cash tied up in accounts receivable, collects this cash faster, and usually has fewer customers who don't pay.

The amount of net *credit* sales is the best amount to use as the numerator. This is the number that managers use when making this calculation. However, since companies don't give a breakdown between credit and cash sales on their income statements, external users must calculate the ratio using total net sales. Because using total net sales increases the numerator (unless all sales are credit sales), this calculation will overestimate the number of times a company's accounts receivable turns over.

Users often divide a company's accounts receivable turnover into the number of days in the business year to show the **number of days in the collection period,** as follows:

$$\text{Number of Days in Collection Period} = \frac{\text{Number of Days in Business Year}}{\text{Accounts Receivable Turnover}}$$

The number of days in a company's collection period is the average time it takes the company to collect its accounts receivable. By comparing a company's average collection period with the days in its credit terms (i.e., 2/10, n/30), a user can see how aggressive the company is in collecting overdue accounts. The user can also compare this number with the ratios for past years and with those of other companies. Because Sweet Temptations made only one credit sale in January, we do not calculate these accounts receivable ratios here. To estimate the number of days in a company's operating cycle, a user can add together the number of days in the company's selling period and the number of days in its collection period.

 Is a company's operating cycle always the same, or can a company control the length of its operating cycle? What, if anything, can be done to control the length of the operating cycle?

Operating Capability Ratios of Actual Companies

The ratios used to evaluate the operating capability of JCPenney and Sears for 2001 were as follows:

	JCPenney	Sears
Inventory Turnover	4.5	5.0
Accounts Receivable Turnover	40.2	1.6

Sears' higher inventory turnover shows that it took less time to sell its inventory than did JCPenney, indicating that Sears was more efficient in managing its inventory. However, JCPenney was more efficient in collecting its cash from credit customers than was Sears, as shown by its higher accounts receivable turnover.[4]

LIMITATIONS OF THE INCOME STATEMENT AND THE BALANCE SHEET

In Chapters 5 and 6 you saw that a company's accounting system is based on several important accounting concepts and principles. The concepts and principles were created to ensure that companies' accounting systems provide useful, reliable, and relevant information to their managers and other interested parties.

 The key concepts and principles are the entity concept, the monetary unit concept, the historical cost concept, the accounting period concept, the matching principle, and accrual accounting. In your own words, describe why each of these is important.

These concepts and principles guide accountants as they analyze company activities, record transactions, make adjustments, and prepare a company's income statement and balance sheet. Yet even though we have seen how concepts and principles help to build a useful accounting system, they also set limits on the types of information the financial statements provide. These limits restrict the usefulness of the information. For example, the historical cost concept requires that the asset Land be reported on a company's balance sheet at its original cost. So if a company purchased land in 1979 for $10,000, its

---

[4]However, care must be taken when comparing the two companies' ratios. JCPenney "sells" a significant amount of its accounts receivable to a finance company, which results in lower average accounts receivable and therefore a higher accounts receivable turnover. Furthermore, Sears' accounts receivable turnover may have been lower in part because it encourages its credit customers to extend the payments on their "revolving charge" accounts, for which Sears charges these customers interest.

2004 ending balance sheet will list "Land $10,000," no matter how much the land is currently worth. In 2004, the land may be worth much more than $10,000. Thus, the balance sheet doesn't always show each asset's current value. But if the company has no intention of selling an asset, the current value may not be relevant.

Another limitation of the income statement and the balance sheet is that they do not provide much information about a company's cash management because they are based on accrual accounting. Hence, investors and creditors also need a financial statement that provides a summary of a company's cash flows during an accounting period. Thus, a company prepares and reports a third financial statement—the cash flow statement, which we will discuss in Chapter 8.

 What do a company's balance sheet and income statement reveal about its management of cash? What else would an investor or creditor want to know?

Business Issues and Values

Recall that assets are a company's economic resources that it expects will provide future benefits to the company. As we mentioned above, in accordance with GAAP, companies record assets at their historical cost and do not change these amounts for changes in their values. One of the major economic resources of many companies is their employees. A company that has a loyal, well-trained employee group has a valuable "asset" that may increase in value over time because of additional training and job satisfaction. This employee group makes very important contributions to the company's ability to earn profits. (Do you continue to shop at a store where the employees are rude?) But because of the historical cost concept, the company cannot report this economic resource as an asset on the balance sheet it issues to external users. However, some companies that take pride in the quality of their employees do prepare internal reports that include measures of their employees' values; their managers use these reports for internal decision making.

 Does "investing" in its employees worsen a company's reported performance in the current year?

SUMMARY

At the beginning of the chapter we asked you several questions. During the chapter, we asked you to STOP and answer some additional questions to build your knowledge about specific issues. Be sure you answered these additional questions. Below are the questions from the beginning of the chapter, with a brief summary of the key points relating to the answers. Use your creative and critical thinking skills to expand on these key points to develop more complete answers to the questions and to determine what other questions you have that might lead you to learn more about the issues.

1 Why is a company's balance sheet important?

A company's balance sheet is important because this statement provides internal and external users with information to help evaluate the company's ability to achieve its primary goals of earning a satisfactory profit and remaining solvent. A balance sheet provides information about a company's economic resources and the claims on those resources (its financial position) on a specific date.

2 What do users need to know about a company's classified balance sheet?

Users need to know that a company's classified balance sheet shows important subtotals, in related groupings, for the assets, liabilities, and owner's equity of the company. The groupings include current assets and noncurrent assets, as well as current liabilities and noncurrent liabilities.

Current assets are cash and other assets that a company expects to convert into cash, sell, or use up within one year. Current assets include cash, marketable securities, receivables, inventory, and prepaid items. Noncurrent assets are assets other than current assets; these include items such as long-term investments, as well as property and equipment. Current liabilities are obligations that a company expects to pay within one year by using current assets. Current liabilities include accounts payable and salaries payable, unearned revenues, and short-term notes (and interest) payable. Noncurrent liabilities are obligations that a company does not expect to pay within the next year; these include items such as long-term notes payable, mortgages payable, and bonds payable.

3 What is a company's liquidity, and how do users evaluate it?

A company's liquidity is a measure of how quickly it can convert its current assets into cash to pay its current liabilities as they become due. Users evaluate a company's liquidity by studying its working capital (current assets minus current liabilities), current ratio (current assets divided by current liabilities), and quick (acid-test) ratio (quick assets divided by current liabilities).

4 What is a company's financial flexibility, and how do users evaluate it?

A company's financial flexibility is its ability to adapt to change. Measures of a company's financial flexibility are used to assess whether the company can increase or reduce its operating activities as needed. Users study a company's current ratio and quick ratio to evaluate its short-term financial flexibility. They study a company's debt ratio (total liabilities divided by total assets) to evaluate its long-term financial flexibility.

5 Why and how do users evaluate a company's profitability?

Users evaluate a company's profitability to determine how well it has met its profit objectives in relation to the resources invested. They study a company's return on total assets [(net income plus interest expense) divided by average total assets] and return on owner's equity (net income divided by average owner's equity) ratios to evaluate a company's profitability.

6 What is a company's operating capability, and how do users evaluate it?

A company's operating capability is its ability to sustain a given level of operations. Measures of a company's operating capability are used to assess how well the company is maintaining its operating level and to predict future changes in its operating activity. Users study a company's activity ratios to determine the length of the parts of the company's operating cycle. These ratios include the inventory turnover (cost of goods sold divided by average inventory) and the accounts receivable turnover (net credit sales divided by average accounts receivable).

KEY TERMS

accounting equation (p. 200)
accounts receivable turnover (p. 215)
accumulated depreciation (p. 203)
assets (p. 202)
balance sheet (p. 200)
book value (p. 203)
classified balance sheet (p. 202)
current assets (p. 202)
current liabilities (p. 204)
current ratio (p. 207)
debt ratio (p. 209)
financial flexibility (p. 208)
inventory turnover (p. 214)
liabilities (p. 204)

liquidity (p. 206)
long-term investments (p. 203)
noncurrent liabilities (p. 204)
number of days in the collection period
    (p. 216)
number of days in the selling period
    (p. 215)
operating capability (p. 214)
owner's equity (p. 205)
property and equipment (p. 203)
quick ratio (p. 208)
return on owner's equity (p. 213)
return on total assets (p. 212)
working capital (p. 206)

SUMMARY SURFING

Here is an opportunity to gather information on the Internet about real-world issues related to the topics in this chapter. Go to http://www.cunningham.swlearning.com and click on the Interactive Study Center. Click on this chapter number then click on Summary Surfing and answer the following questions.

- Click on JCPenney. Find the appropriate financial statement(s). Compute the current ratio, quick ratio, debt ratio, return on owners' equity, and inventory turnover for the most current year. How do these results compare with the 2001 ratios we discussed in this chapter?

- Click on Sears. Find the appropriate financial statements. Compute the current ratio, quick ratio, debt ratio, return on owners' equity, and inventory turnover for the most current year. How do these results compare with the 2001 ratios we discussed in this chapter?

INTEGRATED BUSINESS AND ACCOUNTING SITUATIONS

Answer the Following Questions in Your Own Words.

Testing Your Knowledge

7-1   What is a balance sheet, and what types of questions can a user answer by studying the balance sheet?

7-2   What is the accounting equation, and how does it relate to the balance sheet of a company?

7-3   What is the difference between an account form and a report form of balance sheet?

7-4   Explain what is meant by a *classified* balance sheet, and identify the major groupings of assets and liabilities.

7-5   Explain the meaning of the term *current assets*.

7-6   Identify and briefly explain the major current assets.

7-7   What are long-term investments? Give several examples.

7-8   What is property and equipment? At what amount is each item of property and equipment listed on the balance sheet?

7-9   Explain the meaning of the term *current liabilities*.

7-10   Identify and briefly explain the major current liabilities.

7-11   What are noncurrent liabilities? Give several examples.

7-12   What is owner's equity, and why is it sometimes called *residual equity*?

7-13   What is meant by the term *liquidity*, and why is it important?

7-14   Explain how to compute the current ratio and what it is used for.

7-15   Explain how to compute the quick ratio and what it is used for.

7-16   What is meant by the term *financial flexibility*, and why is it important?

7-17   Explain how to compute the debt ratio and what it is used for.

7-18   Explain how to compute a company's return on total assets and what it is used for.

7-19   Explain how to compute a company's return on owner's equity and how it relates to the return on total assets.

7-20   What is meant by the term *operating capability*, and why is information about it important?

7-21   Explain how to compute a company's inventory turnover. Is a high inventory turnover good or bad? Why?

7-22   Explain how to compute a company's accounts receivable turnover. What is a "good" accounts receivable turnover? Why?

Applying Your Knowledge

7-23   In each of the following situations, the total increase or decrease for one component of the accounting equation is missing:
(a) Assets increased by $10,400; liabilities increased by $3,200.
(b) Liabilities decreased by $2,000; owner's equity increased by $10,000.
(c) Assets decreased by $6,200; owner's equity decreased by $13,500.
(d) Owner's equity increased by $27,500; liabilities decreased by $5,715.
(e) Assets increased by $12,600; owner's equity decreased by $25,750.

Required: Using Assets: $60,000 = Liabilities: $20,000 + Owner's Equity: $40,000 as the beginning accounting equation for each of the preceding situations, determine (1) the total increase or decrease for the missing component of the equation and (2) the amount of each component in the *ending* accounting equation. Treat each situation independently.

7-24   The total increase or decrease for one component of the accounting equation is missing in each situation that follows:
(a) Assets decreased by $10,000; liabilities decreased by $6,500.
(b) Owner's equity decreased by $15,750; assets decreased by $7,500.
(c) Liabilities increased by $1,000; owner's equity decreased by $5,000.
(d) Owner's equity increased by $18,000; assets increased by $9,650.

Required: Using Assets: $45,000 = Liabilities: $15,000 + Owner's Equity: $30,000 as the beginning accounting equation for each of the preceding situations, determine (1) the total increase or decrease for the missing component of the equation and (2) the amount of each component in the *ending* accounting equation. Treat each situation separately.

7-25   Listed below are the balances of selected accounts of the Watson Company at the end of the current year:

Equipment	$18,500
Prepaid insurance	2,600
Notes payable (due in 30 days)	7,100
Cash	3,900
Land	11,700
Accounts receivable (net)	10,200
Inventory	24,400
Mortgage payable (due next year)	33,000
Notes receivable (due in 60 days)	4,000
Marketable securities (short-term)	6,300
Buildings (net)	74,000
Notes receivable (due in 2 years)	5,600

Required: Prepare the current assets section of the Watson Company's balance sheet.

7-26   Listed below are the balances of selected accounts of the Chriswat Company at the end of the current year:

Notes receivable (due in 3 years)	$14,200
Accounts payable	18,300
Bonds payable (due in 5 years)	46,000
Land	13,500
Marketable securities (short-term)	6,400
Salaries payable	5,700

Notes payable (due in 6 months)	8,000
Mortgage payable (due next year)	4,600
Unearned rent revenue (6 months)	2,400
Notes payable (due in 2 years)	10,000
Mortgage payable (due in 5 years)	18,000

Required: Prepare the current liabilities section of the Chriswat Company at the end of the current year.

7-27 A classified balance sheet contains the following sections:

A. Current assets   D. Current liabilities
B. Long-term investments   E. Noncurrent liabilities
C. Property and equipment F. Owner's equity

Required: The following is a list of accounts. Using the letters A through F, indicate in which section each account is shown.

_____ 1. Land
_____ 2. Accounts payable
_____ 3. A. Smith, capital
_____ 4. Cash
_____ 5. Bonds payable
_____ 6. Equipment
_____ 7. Accounts receivable
_____ 8. Unearned revenue
_____ 9. Mortgage payable (due in 4 years)
_____ 10. Salaries payable
_____ 11. Marketable securities (short-term)
_____ 12. Notes receivable (due in 2 years)
_____ 13. Buildings
_____ 14. Notes payable (due in 9 months)
_____ 15. Prepaid insurance
_____ 16. Inventory

7-28 The following is an alphabetical list of the accounts of Swenson Stores on December 31, 2004:

Accounts payable
Accounts receivable
Administrative expenses
Bonds payable (due 2014)
Buildings (net)
Cash
Cost of goods sold
Equipment (net)
Mortgage payable (10 equal annual payments)
Notes payable (due in 6 months)
Notes payable (due in 4 years)
Notes receivable (due in 8 months)
Notes receivable (due in 3 years)
Office supplies

General expenses
Interest expense
Interest payable (current)
Interest receivable (current)
Interest revenue
Inventory
Investment in government bonds (due 2015)
Land
Prepaid insurance
Salaries payable
Sales
Selling expenses
T. Swenson, capital
Temporary investments in securities

Required: Prepare a December 31, 2004 classified balance sheet (without amounts) for Swenson Stores.

7-29 The financial statement information of the Leon Appraisal Company for 2004 and 2005 is as follows:

	2004	2005
Assets, 12/31	$ (a)	$308,900
Expenses	47,400	51,600
Net income	(b)	39,700
Liabilities, 12/31	153,500	(e)
Leon, capital, 1/1	(c)	115,200
Revenues	83,600	(f)
Leon, withdrawals	24,000	(g)
Leon, capital, 12/31	(d)	124,900

Required: Fill in the blanks lettered (a) through (g). All the information is listed. (*Hint:* It is not necessary to calculate your answers in alphabetical order.)

7-30 The financial statement information of the Charles Adjusting Company for 2004 and 2005 is shown on the following page.

	2004	2005
Charles, capital, 12/31	$ 83,500	$ (d)
Charles, withdrawals	(a)	24,000
Revenues	(b)	65,000
Charles, capital, 1/1	69,400	(e)
Liabilities, 12/31	(c)	116,800
Net income	24,100	(f)
Charles, additional investments	8,000	(g)
Expenses	35,200	39,800
Assets, 12/31	184,500	211,500

Required: Fill in the blanks lettered (a) through (g). All the information is listed. (*Hint:* It is not necessary to calculate your answers in alphabetical order.)

7-31   The balance sheet information at the end of 2004 and 2005 for the Decatur Medical Equipment Company is as follows:

	2004	2005
Current assets	$ (a)	$ 27,000
Noncurrent liabilities	(b)	34,900
Long-term investments	19,200	22,500
Davis, capital	81,900	(d)
Total liabilities	(c)	(e)
Current liabilities	14,500	12,300
Total assets	130,200	(f)
Property and equipment (net)	85,700	93,100

Required: Fill in the blanks labeled (a) through (f). All the necessary information is provided. (*Hint:* It is not necessary to calculate your answers in alphabetical order.)

7-32   The balance sheet information at the end of 2004 and 2005 for Columbia Electronics is as follows:

	2004	2005
Bevis, capital	$ 83,500	$ 88,700
Current liabilities	(a)	9,800
Property and equipment (net)	(b)	87,500
Current assets	18,500	(e)
Long-term liabilities	(c)	30,200
Total assets	(d)	(f)
Working capital	9,300	10,200
Long-term investments	23,700	(g)
Total liabilities	38,100	(h)

Required: Fill in the blanks labeled (a) through (h). All the necessary information is provided. (*Hint:* It is not necessary to calculate your answers in alphabetical order.)

7-33   The following items and their corresponding amounts appeared in the accounting records of the Office Equipment Specialists Company on December 31, 2004:

Accounts receivable	$ 4,900
Accounts payable	2,900
Building (net)	24,000
Cash	1,400
Delivery equipment (net)	10,000
Inventory	7,500
J. Jenlon, capital	35,400
Mortgage payable (due 9/1/2006)	29,000
Marketable securities	2,000
Notes payable (due 10/1/2005)	10,000
Office supplies	2,600
Land	6,000
Notes receivable (due 12/31/2006)	7,000
Office equipment (net)	6,400
Prepaid insurance	1,700

Notes payable (due 12/31/2008)	11,000
Interest payable (due 10/1/2005)	1,000
Unearned revenue	3,000
Investment in government bonds (due 12/31/2013)	20,000
Salaries payable	1,200

Required: (1) Prepare a classified balance sheet for the Office Equipment Specialists Company on December 31, 2004.

(2) The Office Equipment Specialists Company is applying for a short-term loan at a local bank. If you were the banker, would you grant the company a loan? Explain your decision using what you learned in this chapter about evaluating a company's liquidity.

7-34 The following accounts and account balances were listed in the accounting records of the Rigons Lighting Company on December 31, 2004:

Salaries payable	$ 1,100
Accounts receivable	11,300
Investment in government bonds (due 12/31/2008)	30,000
Accounts payable	7,700
Unearned revenue	1,000
Building (net)	37,000
Interest payable (due 9/1/2005)	200
Cash	6,100
Notes payable (due 12/31/2006)	15,000
Store equipment (net)	14,000
Prepaid insurance	900
Office equipment (net)	9,600
Inventory	13,200
Notes receivable (due 12/31/2007)	8,000
P. Rigons, capital	85,300
Land	4,000
Mortgage payable (due 7/1/2006)	22,500
Office and store supplies	2,700
Marketable securities	3,000
Notes payable (due 9/1/2005)	7,000

Required: (1) Prepare a classified balance sheet for the Rigons Lighting Company on December 31, 2004.

(2) The Rigons Lighting Company is applying for a $2,000 short-term loan at a local bank. If you were the banker, would you grant a loan to the company? Explain your decision using what you learned in this chapter about evaluating a company's liquidity.

7-35 Taylor Machines Company has the following condensed balance sheet on December 31, 2004:

Current assets..........................	$ 13,400	Current liabilities..........................	$ 6,800
Noncurrent assets.................	91,200	Noncurrent liabilities.................	36,700
		Total Liabilities..............................	$ 43,500
		T. Taylor, capital..........................	61,100
		Total Liabilities and	
Total Assets.............................	$104,600	Owner's Equity..........................	$104,600

The company's quick assets are 60% of its current assets.

Required: Compute the company's working capital and its current, quick, and debt ratios at the end of 2004.

7-36 Simpson Company reported net income of $78,200 for 2005. Interest expense of $4,800 was deducted in the calculation of this net income. The following schedule shows other information about the company's capital structure:

	12/31/2004	12/31/2005
Total Assets	$670,000	$730,000
Total Owner's Equity	415,000	465,000

Required: (1) Compute the return on total assets for 2005.

(2) Compute the return on owner's equity for 2005.

(3) Compute the debt ratio at the end of 2005. How does this compare with the debt ratio at the end of 2004?

7-37 Parket Company began 2004 with accounts receivable of $32,000 and inventory of $40,000. During 2004, the company made total net sales of $600,000, of which 70% were credit sales. The company's cost of goods sold averaged 60% of total net sales during 2004. Parket was open for business each day of the year, and at the end of the year it had accounts receivable of $36,000 and inventory of $60,000.

Required: (1) Compute the inventory turnover and the number of days in the selling period for 2004.

(2) Compute the accounts receivable turnover and the number of days in the collection period for 2004.

(3) What is your estimate of the number of days in the company's operating cycle during 2004?

Making Evaluations

7-38 A friend of yours makes this statement: "Accumulated depreciation and depreciation expense are the same thing, since they both measure the portion of the cost of an asset that has been 'used up' to earn revenues."

Required: Do you agree or disagree with your friend's statement? Support your answer.

7-39 Many long-term loans are payable over a period of time. For example, when a company takes out a mortgage to finance a building, it pays off a fraction of that mortgage every month.

Required: What criteria would you use to decide whether to classify the mortgage as a current liability or a long-term liability, and how would you classify the mortgage?

7-40 In this chapter, we said that the quick ratio is a better measure of liquidity than is the current ratio because the quick ratio includes only those current assets that may be easily converted to cash.

Required: What is the quick ratio? Do you think this is the best possible measure of liquidity? If so, defend your answer. If not, design a better measure and defend it.

7-41 On March 8, 2004, Peter Bailey started his own company by depositing $10,000 in the Bailey Company checking account at the local bank. On March 14, 2004, the Bailey Company checkbook was stolen. During that period of time, the Bailey Company had entered into several transactions, but unfortunately, it had not set up an accounting system for recording the transactions. Bailey did save numerous source documents, however, which had been put into an old shoebox.

In the shoebox is a fire insurance policy dated March 13, 2004, on a building owned by the Bailey Company. Listed on the policy was an amount of $300 for one year of insurance. "Paid in Full" had been stamped on the policy by the insurance agent. Also included in the box was a deed for land and a building at 800 East Main. The deed was dated March 10, 2004, and showed an amount of $40,000 (of which $8,000 was for the land). The deed indicated that a down payment had been made by the Bailey Company and that a mortgage was signed by the company for the balance owed.

The shoebox also contained an invoice dated March 12, 2004, from the Ace Office Equipment Company for $600 of office equipment sold to the Bailey Company. The invoice indicates that the amount is to be paid at the end of the month. A $34,000 mortgage, dated March 10, 2004, and signed by the Bailey Company, for the purchase of land and a building is also included in the shoebox. Finally, a 30-day, $4,000 note receivable is in the shoebox. It is dated March 15, 2004, and is issued to the Bailey Company by the Ret Company for "one-half of the land located at 800 East Main."

The Bailey Company has asked for your help in preparing a classified balance sheet as of March 15, 2004. Peter Bailey indicates that company checks have been issued for all cash payments. Bailey has called its bank. The bank's records indicate that the Bailey Company's checking account balance is $9,500, consisting of a $10,000 deposit, a $200 canceled check made out to the Finley Office Supply Company, and a $300 canceled check made out to the Patz Insurance Agency.

You notice that the Bailey Company has numerous office supplies on hand. Peter Bailey states that a company check was issued on March 8, 2004, to purchase the supplies but that none of the supplies had been used.

Required: Based on the preceding information, prepare a classified balance sheet for the Bailey Company on March 15, 2004. Show supporting calculations.

7-42 The following items appear (in thousands) on the February 3, 2001 and January 29, 2000, financial statements of **Dillard's Department Stores Inc.:**

http://www.dillards.com

	February 3, 2001	January 29, 2000
Accounts receivable	$ 979,241	$1,104,925
Inventory	1,616,186	2,047,830
Sales (net)	8,566,560	8,676,711
Cost of goods sold	5,802,147	5,762,431

Suppose that in February 2001, Dillard's wants to arrange with its supplier to pay for merchandise 90 days after the purchase.

Required: Based on the above information, would you, as a supplier, feel confident about the ability of Dillard's to pay you in 90 days? Justify your answer. If you were making the decision to grant Dillard's credit, what other information would you like to know about Dillard's Department Stores Inc.?

7-43 Bart Brock is thinking about starting his own company, BB's. At the beginning of October 2004, he plans to invest $16,000 into the business. During October the company will purchase land, a small building to house the business, some office equipment, and some supplies. Bart has found land and a building that would be suitable for the company. The purchase price of both the land and the building is $60,000. Bart estimates that the cost of the land is 15% of the total price and the building is 85% of the total price. Bart wants the company to "finance" this purchase through its bank. The bank would require BB's to make a 20% down payment and would also require the company to sign a mortgage for the balance. Bart has determined that there is too much land, however, so that if BB's purchased the land and building, it would sell one-quarter of the land to another company to use as a parking lot. The other company has agreed to buy the land at a price equal to the cost paid by BB's and to sign a note requiring payment of this cost at the end of two years. Bart has found some used office equipment that could be purchased by BB's for $1,800 on credit, to be paid in 60 days. He also expects that BB's will need $800 of office supplies, which the company would purchase with cash. Before the bank will lend BB's the money to buy the land and building, it has requested a "projected" balance sheet for the company, along with a "projected" current ratio and debt ratio as of October 31, 2004, based on the preceding plans. Bart Brock has asked for your help.

Required: (1) Using the preceding information, prepare a projected balance sheet, current ratio, and debt ratio for BB's as of October 31, 2004. Show supporting calculations.

(2) Basing your decision solely on this information, if you were the banker would you give BB's the loan? What other information would help you make your decision?

7-44 Today is January 1, 2005. Last night you were at a New Year's Eve party at which you ran into a long-lost friend, Art Washet, who is the owner of Washet Company. In a conversation, Art mentioned that his company would like to borrow $5,000 from you now and repay you $6,000 at the end of two years. You told him to stop by your house today with his financial records. He has just dropped off the following balance sheet, along with his accounting records:

WASHET COMPANY
Balance Sheet
For Year Ended December 31, 2004

Working capital.................	$ 11,100		Noncurrent liabilities.................	$ 9,400
Other assets	93,900		Owner's equity	95,600
Total..................................	$105,000		Total	$105,000

Your analysis of these items and the accounting records reveals the following information (the amounts in parentheses indicate deductions from each item):

(a) Working capital consists of the following:

Equipment (net)..	$ 14,000
Land..	10,000
Accounts due to suppliers	(28,000)
Inventory, including office supplies of $3,700	34,700
Salaries owed to employees	(2,600)
Note owed to bank (due June 1, 2005).............................	(17,000)
	$ 11,100

(b) Other assets include the following:

Cash..	$ 6,000
Prepaid insurance..	1,900
Buildings (net)...	46,000
Long-term investment in government bonds....................	30,000
A. Washet, withdrawals...	10,000
	$ 93,900

(c) Noncurrent liabilities consist of the following:

Mortgage payable (due March 1, 2009).............................	33,000
Accounts due from customers..	(16,600)
Notes receivable (due December 31, 2007)......................	(7,000)
	$ 9,400

(c) Owner's equity includes the following:

A. Washet, capital	$104,900
Securities held as a temporary investment........................	(11,000)
Interest payable (due with note on June 1, 2005)............	1,700
	$ 95,600

Required: (1) Using your analysis, prepare a properly classified December 31, 2004 balance sheet (report form) for Washet Company.

(2) Compute the current ratio and the quick ratio for the company on December 31, 2004. Basing your decision solely on this information, would you loan $5,000 to the company?

7-45   Ray Young owns and operates a repair service called Ray's Rapid Repairs. It is the end of the year, and his bookkeeper has recently resigned to move to a warmer climate. Knowing only a little about accounting, Ray prepared the following financial statements, based on the ending balances in the company's accounts on December 31, 2004:

RAY'S RAPID REPAIRS
Income Statement
For Year Ended December 31, 2004

Repair service revenues		$ 29,000
Operating expenses:		
Rent expense...	$ 3,200	
Salaries expense......................................	9,900	
Utilities expense	1,100	
R. Young, withdrawals...........................	16,000	
Total operating expenses		(30,200)
Net Loss..		$ (1,200)

RAY'S RAPID REPAIRS
Balance Sheet
December 31, 2004

Assets		Liabilities and Owner's Equity	
Cash	$ 1,500	Accounts payable	$ 2,600
Repair supplies	2,400	Note payable (due 1/1/07)	10,000
Repair equipment	15,000	Total Liabilities	$12,600
		R. Young, capital[a]	6,300
		Total Liabilities and	
Total Assets	$18,900	Owner's Equity	$18,900

[a]Beginning capital − net loss

Ray is upset and says to you, "I don't know how I could have had a net loss in 2004. Maybe I did something wrong when I made out these financial statements. Could you help me? My business has been good in 2004. In these times of high prices, people have been getting their appliances and other items repaired by me instead of buying new ones. I used to have to rent my repair equipment, but business was so good that I purchased $15,000 of repair equipment at the beginning of the year. I know this equipment will last 10 years even though it won't be worth anything at the end of that time. I did have to sign a note for $10,000 of the purchase price, but the amount (plus $1,200 annual interest) will not be due until the beginning of 2007. I still have to rent my repair shop, but I paid $3,200 for two years of rent in advance at the beginning of 2004, so I am OK there. And besides, I just counted my repair supplies, and I have $1,100 of supplies left from 2004 which I can use in 2005."

He continues: "I'm not too worried about my cash balance. I know that customers owe me $700 for repair work I just completed in 2004. These are good customers and always pay, but I never tell my bookkeeper about this until I collect the cash. I am sure I will collect in 2005, and that will also make 2005 revenues look good. In fact, it will almost offset the $600 I just collected in advance (and recorded as a revenue) from a customer for repair work I said I would do in 2005. I still have to write a check to pay my bookkeeper for his last month's salary, but he was my only employee in 2004. In 2005 I am going to hire someone only on a part-time basis to keep my accounting records. You can have the job, if you can determine whether the net loss is correct and, if not, what it should be and what I am doing wrong."

Required: (1) Set up the following account columns: under *Assets*: Accounts Receivable, Repair Supplies, Prepaid Rent, and Repair Equipment; under *Liabilities*: Unearned Revenues, Salaries Payable, and Note Payable; under *Owner's Equity* (*Revenues*): Repair Service Revenues; and (*Expenses*): Depreciation Expense, Interest Expense, Rent Expense, Supplies Expense, and Salaries Expense. Enter any balances for these accounts shown on the financial statements.

(2) Using the accounts from (1), prepare any year-end adjustments you think are appropriate for 2004. Show any supporting calculations. Compute the ending balance of each account.

(3) Prepare a corrected 2004 income statement, statement of changes in owner's equity, and ending classified balance sheet (report form).

(4) Write a brief report to Ray Young, summarizing your suggestions for improving his accounting practices.

7-46 The following are a condensed 2004 income statement and a December 31, 2004 balance sheet for Murf Company:

MURF COMPANY
Income Statement
For Year Ended December 31, 2004

Sales (net)	$154,000
Cost of goods sold	(91,300)
Gross profit	$ 62,700
Operating expenses	(47,300)
Interest expense	(2,800)
Net Income	$ 12,600

MURF COMPANY
Balance Sheet
December 31, 2004

Cash	$ 3,200
Marketable securities (short-term)	2,100
Accounts receivable	7,370
Inventory	9,650
Property and equipment (net)	97,680
Total Assets	$120,000
Current liabilities	$ 12,400
Note payable (due 12/31/09)	35,000
Total Liabilities	$ 47,400
S. Murf, capital	72,600
Total Liabilities and Owner's Equity	$120,000

On January 1, 2004, the accounts receivable were $6,050, the inventory was $10,950, the total assets were $110,000, and the owner's capital was $62,600. The company makes 60% of its net sales on credit and operates on a 300-day business year. At the end of 2003, the following ratio results were computed, based on the company's financial statements for 2003:

(a) Current	2.0
(b) Quick	1.3
(c) Debt	43.3%
(d) Inventory Turnover	8.5 times (35.3 days)
(e) Accounts Receivable Turnover	13.6 times (22.1 days)
(f) Gross Profit Percentage	39.2%
(g) Profit Margin	7.8%
(h) Return on Total Assets	12.4%
(i) Return on Owner's Equity	17.0%

The company has hired you to update its ratio results and compare its performance in 2004 with that in 2003.

Required: (1) Compute the preceding ratios for 2004.

(2) Write a short report that compares the company's performance in 2004 with that in 2003 regarding its liquidity, financial flexibility, operating capability, and profitability.

7-47 Yesterday, you received the letter shown below for your advice column in the local paper:

DR. DECISIVE

Dear Dr. Decisive:

I always read your column and think you do a good job settling squabbles. Here's one for you. I took my accounting book home over the break, and one of my parents (let's just call him "Dad") started to look through it. Soon he encountered a statement he didn't agree with, and the squabble began. Here's the statement: "If a company has a higher return on owner's equity than its return on total assets, this shows users that the company has benefited from using debt to help finance its assets." Dad says that a company will always have a higher return on owner's equity than it will have on total assets because owner's equity is always going to be smaller than total assets. I say that Dad is not correct. I think that sometimes a company can have a return on owner's equity that is lower than its return on total assets. His logic doesn't take into account the fact that when a company has debt, it also has interest, and that interest expense affects the company's return on assets but not its return on owner's equity. But when I challenge Dad, we always end up arguing. And every time we talk, the same subject comes up. Please help, and please show us with numbers! I need some peace and quiet.

"Enough Already"

Required: Meet with your Dr. Decisive team and write a response to "Enough Already."

CHAPTER 9

MANAGING AND REPORTING WORKING CAPITAL

"MOST PEOPLE HAVE WRONG-MINDED IDEAS ABOUT WHY COMPANIES FAIL. THEY THINK IT'S BECAUSE OF A LACK OF MONEY. IN MOST CASES, IT HAS VERY LITTLE TO DO WITH THAT."

—MICHAEL E. GERBER, AUTHOR AND BUSINESS CONSULTANT

1. What is working capital, and why is its management important?

2. How can managers control cash receipts in a small company?

3. How can managers control cash payments in a small company?

4. What is a bank reconciliation, and what are the causes of the difference between a company's cash balance in its accounting records and its cash balance on its bank statement?

5. How can managers control accounts receivable in a small company?

6. How can managers control inventory in a small company?

7. How can managers control accounts payable in a small company?

How do you protect your cash? Do you organize your money in your wallet so that 1-dollar bills are in front and larger denominations are behind? Do you keep a jar of "change" to pay for small items? At your apartment or dorm room, do you store your wallet in a "secret" place? When you get your paycheck, do you examine the "pay stub" to be sure you were paid the right amount and that the deductions are correct? Do you have a checking account or savings account? Do you earn interest on the balance you keep in the account? Do you immediately deposit your paycheck in this checking or savings account? Do you record each check that you write in your check register? Do you reconcile your checkbook each time you get a bank statement? Have any of your friends asked you to lend them money? Did you consider the likelihood they would pay you back before you decided whether to lend them the money? If you lent them money, did you make them sign an agreement to pay you back? When you pay for several items at a store by using your credit card, do you examine the receipt to make sure the store has not over-charged you? These are all ways that individuals might keep "control" over their cash, amounts owed to them, and amounts they owe. To be successful, companies also need sound controls over their cash, accounts receivable, and accounts payable.

According to Michael E. Gerber, accounting issues are the basis for three of the top ten reasons why small companies fail. These three are (1) a lack of management systems, such as financial controls, (2) a lack of financial planning and review, and (3) an inadequate level of financial resources.[1] The third reason can be interpreted to mean "a lack of money." The first two reasons, however, focus on managing a company's financial resources.

Often, it is difficult for a new company to keep an adequate amount of financial re-sources. Because so many companies start with very little cash, a relatively new term, *boot-strapping*, describes how these companies operate under such tight financial constraints. The term is taken from an old phrase about being self-reliant: these companies are "pulling them-selves up by their bootstraps." Some very successful companies, for example **Dell Computer Corporation** and **Joe Boxer Company**, started business with very limited resources. These companies were able to manage their resources effectively and grow into sizable businesses.

http://www.dell.com
http://www.joeboxer.com

In this chapter, we build on the knowledge you gained in previous chapters. We take a closer look at how companies manage and report four important balance sheet items: cash, accounts receivable, inventory, and accounts payable. As you will learn, how a com-pany manages these items affects its cash flows, financial performance, and financial re-porting. More specifically, we define *working capital,* discuss its importance, examine its major components, and explain how managers and external users evaluate it. We believe that through proper short-term financial management, many small companies can increase their likelihood of success.

WORKING CAPITAL

A company's **working capital** is the excess of its current assets over its current liabili-ties. That is, working capital is current assets minus current liabilities.

1   What is working capital, and why is its manage-ment important?

Why do you think this amount is called "working" capital? Why do you think a company is concerned with its working capital?

Recall that the current asset section of a company's balance sheet includes assets that the company expects to convert into cash, sell, or use up within one year. The current lia-bility section includes liabilities that it expects to pay within one year by using current as-sets. The term *working capital* represents the *net* resources that managers have to *work* with (manage) in the company's day-to-day operations. Exhibit 9-1 shows Sweet Temptations' balance sheets for December 31, 2005 and December 31, 2004. We have assumed that Sweet Temptations has operated for two years and has recorded all its transactions correctly. We

---

[1]Interview with Michael E. Gerber by Maria Shao, *Des Moines Register,* February 14, 1994, 15-B.

EXHIBIT 9-1	SWEET TEMPTATIONS' BALANCE SHEETS

SWEET TEMPTATIONS
Comparative Balance Sheets
December 31, 2005 and 2004

Assets	December 31, 2005		December 31, 2004	
Current Assets				
Cash	$ 5,818		$ 5,014	
Accounts receivable (net)	7,340		8,808	
Inventory	1,570		1,300	
Total current assets		$14,728		$15,122
Property and Equipment				
Store equipment (net)	$13,500		$10,420	
Total property and equipment		13,500		10,420
Total Assets		$28,228		$25,542
Liabilities				
Current Liabilities				
Accounts payable	$ 7,540		$ 7,731	
Total current liabilities		$ 7,540		$ 7,731
Noncurrent Liabilities				
Notes payable	$ 5,000		$ 5,000	
Total noncurrent liabilities		5,000		5,000
Total Liabilities		$12,540		$12,731
Owner's Equity				
A. Cox, capital		$15,688		$12,811
Total Liabilities and Owner's Equity		$28,228		$25,542

Working Capital = Current Assets − Current Liabilities
12/31/2005 Working Capital = $14,728 − $7,540 = $7,188
12/31/2004 Working Capital = $15,122 − $7,731 = $7,391

highlight the current sections of the balance sheet and calculate Sweet Temptations' working capital at the bottom of Exhibit 9-1. Note that the amount of the current liabilities (accounts payable) is subtracted from the total amount for the current assets (cash, accounts receivable, and inventory) to calculate working capital. Changes in any of these four items affect Sweet Temptations' working capital. Decisions that managers make regarding any of these items are considered part of *working capital management*. Other terms for working capital management are *operating capital management* and *short-term financial management*.

 How much working capital do you need in your personal life? Why?

Companies manage working capital because they want to keep an appropriate amount on hand. But what is an *appropriate* amount of working capital for a company? An appropriate amount is enough working capital to finance its day-to-day operating activities plus an extra amount in case something unexpected happens. For instance, the extra amount may enable the company to buy inventory when it is offered at a reduced price or to cover the lost cash when a customer doesn't pay its account. If a company has too little working capital, it risks not having enough liquidity. If it has too much, the company risks not putting its resources to their best use. In summary, companies manage working capital to keep an appropriate balance between (1) having enough working capital to operate and to handle unexpected needs for cash, inventory, or short-term credit and (2) having so much excess cash, inventory, or available credit that profitability is reduced.

Keeping the right amount of working capital requires careful planning and monitoring. For instance, the timing of inventory purchases usually does not coincide with the timing of sales. Thus, at any given time a company may find itself with either too little or too much inventory. Cash receipts usually do not coincide with the company's need to use its cash. So, a company may have excess cash sitting idly in its checking account, or it may need additional short-term financing.

The fact that customers have some control over when they make their payments affects a company's management of cash collections from its accounts receivables. The longer customers take to pay, the longer the company must wait between the time when it purchased inventory and the time when it receives cash from the sale. The company can manage this aspect of working capital by setting policies that encourage early payment of accounts receivable.

On the other hand, the company must also manage the payments of its obligations. It should make these payments on time, as well as take advantage of purchases discounts available from its suppliers.

Managing working capital affects all aspects of a company's operating activities. Exhibit 9-2 shows a time line of Sweet Temptations' (ST) operating activities. The top half of the exhibit shows Sweet Temptations' transactions with its supplier, Unlimited Decadence (UD). These consist of ST purchasing chocolates (increasing inventory) on credit (increasing accounts payable), paying invoices as they come due (decreasing cash and accounts payable), and monitoring inventory to determine when to restock (not shown).

Purchasing and Cash Payments

1. ST purchases boxes of chocolates from UD on credit.

2. ST waits until almost the invoice due date to process cash payment.

3. ST sends check for payment.

4. UD processes receipt and deposits ST's check; ST's bank deducts check from its checking account.

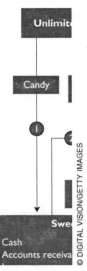

© DIGITAL VISION/GETTY IMAGES

How can a company encourage its customers to pay their bills more quickly?

EXHIBIT 9-2	**WORKING CAPITAL FLOWS**

Unlimited Decadence (UD)

Purchasing and Cash Payments

1. ST purchases boxes of chocolates from UD on credit.

2. ST waits until almost the invoice due date to process cash payment.

3. ST sends check for payment.

4. UD processes receipt and deposits ST's check; ST's bank deducts check from its checking account.

Candy   ST's Check   Deposit ST's Check

Banking System

Deposit Customer's Check

Sweet Temptations (ST)

Cash
Accounts receivable   Accounts payable
Inventory

Candy   Customer's Check

CUSTOMER

Sales and Cash Receipts

5. ST sells boxes of chocolates to a customer on credit.

6. Customer waits until almost the due date to process cash payment.

7. Customer sends check for payment to ST.

8. ST processes receipt and deposits the customer's check with other daily receipts in its checking account.

Why do you think Sweet Temptations waits until almost the invoice due date to process its cash payment to Unlimited Decadence?

The bottom half of Exhibit 9-2 shows Sweet Temptations' transactions with its customers. These consist of Sweet Temptations making candy sales (decreasing inventory) on credit (increasing accounts receivable) or for cash (not shown), and of credit customers mailing payments based on the credit terms of the sales (decreasing accounts receivable). When Sweet Temptations receives customers' payments, it deposits the checks in the bank (increasing cash). These deposits are then available to make cash payments.

Managers control each aspect of the operating cycle to ensure that operating activities are performed in accordance with company objectives. As you will see, they do this by establishing an internal control structure. An **internal control structure** is a set of policies and procedures that directs how employees should perform a company's activities.

The reporting by a company of the amount of each current asset and current liability on its balance sheet provides external users with information about the company's ability to keep an appropriate level and mix of working capital. In turn, this reporting helps managers and users evaluate the company's liquidity.

To put this another way, working capital is to a company what water is to a plant. If the plant does not have enough water, it will not grow. Eventually, it will wither and die. If the plant receives too much water, it will drown. Just as plants need the right amount of water in order to grow, companies need the right amount of working capital to achieve desired levels of profitability and liquidity. In the following sections we will discuss how a company manages and reports its working capital items—cash, accounts receivable, inventory, and accounts payable.

CASH

A company's **cash** includes money on hand, deposits in checking and savings accounts, and checks and credit card invoices that it has received from customers but not yet deposited. A simple rule is that cash includes anything that a bank will accept as a deposit.

In addition to being an integral part of a business, cash is also the most likely asset for employees and others to steal or for the company to misplace. For example, cash received from customers in a retail store has no identification marks that have been recorded by the store. Therefore, when cash is "missing" it is very difficult to prove the cash was stolen or who stole it. Also, cash that is illegally transferred from a company bank account involves no physical possession of the cash by the thief, and if the thief can conceal or destroy the records, the theft of the money may not be traceable. Although internal control procedures are necessary for all phases of a company's business, they are usually most important for cash.

Simple Cash Controls

For any size company, the best way to prevent both intentional and unintentional losses is to hire competent and trustworthy personnel and to establish cash controls. Next, we discuss two categories of simple cash controls. These are internal controls over (1) cash receipts and (2) cash payments. These controls apply to all cash transactions except those dealing with a company's petty cash fund, which we will discuss later in the chapter.

Controls over Cash Receipts

2 How can managers control cash receipts in a small company?

A company uses internal control procedures for cash receipts to ensure that it properly records the amounts of all cash receipts in the accounting system and to protect them from being lost or stolen. Cash receipts from a company's operating activities result from cash sales and from collections of accounts receivable mailed in by its customers.

How might an employee steal from his or her employer when working at a company's cash register?

For cash sales, a company should use three control procedures. The most important control procedure is the proper use of a cash register. Managers should make sure that a prenumbered sales receipt is completed for every sale and that the salespeople ring up each sale on the register. In most companies, the cash register produces the receipt as well as a tape containing a chronological list of all sales transactions rung up on the register. This step is important because it is the first place that sales get entered into the accounting system. The fact that customers expect to receive a copy of the receipt helps ensure that each sale is entered. As a customer, you may have been part of a company's cash controls without even knowing it. At many **Sbarro Pizza** franchises, for example, there is a sign near the cash register that reads,

http://www.sbarro.com

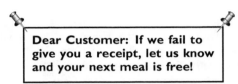

Dear Customer: If we fail to give you a receipt, let us know and your next meal is free!

This added control increases the likelihood that salespeople will enter all sales into the cash register, and it signals to company employees the importance of this activity.

 Could an employee at Sbarro's still take money from the company? How?

Second, when a check is accepted for payment, the salesperson should make sure that the customer has proper identification in order to minimize the likelihood that the check will "bounce." Even this procedure is not always adequate. For instance, in Vail, Colorado (and in other ski areas), retail and service companies must be especially careful in accepting checks at the end of the ski season because some customers close out their accounts when they leave.

Third, at the end of each salesperson's work shift, the employee should match the total of the amounts collected (cash plus checks and credit card sales) against the total of the cash register tape and report any difference between the two totals to a supervisor.

Some companies (e.g., **QuikTrip**) use a fourth control procedure for cash sales. These companies remove the "big bills" from cash registers even during one employee's shift. For example, if more than five 20-dollar bills are in the register, the employee inserts the excess bills through a slot into a locked safe that is kept behind the counter. Only the store manager knows the combination to the safe.

A company should use three control procedures to safeguard collections of cash from accounts receivable. First, either the owner-manager or an employee who does not handle accounting records should open the mail. This control procedure is called *separation of duties.* Separating the duties of handling accounting records and opening the mail prevents an employee from stealing undeposited checks *and* covering up the theft by making a fictitious entry in the accounting records. Second, immediately after opening the mail, the employee should list all of the checks received. Later, if a customer claims to have previously paid a bill, the company can review the list. You may be wondering what happens if the customer *did* pay the bill but the receipt is not listed because the employee stole the undeposited check. In this case, the customer's canceled check (from the customer's bank) may help the company discover that its employee stole the check. Third, while opening the mail, the employee should restrictively endorse each incoming check for deposit in the company's bank account. At Sweet Temptations, this is done by stamping "for deposit only—Sweet Temptations" on the back of each check. If a check is lost or stolen, the endorsement makes it more difficult to cash the check illegally.

How can this customer help QuickTrip keep control of the cash in this register?

http://www.quiktrip.com

Why do you think an owner might be interested in not recording all cash receipts?

Finally, a company should adopt one additional procedure to help it safeguard the cash collected from both its cash sales and its accounts receivable. It should deposit all cash receipts intact daily. This means that at the end of the day, the company should take all of its cash (everything included in our definition of cash), fill out a deposit slip, and make a bank deposit. These daily bank deposits aid in two ways. First, keeping a substantial amount of cash at the company overnight is taking an unnecessary risk of theft. By depositing all cash receipts on a daily basis, the company does not leave cash unattended overnight. Second, the bank's deposit records show the company's cash receipts for each day. When the company receives its monthly bank statement, the company can check the daily bank deposits listed in the bank statement against its Cash account to determine that it deposited all its recorded cash receipts in the bank and that it properly recorded all bank deposits in its Cash account.

Controls over Cash Payments

 How can managers control cash payments in a small company?

The basic rule for good internal control over cash payments is to have all payments made by check. A very small company that the owner operates may have little need for any additional internal control procedures. The owner purchases items, signs checks for payment, and pays employees by check. As the company grows, two more controls over cash payments can provide added security over cash. First, the company should pay only for approved purchases that are supported by proper documents. The proper documents generally include an approved copy of the company's purchase order providing evidence that the company actually ordered the items (which we will discuss in detail later in the chapter), a freight receipt showing evidence that the company received the items it ordered, and the supplier's invoice. This procedure reduces the chances that the company will pay either for items that it did not want to purchase or for items that it has not received. Second, immediately after writing the check for payment, the owner should stamp "PAID" on the supporting documents. Canceling the documents in this way prevents the company from paying for items more than once.

 Why do you think an employee might want to deceive the company about its cash payments? Why do you think an owner might be interested in not recording all cash payments?

Bank Reconciliation

Despite all of the procedures used to control the receipts and payments of cash, errors in a company's records can still occur. Since the bank also keeps a record of the company's cash balance, the company can use both sets of records to determine what its correct cash balance should be. However, the time when the company records its receipts and payments differs from the time when the bank records them. Therefore, a company uses a bank reconciliation to determine the accuracy of the balance in its Cash account. In this section, we discuss what a bank reconciliation is, why it is necessary, and how it is performed.

Do you reconcile your bank statement every month? What risks do you take if you don't reconcile the statement?

A company's bank independently keeps track of the company's cash balance. Each month the bank sends the company a **bank statement** that summarizes the company's banking activities (e.g., deposits, paid checks) during the month. A company uses its bank statement, along with its cash records, to prepare a bank reconciliation.

When a company uses the internal control procedures of depositing daily receipts and paying only by check, the ending balance in its Cash account should be the same as the bank's ending cash balance for the company's checking account, except for a few items.

(We will discuss the various causes of the difference between the two balances below.) A company prepares a **bank reconciliation** to analyze the difference between the ending cash balance in its accounting records and the ending cash balance reported by the bank in the bank statement. Through this process, the company learns what changes, if any, it needs to make in its Cash account balance. This enables the company to report the correct cash balance on its balance sheet.

Exhibit 9-3 summarizes the causes of the difference between the ending cash balance listed on the bank statement and the ending cash balance listed in the company's records. The causes include (1) deposits in transit, (2) outstanding checks, (3) deposits made directly by the bank, (4) charges made directly by the bank, and (5) errors.

4 What is a bank reconciliation, and what are the causes of the difference between a company's cash balance in its accounting records and its cash balance on its bank statement?

 Which of the five listed items are most important when you reconcile your bank account?

EXHIBIT 9-3	CAUSES OF DIFFERENCE IN CASH BALANCES

1. *Deposits in Transit.* A **deposit in transit** is a cash receipt that the company has added to its Cash account but that the bank has not included in the cash balance reported on the bank statement. When a company receives a check, it records an increase to its Cash account. As illustrated in Exhibit 9-2, a short period of time may pass before the company deposits the check and the bank records it. At the end of each month the company may have deposits in transit (either cash or checks) that cause the deposits recorded in the company's Cash account to be greater than deposits reported on the bank statement.

2. *Outstanding Checks.* An **outstanding check** is a check that the company has written and deducted from its Cash account but that the bank has not deducted from the cash balance reported on the bank statement because the check has not yet "cleared" the bank. As illustrated in Exhibit 9-2, a period of time is necessary for the check to be received by the payee (the company to whom the check is written), deposited in the payee's bank, and forwarded to the company's bank for subtraction from the company's bank balance. Therefore, at the end of each month a company usually has some outstanding checks that cause the cash payments recorded in its Cash account to be more than the cancelled checks itemized on the bank statement.

3. *Deposits Made Directly by the Bank.* Many checking accounts earn interest on the balance in the account. For these accounts, the bank increases the company's cash balance in the bank's records by the amount of interest the company earned on its checking account; the bank lists this amount on the bank statement. This causes deposits listed on the bank statement to be greater than the deposits listed in the company's Cash account. The company is informed of the amount of interest when it receives the bank statement.

4. *Charges Made Directly by the Bank.* A bank frequently imposes a service charge for a depositor's checking account and deducts this charge directly from the bank account. Banks also charge for the cost of printing checks, according to an agreed price, and for the cost of stopping payment on checks. The company is informed of the amount of the charge when it receives the bank statement showing the amount of the deduction.

When the company receives a customer's check, it adds the amount to its Cash account and deposits the check in its bank account for collection. The company's bank occasionally is unable to collect the amount of the customer's check. That is, the customer's check has "bounced." A customer's check that has "bounced" is called an **NSF (not sufficient funds) check**. Because the bank did not receive money for the customer's check, it lists the check as an NSF check on the bank statement. Although the bank usually informs the company immediately of each NSF check, there may be some NSF checks that are included in the bank statement and that the company has not recorded.

At the end of the month, the bank lists any service charges and NSF checks as deductions on the bank statement—deductions not yet listed in the company's Cash account.

5. *Errors.* Despite the internal control procedures established by the bank and the company, errors may arise in either the bank's records or the company's records. The company may not discover these errors until it prepares the bank reconciliation. For example, a bank may include a deposit or a check in the wrong depositor's account or may make an error in recording an amount. Or, a company may record a check for an incorrect amount or may forget to record a check.

The Structure of a Bank Reconciliation

Exhibit 9-4 shows a common way to structure a bank reconciliation. Notice that the reconciliation has two sections: an upper section starting with the bank's record of the company's ending cash balance, and a lower section starting with the company's record of its cash balance. It is logical to set up these two sections because the purpose of the bank reconciliation is to determine the company's correct ending cash balance. By adjusting the cash balance in each section for the amounts that either are missing or are made in error, the company is able to determine its reconciled (correct) ending cash balance.

For example, in the upper section, a deposit in transit is added to the ending cash balance from the bank statement because this deposit represents a cash increase that the bank has not yet added to the company's checking account. In the lower section, a service charge made by the bank is subtracted from the ending balance in the company's Cash account because this charge represents a cash decrease that the company has not yet recorded.

The bank reconciliation is complete when the ending reconciled cash balances calculated in these two sections are the same. This ending reconciled cash balance is the correct cash balance that the company includes in its ending balance sheet. This form of bank reconciliation acts as another type of internal control over cash because it enables a company to identify errors in its cash-recording process and to know its correct cash balance at the end of each month.

Preparing a Bank Reconciliation

When you prepare a bank reconciliation, keep in mind that you are doing it to determine the correct ending cash balance to be shown on the company's balance sheet. The cash balance at the end of the month is correct if it includes *all* of the company's transactions and events that affected cash. As you work through the upper section of the reconciliation, ask yourself, "What cash transactions (e.g., checks written and deposits made) have taken place that the bank doesn't know about?" In the lower section ask yourself, "What is not included in calculating the company's ending cash balance but should be (e.g., bank service charges, interest earned)?" Keep these questions in mind as you work through the reconciliation until the reconciled balances are the same.

To prepare a bank reconciliation, you need two sets of items: (1) the bank statement for the month being reconciled, along with all of the items returned with the statement, and (2) the company's cash records. With these items, you can work through a reconciliation in a step-by-step manner. Exhibit 9-5 summarizes the eight steps to follow in preparing a bank reconciliation.

In the following section, we illustrate the reconciliation process by preparing Sweet Temptations' December 31, 2005 bank reconciliation.

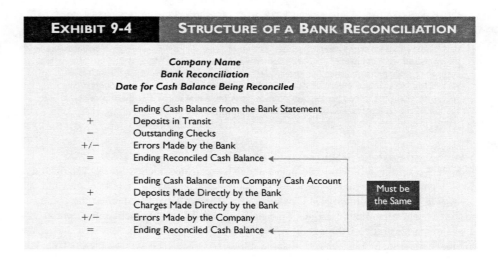

EXHIBIT 9-4	STRUCTURE OF A BANK RECONCILIATION

Company Name
Bank Reconciliation
Date for Cash Balance Being Reconciled

	Ending Cash Balance from the Bank Statement	
+	Deposits in Transit	
−	Outstanding Checks	
+/−	Errors Made by the Bank	
=	Ending Reconciled Cash Balance	
	Ending Cash Balance from Company Cash Account	
+	Deposits Made Directly by the Bank	Must be the Same
−	Charges Made Directly by the Bank	
+/−	Errors Made by the Company	
=	Ending Reconciled Cash Balance	

EXHIBIT 9-5	STEPS IN PREPARING A BANK RECONCILIATION

1. *Set up the proper form for the bank reconciliation.* Fill in the information you already know (e.g., ending unadjusted cash balances from the bank statement and the Cash account).

2. *Look for deposits in transit.* Compare the increases in cash listed in the company's Cash account with the deposits shown on the bank statement. Check to see if any increase in the company's Cash account is not listed as a deposit on the bank statement. For any deposit in transit, add the amount to the ending cash balance from the bank statement listed on the reconciliation.

3. *Look for outstanding checks.* Compare the decreases in cash listed on the company's Cash account with the paid checks shown on the bank statement. Identify any decrease that is shown in the company's Cash account during the month but that is not matched with a corresponding check deduction on the bank statement. Starting from the company's records, trace each decrease to its check listing on the bank statement. Subtract the amounts of the outstanding checks from the ending cash balance from the bank statement listed on the bank reconciliation.

4. *Identify any deposits that were made directly by the bank but that are not included as increases in the company's Cash account.* Look through the bank statement for bank deposits that the company has not recorded as increases in its Cash account. Usually, these deposits are for interest earned on the company's checking account balance. Add these deposits to the balance of the company's Cash account listed on the bank reconciliation.

5. *Identify any charges that were made directly by the bank but that are not included as decreases in cash on the company's records.* Look through the bank statement for bank charges that the company has not recorded as decreases in its Cash account. Usually, these charges result from bank services such as printing checks or handling the company's own NSF checks. Deduct these charges from the balance of the company's Cash account listed on the bank reconciliation.

6. *Determine the effect of any errors.* While completing steps 1 through 5, you may discover that the bank or the company (or both) made an error during the processing of the cash transactions. If you find a bank error, contact the bank to get the error corrected in the company's checking account, and correct the amount of the error in the upper section of the bank reconciliation. If the company made an error, correct the amount of the error in the lower section of the reconciliation.

7. *Complete the bank reconciliation.* After you have finished steps 1 through 6, complete the reconciliation. Include the date and amount for any deposit in transit, and list the check numbers for any outstanding checks. This improves documentation and makes the reconciliation easier for others to understand. Describe any bank charges or error corrections in sufficient detail so that these activities can be recorded properly in the company's accounting records. At this point, the reconciled (correct) cash balances in both sections of the reconciliation should be the same. If not, trace back through the process carefully to locate any mistakes (e.g., outstanding checks you failed to include, math errors, etc.).

8. *Adjust the balance of the company's Cash account to agree with the corrected cash balance.* The lower section of a completed reconciliation answers the question "What is not included in the company's ending cash balance but should be?" The last step in preparing a reconciliation is to record these items in the company's Cash account (and the other related accounts). This recording changes the company's cash balance from the amount listed at the top of the lower section of the bank reconciliation to the correct ending amount.

Sweet Temptations' Bank Reconciliation

Exhibit 9-6 on the following two pages shows several documents: Sweet Temptations' Cash account (for illustrative purposes, we show the increases in the left column and decreases in the right column) for December, the December bank statement the company received from First National Bank, and the completed bank reconciliation. Exhibit 9-6 also summarizes Steps 1 through 7 from Exhibit 9-5 that Anna Cox followed to prepare the reconciliation. We use an arrow and a number to trace each step on the documents.

EXHIBIT 9-6 — SWEET TEMPTATIONS' BANK RECONCILIATION

Cash

Beg Bal. 12/1/2005 $3,238.48

Increases		Decreases		
Date	Amount	Date	Ch#	Amount
Dec. 1	$142.25	Dec. 1	939	$287.94
Dec. 3	155.21	Dec. 3	940	34.51
Dec. 3	142.15	Dec. 3	941	26.79
Dec. 4	154.45	Dec. 8	942	136.00
Dec. 5	198.00	Dec. 8	943	593.15
Dec. 6	98.66	Dec. 10	944	385.00
Dec. 8	190.23	Dec. 10	945	190.12
Dec. 10	163.65	Dec. 14	946	489.57
Dec. 10	187.04	Dec. 24	947	452.18
Dec. 11	156.55	Dec. 24	948	347.00
Dec. 12	177.91	Dec. 28	949	1,904.78
Dec. 13	217.87	Dec. 29	950	121.00
Dec. 15	313.57			
Dec. 15	293.32			
Dec. 17	336.58			
Dec. 18	387.22			
Dec. 19	441.10			
Dec. 20	457.16			
Dec. 22	451.82			
Dec. 22	591.78			
Dec. 24	458.25			
Dec. 27	287.35			
Dec. 28	335.76			
Dec. 29	418.43			
Dec. 31	786.00			

End Bal. 12/31/2005
Before Reconciliation $5,812.75

First National Bank
7th and Grand
Sachse, TX 75662-3443
Phone (214) 555-9800
Member FDIC

SWEET TEMPTATIONS
WESTWOOD MALL #117
SACHSE, TX 75665-0117

NO. 137-187-8
Beginning balance December 1, 2005 $ 3,158.48
Deposits and other additions:
Deposits:

Dec. 3	$101.00		
Dec. 3	142.25	Dec. 17	$313.57
Dec. 3	155.21	Dec. 17	293.32
Dec. 3	142.15	Dec. 17	336.58
Dec. 4	154.45	Dec. 19	387.22

Steps to Complete Bank Reconciliation

Step 1. Anna transferred the $5,465 cash balance from the bank statement and the $5,812.75 balance from Sweet Temptations' Cash account to the reconciliation. Next, she completed the upper section of the reconciliation.

Step 2. Anna compared the increases in the Cash account with the bank deposits and found that the December 31 increase of $786.00 was not listed on the bank statement. She entered $786.00 on the reconciliation as a deposit in transit.

Step 3. Anna compared the bank statement's listing of checks and Sweet Temptations' record of decreases in its Cash account and found that all but two of the decreases (check #948 for $347.00 and check #950 for $121.00) were deducted on the current month's bank statement. She subtracted these outstanding checks from the ending cash balance of the bank statement in the reconciliation. After completing the upper section of the bank reconciliation, Anna calculated the reconciled ending cash balance to be $5,783.00.

Step 4. Anna began completing the lower section of Sweet Temptation's reconciliation. Anna reviewed the deposits listed in the bank statement and found that Sweet Temptations had not recorded a $12.25 bank deposit for interest earned as an increase in its Cash account. She added the $12.25 deposit to the company's ending balance on the reconciliation.

Step 5. Anna reviewed the charges on the bank statement and found that Sweet Temptations had not recorded a $15.00 bank service charge (for printed checks) as a decrease in its Cash account. She subtracted the $15.00 charge from the company's ending cash balance on the reconciliation.

Step 6. When Anna compared the decreases in Sweet Temptations' Cash account with the bank statement in Step 3, she also found that Sweet Temptations had incorrectly recorded check #942 (in payment of an account payable) in its records for $136 instead of $163. Because the amount that Sweet Temptations should have recorded is $27.00 more than the amount that it did record ($163.00 – $136.00), Anna subtracted $27.00 from the company's ending cash balance on the reconciliation.

Step 7. After completing the lower section of the reconciliation, Anna calculated the reconciled ending cash balance to be $5,783.00. Anna also observed that the reconciled balances shown in the upper and lower sections of the bank reconciliation are the same. This indicates that she completed the bank reconciliation properly.

EXHIBIT 9-6 CONTINUED

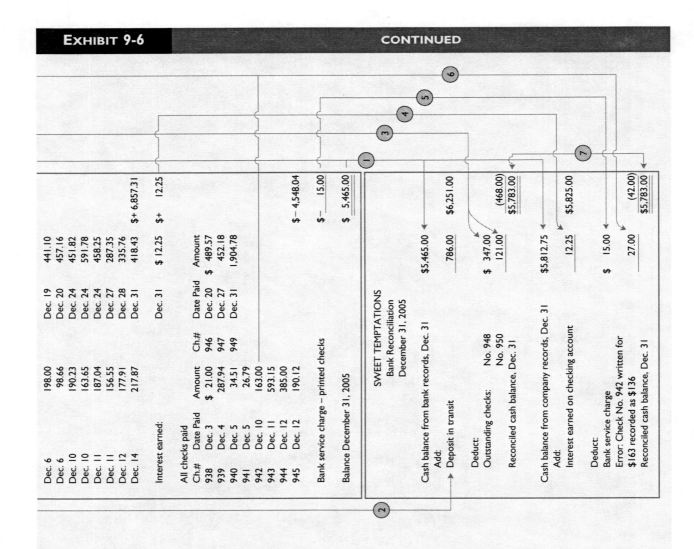

Date	Amount		Date	Amount
Dec. 6	198.00		Dec. 19	441.10
Dec. 6	98.66		Dec. 20	457.16
Dec. 10	190.23		Dec. 24	451.82
Dec. 10	163.65		Dec. 24	591.78
Dec. 11	187.04		Dec. 24	458.25
Dec. 11	156.55		Dec. 27	287.35
Dec. 12	177.91		Dec. 28	335.76
Dec. 14	217.87		Dec. 31	418.43	$+ 6,857.31

Interest earned: Dec. 31 $ 12.25 $+ 12.25

All checks paid

Ch.#	Date Paid	Amount		Ch.#	Date Paid	Amount
938	Dec. 3	$ 21.00		946	Dec. 20	$ 489.57
939	Dec. 4	287.94		947	Dec. 27	452.18
940	Dec. 5	34.51		949	Dec. 31	1,904.78
941	Dec. 5	26.79				
942	Dec. 10	163.00				
943	Dec. 11	593.15				
944	Dec. 12	385.00				
945	Dec. 12	190.12				$– 4,548.04

Bank service charge – printed checks Dec. 31 $– 15.00

Balance December 31, 2005 $ 5,465.00

SWEET TEMPTATIONS
Bank Reconciliation
December 31, 2005

Cash balance from bank records, Dec. 31			$5,465.00
Add:			
Deposit in transit			786.00
			$6,251.00
Deduct:			
Outstanding checks:	No. 948	$ 347.00	
	No. 950	121.00	(468.00)
Reconciled cash balance, Dec. 31			$5,783.00
Cash balance from company records, Dec. 31			$5,812.75
Add:			
Interest earned on checking account			12.25
			$5,825.00
Deduct:			
Bank service charge		$ 15.00	
Error: Check No. 942 written for			
$163 recorded as $136		27.00	(42.00)
Reconciled cash balance, Dec. 31			$5,783.00

In Step 8, Anna entered the reconciling items listed in the lower section of Sweet Temptations' bank reconciliation (interest earned, bank service charge, and correction of error) into the company's accounting records on December 31, 2005, as follows:

Assets	=	Liabilities	+	Owner's Equity		
					Net Income	
				Revenues	−	Expenses
		Accounts		Interest		Banking
Cash		Payable		Revenue		Expense
+$12.25				+$12.25		
−$15.00					−	+$15.00
−$27.00		−$27.00				
−$29.75		−$27.00		+$12.25	−	+$15.00

Exhibit 9-7 shows that after Anna recorded the reconciling items, Sweet Temptations' Cash account balance is the correct amount: $5,783. Anna will add this amount to the total amount in the petty cash fund (discussed later) and will show the combined total as "Cash" on Sweet Temptations' December 31, 2005 balance sheet.

Additional Controls over Cash

Two other steps in preparing a bank reconciliation help a company keep control over its cash. First, the company should make sure that any deposit in transit listed on the bank reconciliation for the *previous* month is listed as a deposit on the current bank statement. If it is not listed, the company should investigate to determine what happened to the deposit. It may be that the deposit was misplaced or even stolen. Second, the company should investigate any outstanding check from the *previous* month that is still outstanding for the current month. It may be that the check was misplaced or was lost in the mail. Sweet Temptations has neither of these situations. The $101 deposit recorded by the bank on December 1 was a deposit in transit at the end of November. And check #938 for $21 paid by the bank on December 1 was an outstanding check at the end of November.

Petty Cash Fund

Although paying for all items by check is excellent internal control, it can be inconvenient. So, to make it easier for employees to make small, but necessary, purchases, a company may set up a petty cash fund. A **petty cash fund** is a specified amount of money that is under the control of one employee and that is used for making small cash payments for the company. A company uses a petty cash fund because some payments can be made only with "currency" or because writing checks for small amounts (e.g., for postage) would be cumbersome. There is less control over these expenditures, but the amounts involved are so small that an employee probably will not be tempted to steal.

 What items do you always pay for with "currency"?

To start a petty cash fund, a company gives an employee an amount of money, say $50, to be kept at the company. Usually the employee keeps the money locked in his or her desk drawer. Each time a payment is made from the fund, the employee makes a record of the payment and keeps a written receipt. At any time, the total of the receipts plus the remaining cash should equal the amount (in this case, $50) that was originally given to the employee. When the fund gets low or on the date of its balance sheet, the company replenishes the fund to the original amount and uses the receipts to record the various cash transactions in its accounting system. For each receipt, the company records an increase in the related expense (or asset) account (e.g., postage expense, office supplies) and a decrease in its Cash account. This ensures that all of the petty cash payments are included in the amounts reported in the company's financial statements.

EXHIBIT 9-7   ADJUSTMENTS FROM BANK RECONCILIATION

Cash

Beg Bal. 12/1/2005	$3,238.48			
Increases		**Decreases**		
Date	Amount	Date	Ch#	Amount
Dec. 1	$142.25	Dec. 1	939	$ 287.94
Dec. 3	155.21	Dec. 3	940	34.51
Dec. 3	142.15	Dec. 3	941	26.79
Dec. 4	154.45	Dec. 8	942	136.00
Dec. 5	198.00	Dec. 8	943	593.15
Dec. 6	98.66	Dec. 10	944	385.00
Dec. 8	190.23	Dec. 10	945	190.12
Dec. 10	163.65	Dec. 14	946	489.57
Dec. 10	187.04	Dec. 24	947	452.18
Dec. 11	156.55	Dec. 24	948	347.00
Dec. 12	177.91	Dec. 28	949	1,904.78
Dec. 13	217.87	Dec. 29	950	121.00
Dec. 15	313.57			
Dec. 15	293.32			
Dec. 17	336.58			
Dec. 18	387.22			
Dec. 19	441.10			
Dec. 20	457.16			
Dec. 22	451.82			
Dec. 22	591.78			
Dec. 24	458.25			
Dec. 27	287.35			
Dec. 28	335.76			
Dec. 29	418.43			
Dec. 31	786.00			

Ending Bal.12/31/2005 Before Reconciliation	$5,812.75
Adjustments to company records based on bank reconciliation:	
Interest earned	+12.25
Bank service charge	−15.00
Correct error for check no. 942	−27.00
Reconciled Cash Bal 12/31/2005	$5,783.00

    Sweet Temptations keeps a petty cash fund totaling $35. As we will discuss next, Anna adds this amount to Sweet Temptations' ending reconciled cash balance for its checking account so that Sweet Temptations shows its total cash on its balance sheet.

Reporting the Cash Balance on the Balance Sheet

Cash is usually the first asset listed on the balance sheet because it is the most liquid current asset. Recall that Cash includes money on hand, deposits in checking and savings accounts, and checks and credit card receipts that a company has received but not yet deposited. As we discussed in the previous sections, a company's accounting system keeps track of these items separately. When reporting Cash on its balance sheet, a company must combine the balances of each of these items.

Sweet Temptations' total cash balance at December 31, 2005 consists of two items:

1. The December 31 reconciled cash balance
   in its checking account... $5,783.00
2. The amount in its petty cash fund 35.00

Total cash balance on December 31, 2005......................... $5,818.00

Notice that Sweet Temptations shows this amount as Cash on its December 31, 2005 balance sheet in Exhibit 9-1.

A company that sells on credit cannot manage its cash without managing its accounts receivable. We will discuss the management of accounts receivable next.

Have you ever applied for credit? What steps did you have to go through? Why do you think you had to go through that process in order to get credit?

ACCOUNTS RECEIVABLE

Accounts receivable are the amounts owed to a company by customers from previous credit sales. The company intends to collect these amounts in cash. Companies make sales on credit for three basic reasons. The first reason is that selling on credit may be more convenient than selling for cash. For example, when a company is selling goods that must be shipped, it is common for the purchaser to pay for the goods after receiving them. Between the time that the purchaser receives the goods and the time that the seller collects the payment, the seller has extended credit to the purchaser. The second reason a company makes credit sales is that managers believe that offering credit will encourage customers to buy items that they might not otherwise purchase. This is common in retail sales, when the customer may not have enough cash to make the purchase. The third reason a company makes credit sales is to signal product quality. By allowing customers to pay after receiving, seeing, and using the goods, a company shows that it is confident in the quality of the goods.

Do credit card sales result in accounts receivable? Some? All? None?

http://www.visa.com
http://www.mastercard.com
http://www.discovercard.com

Credit sales using accounts receivable are not the same as "credit card sales." If you use a credit card to pay for goods that are sold to you, it is the credit card company (e.g., **VISA, Mastercard, Discover**) that is extending credit to you, not the company that sold you the goods. A retail store deposits its credit card receipts into its checking account just as it does its cash receipts. Because of this, credit card sales receipts are sometimes referred to as "instant cash."

Who is extending credit in this transaction?

The Decision to Extend Credit

If accounts receivable increase a company's sales, why not automatically decide to grant all customers credit? The decision is not automatic because accounts receivable also have two disadvantages. One disadvantage of credit sales is that having accounts receivable requires significant management effort. Managers must make credit investigations, prepare and send bills, and encourage payments from the customers. All of these activities involve a cost to the company in money and in employee time. The second disadvantage is that when a company makes credit sales, there is always the chance that the purchaser will not pay. However, just because a company has some uncollectible accounts ("bad debts"), it does not mean that the company should not make credit sales. If, given the additional revenues and costs of managing accounts receivable, the company's profits are increased by extending credit, then credit sales help the company achieve its goals. A company uses a form of cost-volume-profit (C-V-P) analysis for this evaluation. We will explain more about a company's decision to extend credit in Chapter 13. In this chapter, we focus on managing and reporting the accounts receivable for a small company.

Simple Controls over Accounts Receivable

 Has anyone ever asked to borrow money from you? If they did, what factors affected your decision? Will you deal with the next situation in the same way? Why or why not?

Accounts receivable provide a greater increase in profit if credit sales are monitored properly. Internal controls over accounts receivable focus on procedures that help maximize the increase in profit from granting credit. For a small company, such as Sweet Temptations, three control procedures should be used with accounts receivable.

5 How can managers control accounts receivable in a small company?

First, before extending credit, a company should determine that a customer is likely to pay. The risk of not collecting customers' accounts is greatly reduced if a company extends credit only to customers who have a history of being financially responsible. But how does a company decide if a customer is creditworthy? And how much credit should a company extend?

 Some companies grant credit "on the spot" with no credit checks. Why would a company do this? What problems could the company later encounter?

To answer these questions, a company asks each potential credit customer to complete a credit application (similar to the one you would fill out for a car loan). Normally, a credit application requests that the applicant provide the following information: (1) the name of the applicant's employer and the applicant's income, (2) the name of the applicant's bank, his or her bank account numbers, and the balances in his or her accounts, (3) a list of assets, (4) credit card account numbers and amounts owed, and (5) a list of other debts. The company will contact the applicant's employer, bank, and credit card companies to verify the application information and ask questions about the applicant's credit history. If the applicant has been financially responsible (i.e., earns a minimum level of income, has not issued many NSF checks, and has made bank and credit card payments in a timely manner), the company approves the application. The amount of credit that it approves depends on the applicant's income, amounts of other debt, and the specific results of the company's investigation. Credit sales should be made only to customers whose credit it has approved.

Second, a company should monitor the accounts receivable balances of its customers. Recall from Chapter 6 that credit customers agree to accept certain payment terms. Common credit terms are "2/10, net/30." Under this arrangement the customer agrees that a 2 percent cash discount will be granted if it makes payment within 10 days and that, if it does not make payment then, the full amount is due in 30 days. To monitor customer credit effectively, a company needs to have an accounting system that is able to keep track of each customer's credit activity. The company also needs to have an organized collection effort. It should mail monthly statements to customers and should consider payments not received in 30 days (in this case) to be past due. It should send personalized letters to customers whose accounts become past due and should deny them additional credit until it collects the past-due amounts. If accounts become very overdue, say 90 days or more, the company can use telephone calls to encourage payments. At some point, it may consider an overdue account to be uncollectible and may decide not to make any further effort to collect the account or to turn it over to a collection agency.

Third, a company should monitor its total accounts receivable balance. If the balance increases, the company should investigate the reasons for the increase. If the increase resulted from an increase in credit sales from creditworthy customers, the company will continue with its standard collection efforts. However, if the increase resulted from a slow-down in cash collections, the company should reexamine its credit and collection policies to try to solve the problem.

Regardless of the collection effort made by a company, it can expect that some of its accounts receivable will not be collectible. The point of the collection effort is to improve the percentage of accounts receivable that *are* collected. Most financial statement users know that some of a company's accounts receivable are not collectible. In the next section, we will discuss how a company includes this information when reporting the amount of its accounts receivable on its balance sheet.

 Would you rather know the amount you are owed or the amount you expect to receive? Why?

Accounts Receivable Balance

The amount of accounts receivable that a company reports on its balance sheet is the amount of cash it expects to receive from customers as payments for previous credit sales. The words "expects to receive" reflect the fact that the company may not collect all of its accounts receivable. So, the amount a company shows on its balance sheet as accounts receivable is the total owed by customers (the "gross" amount) less an amount that it expects to be uncollectible. GAAP refers to this amount as the *net realizable value* of accounts receivable.

A company shows its accounts receivable at their net realizable value because, as part of an analysis of its liquidity, financial statement users are concerned with the company's ability to turn accounts receivable into cash. As you learned in Chapter 8, predicting a company's cash flows helps external users make business decisions.

The gross amount of the total accounts receivable at year-end is calculated by adding all of the individual customers' balances. However, the dollar amount of accounts receivable that are uncollectible requires an estimate. This is because the company doesn't know which customers won't pay. (If, at the time of the credit sale, the company thought a particular customer would not pay for the goods, it would not have granted the credit!)

Given the uncertainties of collecting accounts receivable, how does a company estimate the amount that it expects will be uncollectible? In general, a company bases this estimate on its past experience with collections. Using the company's history as a guide, it either calculates the estimate as a percentage of credit sales (e.g., 1 percent of credit sales) or bases the estimate on an "aging analysis" of the accounts receivable (i.e., the older a receivable is, the more likely it is to be uncollectible). We will discuss specific types of estimation methods in Chapter 13.

To inform financial statement users that a company is showing its accounts receivable at the net realizable value, the company places the word "net" after accounts receivable on the balance sheet: Accounts receivable (net).

Sweet Temptations shows its accounts receivable on its December 31, 2005 balance sheet (Exhibit 9-1) as follows:

Accounts receivable (net) $7,340

To determine this net amount, Sweet Temptations calculated its dollar estimate of uncollectible accounts receivable and subtracted it from the total amount of accounts receivable listed in the accounting records. Assuming Sweet Temptations' gross accounts receivable are $7,874 (the company's accounting system keeps track of this amount), we can determine that its estimated uncollectible accounts receivable at December 31, 2005 are $534 ($7,874 − $7,340).

Many companies' accounts receivable result from selling inventory on credit. We will discuss inventory management in the next section.

INVENTORY

Why do companies sometimes sell their goods for 50 percent off the retail price? Why do they advertise that they are having an inventory reduction sale? Should this affect the way they account for their inventory?

A company's **inventory** is the merchandise being held for resale. In Chapter 6, we discussed how a company uses either a perpetual inventory system or a periodic inventory system to keep track of its merchandise. That discussion focused on the calculation of a company's cost of goods sold. Remember that the cost of goods sold is the cost a company has incurred for the merchandise (goods) it has sold to customers during the accounting period. The company includes the cost of goods sold as an expense on its income statement. In this section, we focus on the calculation of a company's ending inventory.

Accounting for, controlling, and reporting on inventory are important for several reasons. First, selling inventory is the primary way a retail or manufacturing company gets cash from operating activities (and earns a profit). If the amount of inventory is too low, the company could have future difficulties providing the cash it will need for operations. Second, a company usually expects to turn over its inventory (purchase it, sell it, and replace it with newly purchased inventory) several times during the year. If inventory sales slow down, investors and creditors may become concerned about the company's ability to continue to sell the inventory at a satisfactory profit. Third, storing inventory is expensive due to storage space, utilities, and insurance costs. Finally, inventory can be stolen and/or can become obsolete. For these reasons, a company must effectively account for, control, and report on its inventory.

Simple Inventory Controls

A company should establish several simple internal controls that will help safeguard its inventory and improve record-keeping. First, it should control the ordering and acceptance of inventory deliveries. In a small company, the owner is usually the only person who places orders for inventory. But even in a small company, the owner should place orders using a purchase order. A **purchase order** is a document authorizing a supplier to ship the items listed on the document at a specific price. It is signed by an authorized person in the company. Use of purchase orders helps ensure that purchasing activities are efficient and that no unauthorized person can purchase inventory.

6 How can managers control inventory in a small company?

A company should keep a list of the purchase orders or copies of the purchase orders where employees can have access to them. Employees receiving inventory need to know what has been ordered because they should accept only approved orders. In addition, employees should check the quantity and condition of every order received. If, on further inspection, an employee finds that the order was not filled properly (e.g., the wrong boxes of chocolates are received) or that the goods are damaged (e.g., the chocolates melted), the supplier should be notified immediately.

Think about the last time you went shopping. What physical controls over inventory did you notice?

Second, a company should establish physical controls over inventory while the inventory is being held for sale. One physical control involves restricting access to inventory. You have probably seen signs on certain company doors that state:

FOR EMPLOYEES ONLY

Companies post the signs to help keep customers out of storage areas. Other controls include locked display cases, magnetic security devices, and camera surveillance systems.

Finally, to make sure that inventory records are accurate, a company should periodically take a physical count of its inventory. Whether a company uses a perpetual or a periodic inventory system (discussed in Chapter 6), by physically counting inventory the company can determine the accuracy of its inventory records and can estimate losses from theft, breakage, or spoilage. Almost all companies count inventory at least once a year. Many companies count inventory after closing on the date of their year-end balance sheet. By counting inventory at the end of the fiscal year, a company can use the inventory count to help determine the dollar amount of inventory that it will show on its balance sheet and the cost of goods sold that it will show on its income statement.

Recall that Sweet Temptations uses a perpetual inventory system, so it keeps a running balance of its inventory and cost of goods sold in its accounting records. To verify the accuracy of these balances, Anna Cox and an employee spent two hours counting the boxes of chocolates in Sweet Temptations' inventory after it closed on December 31, 2005. When they were finished, Anna calculated that Sweet Temptations owned 274 boxes of Unlimited Decadence chocolates at year-end.

Determining the Cost of Ending Inventory

A company shows its inventory on its ending balance sheet as a *dollar amount*. So, after the company has counted the number of units in its ending inventory, it must determine the appropriate unit *cost* for each item. How does a company figure out the cost of each inventory item? To answer that question, we need to explain two things: (1) the relationship among cost of goods available for sale, cost of goods sold, and year-end inventory, and (2) the concept of cost flows.

Cost of Goods Available for Sale, Sold, and Held in Inventory

At the start of any month, a company has a certain number of inventory items available for sale—its beginning-of-the-month inventory. For example, say Sweet Temptations starts the month of December 2005 with 590 boxes of chocolates. During the month, a company like Sweet Temptations sells some of its inventory and makes purchases to restock for additional sales. Ideally, at the end of the month, one of two things has happened to all of the items that were available for sale during the month. Either the goods were sold *or* the goods remain in inventory. If Sweet Temptations purchases an additional 300 boxes of chocolates during December and sells 616 boxes, 274 boxes remain in inventory on December 31. Remember that the year-end physical count of inventory was 274 boxes. These calculations for the month of December can be summarized as follows:

	Beginning inventory for December	590 boxes
+	December purchases	300 boxes
=	Goods available for sale during December	890 boxes
−	Goods sold during December	(616) boxes
=	Goods in inventory on December 31, 2005	274 boxes

As we discussed in earlier chapters, when Sweet Temptations prepares its financial statements for the month, it includes the *cost* of the goods sold during the month in the monthly income statement and the *cost* of the ending inventory in the month-end balance sheet, based on its perpetual inventory records. For December, Sweet Temptations' Cost of Goods Sold account shows a balance of $3,399.75, and its Inventory account shows an ending balance of $1,570.25. To arrive at these amounts, Sweet Temptations converted the number of boxes of candy purchased, sold, and on hand to dollar amounts. Sweet Temptations recorded these dollar amounts in its Inventory and Cost of Goods Sold accounts, as we illustrated in Chapter 6.

Exhibit 9-8 shows the December inventory information for Sweet Temptations.[2] Notice that Sweet Temptations' beginning inventory cost $5.50 per box. The purchase it

---

[2]Normally, when a company takes a physical inventory at the end of its fiscal year, it calculates its units and costs of goods available for sale for the entire year. For simplicity, we illustrate Sweet Temptations' inventory information for only one month.

EXHIBIT 9-8	SWEET TEMPTATIONS' INVENTORY INFORMATION		
Beg. Inventory, Dec. 1	590 boxes @ $5.50 per box =	$3,245	
December 13 purchase	300 boxes @ $5.75 per box =	1,725	
Cost of goods available for sale	890 boxes	$4,970	
Sales during December	(616) boxes		
End. Inventory, Dec. 31	274 boxes		

made on December 13 cost $5.75 per box. Since the unit costs of the inventory changed during the month, Sweet Temptations had to decide which cost to assign to the boxes it sold and which cost to assign to the boxes left in inventory. For example, should it assign all of the boxes a cost of $5.50 or all a cost of $5.75? Or should it use both costs and, if so, to which boxes should it assign $5.50 and to which ones $5.75? Or should it use the average of both costs?

A company must have a *method* for deciding how to calculate the dollar amounts for inventory and cost of goods sold, and it may use one of several methods to determine these dollar amounts. However, the company should use its chosen method consistently from year to year, unless a different method would better reflect the company's operations, so that users of its financial statements can compare its performance from year to year. Here, we discuss the specific identification method. We will discuss other methods in Chapter 19.

Specific Identification Method

The **specific identification method** allocates costs to cost of goods sold and to ending inventory by assigning to each unit sold and to each unit in ending inventory the cost to the company of purchasing that particular unit. Under this method a company keeps track of the cost of each inventory item separately. Usually, it does this tracking through a computer system or through an inventory coding system. For example, on every box of chocolates that Sweet Temptations receives from Unlimited Decadence, the date the chocolates were made is stamped on the bottom. Because Unlimited Decadence sends its chocolates out freshly made, Sweet Temptations can use this date to tell which shipment a box came from and the exact cost of the box.[3] Many companies have *point of sale* cash register systems that scan the inventory codes to keep track of the costs of inventory sold and inventory on hand.

For example, say that on December 21 a customer purchases three boxes of chocolates for $12 per box and pays in cash. As Anna rings up the sale, she notes on the sales receipt that two boxes are dated 11-29-2005 and one box is dated 12-13-2005. From Exhibit 9-8 we can see that the boxes dated 11-29-2005 cost $5.50 per box and the box dated 12-13-2005 cost $5.75. When she records this sale, Anna increases the Cash and Sales accounts by $36 ($12 × 3 boxes). She also decreases the Inventory account and increases the separate expense account—Cost of Goods Sold—by $16.75, the exact cost of the items sold ($5.50 + $5.50 + $5.75).

The inventory amount that Sweet Temptations shows on its December 31, 2005 balance sheet is calculated from the results of the physical inventory count. (This amount should be the same as the amount that it shows in its accounting records.) Recall that 274 boxes remained in inventory on December 31. Under the specific identification method, in addition to counting the inventory, Anna and her employee must keep track of the boxes according to the stamped dates. Exhibit 9-9 shows Anna's inventory count instructions, the results of the count, and the year-end inventory and cost of goods sold calculations.

FINANCIAL STATEMENT EFFECTS
Increases current assets and total assets on **balance sheet**. Increases revenues, which increases net income on **income statement** (and therefore increases owner's equity on **balance sheet**). Increases cash flows from operating activities on **cash flow statement**.

FINANCIAL STATEMENT EFFECTS
Decreases current assets and total assets on **balance sheet**. Increases expenses (cost of goods sold), which decreases net income on **income statement** (and therefore decreases owner's equity on **balance sheet**).

[3]Although we did not mention it in Chapter 6, we used the specific identification method in our earlier inventory discussion.

EXHIBIT 9-9	SWEET TEMPTATIONS' YEAR-END INVENTORY CALCULATION

The Inventory Count

After Sweet Temptations closes on the evening of December 31, 2005, Anna and one employee spend two hours counting the company's inventory. Anna tells her employee how the count will work: "You and I will count all of the items independently of each other. I will follow along right behind you. We will count one section of the store at a time. Both of us will mark our findings on inventory count sheets, noting separately the number of boxes dated 11-29-2005 and the number dated 12-13-2005. We have to count these boxes separately because we purchased them at different prices and we value inventory using the specific identification method. After we finish each section of the store, we will compare our results to see that we agree on the count. If the numbers don't match, we will recount the section. After we count all of the inventory, we will compute a total for the number of boxes dated 11-29-2005 and 12-13-2005."

The Results of the Inventory Count

Sweet Temptations' inventory count ran smoothly. After compiling all of the inventory count sheets, Anna concluded that the year-end inventory consisted of the following:

253 boxes of chocolates dated 12-13-2005
21 boxes of chocolates dated 11-29-2005

December 31, 2005 Inventory Calculation

253 boxes @ $5.75 =	$1,454.75
21 boxes @ $5.50 =	115.50
Ending inventory	$1,570.25

December Cost of Goods Sold Calculation

Cost of goods available for sale (Exhibit 9-8)	$4,970.00
− Ending inventory	(1,570.25)
Cost of goods sold	$3,399.75

As a manager, would you "allow" customers to select any box of chocolates from the shelves?

Because Sweet Temptations' physical inventory count of 274 boxes is the same as the calculation of its ending inventory from its inventory records of beginning inventory, purchases, and sales shown earlier, the $1,570.25 cost of the ending inventory calculated in Exhibit 9-9 is the same as the amount in its Inventory account. Furthermore, the $3,399.75 cost of goods sold calculated in Exhibit 9-9 is the same as the amount in its Cost of Goods Sold account. So, by taking a physical count, Sweet Temptations has verified that the amounts in its accounting records are correct.

Now suppose that Anna and her employee counted 253 boxes of chocolates dated 12-13-2005 but only 17 boxes dated 11-29-2005. In this case, 4 boxes of chocolates are missing, and the cost of the ending inventory is $1,548.25 [(253 × $5.75) + (17 × $5.50)]. Anna should try to find out why these boxes are missing. For instance, they may have been given away as "free samples," stolen (or eaten by the employees), or thrown away because they were stale. Whatever the reason, she should adjust the accounting records by increasing the Cost of Goods Sold account and decreasing the Inventory account by $22 (4 × $5.50) for the missing boxes.

Suppose that the year-end count of inventory is less than the accounting records show as ending inventory because Anna threw away stale boxes of chocolates. How might this information affect Anna's future decisions?

FINANCIAL STATEMENT EFFECTS

Decreases current assets and total assets on **balance sheet**. Increases expenses (cost of goods sold), which decreases net income on **income statement** (and therefore decreases owner's equity on **balance sheet**).

ACCOUNTS PAYABLE

As we explained earlier in the chapter, companies often sell on credit to customers. These credit sales result in accounts receivable. Similarly, companies often make purchases on credit, which result in the liability accounts payable. **Accounts payable** are the amounts that a company owes to its suppliers for previous credit purchases of inventory and supplies. The reasons for purchasing on credit are similar to the reasons for selling on credit. The first reason is that purchasing on credit is often more convenient than purchasing with cash. The second reason for purchasing on credit is to delay paying for purchases and, by doing so, to obtain a short-term "loan" from the supplier. Many companies, particularly small companies, are often short of cash and find it difficult to pay for their purchases immediately. Managers of these companies, therefore, try to delay payment until their companies receive the cash from the eventual sale of their products; they then use this cash to pay the amounts their companies owe. This delay is the reason many suppliers offer their customers cash discounts for prompt payment.

Simple Controls over Accounts Payable

A company's accounts payable represent promises to pay the amounts due to other businesses. As is the case with accounts receivable, a company needs controls over accounts payable. Controls over accounts payable should focus on three primary concerns. The first concern involves the ability of employees to obligate the company to an account payable. Giving too many employees the authority to place orders for company purchases makes it more difficult for managers to coordinate and monitor credit purchases, and makes it easier for untrustworthy employees to obligate the company for personal expenditures. In response to this concern, a company should limit the number of employees who have the authority to make company purchases. In a small company, this authority may be given only to the owner. Larger companies usually have a purchasing department that controls all company purchases.

7 How can managers control accounts payable in a small company?

Second, once a company incurs an account payable, the company is concerned that it makes each payment at the appropriate time and that the *supplier* records each payment properly. A company monitors the timeliness of its payments by having an employee keep track of the credit terms of each account payable. If cash discounts are available, the company should take advantage of the cash savings by making the payment within the cash discount period. A company makes sure that the supplier records its payments properly by checking the supplier's monthly statements. If the payment is not recorded properly, an employee should investigate the discrepancy and perhaps contact the supplier.

Finally, managers, investors, and creditors are concerned about a company's total dollar amount of accounts payable because, in the very near future, the company will need to use its cash to pay these liabilities. If the accounts payable are large relative to the company's current assets, the company may experience liquidity problems.

Managers will investigate relatively large increases in accounts payable. If the increase is a result of planned increases in inventory, they assume that increased sales will provide the cash needed to pay the liabilities. If the increase is a result of cash flow problems, managers may postpone purchases of inventory and/or property and equipment, or may contact suppliers to try to arrange an extension of the credit terms.

Accounts Payable Balance

The amount of accounts payable that a company owes on the balance sheet date is listed in the current liabilities section of the ending balance sheet. A company calculates this amount by summing the accounts payable owed to individual suppliers. As Exhibit 9-1 shows, on December 31, 2005, Sweet Temptations' total accounts payable is $7,540.

BUSINESS ISSUES AND VALUES

 Has anyone ever forgotten to repay you for money that he or she borrowed? Has it ever been difficult for you to pay off a debt? How should a company handle these situations? What factors should it consider when developing policies concerning late payments by its customers or to its suppliers?

We started the chapter by stating that managing working capital effectively is an important part of financial management. This is especially true for new companies that have a relatively small amount of capital and may be prone to liquidity problems. But how aggressive should a company be in managing its working capital? When trying to collect accounts receivable payments, some companies repeatedly telephone customers at their offices and homes. On the other hand, when trying to hold off paying their own debts, some companies continue to tell suppliers that "the check is in the mail" when it really is not.

The ethics of aggressive working capital management has been questioned by some business leaders and critics. Instead of being seen as conscientious, a company that uses aggressive collection efforts can be viewed as intimidating and harassing. A company that signs a purchase agreement, even though it knows that it will make suppliers wait an additional 30 or 60 days before paying for the goods, can be viewed as untrustworthy, not as a shrewd financial planner. What do you think? We will continue to discuss these types of issues in future chapters when we examine corporations.

SUMMARY

At the beginning of the chapter we asked you several questions. During the chapter, we asked you to STOP and answer some additional questions to build your knowledge about specific issues. Be sure you answered these additional questions. Below are the questions from the beginning of the chapter, with a brief summary of the key points relating to the answers. Use your creative and critical thinking skills to expand on these key points to develop more complete answers to the questions and to determine what other questions you have that might lead you to learn more about the issues.

1 What is working capital, and why is its management important?

Working capital is current assets minus current liabilities. A company needs to manage its working capital so that it keeps an appropriate balance between having enough to conduct its operations and to handle unexpected needs, and having too much so that profitability is reduced.

2 How can managers control cash receipts in a small company?

Managers can control cash receipts by requiring the proper use of a cash register, separating the duties of receiving and processing collections of accounts receivable, and depositing receipts every day.

3 How can managers control cash payments in a small company?

Managers can control cash payments by paying all bills by check, paying only for approved purchases supported by source documents, and immediately stamping "paid" on the supporting documents after payment.

4 What is a bank reconciliation, and what are the causes of the difference between a company's cash balance in its accounting records and its cash balance on its bank statement?

A bank reconciliation is an analysis that a company uses to resolve the difference between the cash balance in its accounting records and the cash balance reported by the bank on its bank statement. The causes of the difference are deposits in transit, outstanding checks, deposits made directly by the bank, charges made directly by the bank, and errors.

5 How can managers control accounts receivable in a small company?

Managers can control accounts receivable by evaluating a customer's ability to pay before extending credit, monitoring the accounts receivable balance of each customer, and monitoring the total accounts receivable balance.

6 How can managers control inventory in a small company?

Managers can control inventory by establishing policies for ordering and accepting inventory, establishing physical controls over inventory being held for sale, and taking a periodic physical count of the inventory.

7 How can managers control accounts payable in a small company?

Managers can control accounts payable by coordinating and monitoring credit purchases, making payments at the appropriate time, and monitoring the total accounts payable balance.

KEY TERMS

accounts payable *(p. 289)*
accounts receivable *(p. 282)*
bank reconciliation *(p. 275)*
bank statement *(p. 274)*
cash *(p. 272)*
deposit in transit *(p. 275)*
internal control structure *(p. 272)*

inventory *(p. 285)*
NSF (not sufficient funds) check *(p. 275)*
outstanding check *(p. 275)*
petty cash fund *(p. 280)*
purchase order *(p. 285)*
specific identification method *(p. 287)*
working capital *(p. 269)*

SUMMARY SURFING

Here is an opportunity to gather information on the Internet about real-world issues related to the topics in this chapter. Go to http://www.cunningham.swlearning.com and click on the Interactive Study Center. Click on this chapter number then click on Summary Surfing and answer the following questions.

- Click on **SBA** (U.S. Small Business Administration) **On-Line Library**. On left, click on *Publications*. Under *B. Financial Management Series* click on *3. Understanding Cash Flow FM-4* and open the article. How does the article define "working capital cash conversion cycle"? What are some suggestions for (a) more efficient collections, (b) more efficient payments, (c) more efficient management of accounts receivable, (d) more efficient inventory management, and (e) purchasing goods on more favorable terms?

INTEGRATED BUSINESS AND ACCOUNTING SITUATIONS

Answer the Following Questions in Your Own Words.

Testing Your Knowledge

9-1 What is a company's working capital, and what is included in its two components?

9-2 Why does a company manage its working capital?

9-3 Define *cash* for a company.

9-4 Briefly discuss the controls over cash sales.

9-5   Briefly discuss the controls over collections of cash from accounts receivable.

9-6   Briefly discuss the controls over cash payments.

9-7   What is a bank reconciliation?

9-8   Identify the causes of the difference between the ending cash balance in a company's records and the ending cash balance reported on its bank statement.

9-9   Briefly explain what is meant by the terms *deposits in transit* and *outstanding checks*.

9-10   Briefly explain what are included in deposits made directly by the bank and charges made directly by the bank.

9-11   Prepare an outline of a bank reconciliation for a company.

9-12   Briefly explain what a petty cash fund is and how it works.

9-13   Why do companies make sales on credit?

9-14   Briefly discuss the controls over accounts receivable.

9-15   Briefly explain how a company reports its accounts receivable on its ending balance sheet.

9-16   Why is accounting for, controlling, and reporting of inventory important?

9-17   Briefly discuss the controls over inventory.

9-18   Briefly explain how the specific identification method works for determining inventory costs.

9-19   Evaluate this statement: "My company uses a perpetual inventory system, so it doesn't need to take a periodic physical inventory."

9-20   Briefly discuss the controls over accounts payable.

Applying Your Knowledge

9-21   The following are several internal control weaknesses of a small retail company in regard to its cash receipts and accounts receivable:
(a) Sales invoices are not prenumbered.
(b) Receipts from daily sales are deposited every Tuesday and Thursday evening.
(c) One employee is responsible for depositing customer checks from collections of accounts receivable and for recording their receipt in the accounts.
(d) For credit sales on terms of 2/10, net/30, customers are allowed, for convenience, the discount if payment is received within 20 days.
(e) A money box is used instead of a cash register to store both the sales invoices and the cash from the sales.
(f) Credit sales of a large dollar amount can be approved by any sales employee.
(g) When customers write checks for payment, only the identification of customers who look "untrustworthy" is verified.

Required: (1) For each internal control weakness, explain how the weakness might result in a loss of the company's assets.
(2) For each internal control weakness, explain what action should be taken to correct the weakness.

9-22   The following are several internal control weaknesses of a retail company in regard to its cash payments, accounts payable, and inventory:
(a) The inventory of gold jewelry for sale is kept in unlocked display cases.
(b) One employee is responsible for ordering inventory and writing checks.
(c) Some purchases are made by phone, and no purchase order is written up.
(d) The company takes a physical inventory every two years.
(e) Employees are allowed to bring coats, bags, and purses into working areas.
(f) Inventory received at the loading dock is rushed immediately to the sales floor before it is counted.

(g) When inventory is low, any sales employee can prepare a purchase order and mail it to the supplier.

(h) For efficiency, the company pays invoices on credit purchases once a month, even if it has to forgo any cash discounts for prompt payment.

Required: (1) For each internal control weakness, explain how the weakness might result in a loss of the company's assets.

(2) For each internal control weakness, explain what action should be taken to correct the weakness.

9-23 A company is preparing its bank reconciliation and discovers the following items:

(a) Outstanding checks

(b) Deposits in transit

(c) Deposits made directly by the bank

(d) Charges made directly by the bank

(e) The bank's erroneous underrecording of a deposit

(f) The company's erroneous underrecording of a check it wrote

Required: Indicate how each of these items would be used to adjust (1) the company's cash balance or (2) the bank balance to calculate the reconciled cash balance.

9-24 At the end of March, the Elbert Company records showed a cash balance of $7,027. When comparing the March 31 bank statement with the company's Cash account, the company discovered that deposits in transit were $725, outstanding checks totaled $862, bank service charges were $28, and NSF checks totaled $175.

Required: (1) Compute the March 31 reconciled cash balance of the Elbert Company.

(2) Compute the cash balance listed on the March 31 bank statement.

9-25 At the end of September, the Bross Bicycle Company's records showed a cash balance of $3,513. When comparing the September 30 bank statement, which showed a cash balance of $1,860, with the company's Cash account, the company discovered that outstanding checks were $462, bank service charges were $23, and NSF checks totaled $89.

Required: (1) Compute the September 30 reconciled cash balance of the Bross Bicycle Company.

(2) Compute the September deposits in transit.

9-26 The following five situations (columns 1–5) are independent:

	1	2	3	4	5
Ending balance in the company's checking account	(a)	$2,000	$4,000	$12,000	$3,000
Deposits made directly by the bank	$ 200	(b)	500	450	200
Deposits in transit	700	800	(c)	500	900
Outstanding checks	450	1,200	600	(d)	1,000
Ending cash balance from bank statement	6,000	3,000	4,100	12,000	(e)

Required: Compute each of the unknown amounts, items (a) through (e).

9-27 An examination of the accounting records and the bank statement of the Evans Company at March 31, 2004 provides the following information:

(a) The Cash account has a balance of $6,351.98.

(b) The bank statement shows a bank balance of $3,941.83.

(c) The March 31 cash receipts of $3,260.95 were deposited in the bank at the end of that day but were not recorded by the bank until April 1.

(d) Checks issued and mailed in March but not included among the checks listed as paid on the bank statement were as follows:

Check No. 706	$869.38
Check No. 717	212.00

(e) A bank service charge of $30 for March was deducted on the bank statement.

(f) A check received from a customer for $185 in payment of his account and deposited by the Evans Company was returned marked "NSF" with the bank statement.

(g) Interest of $20.42 earned on the company's checking account was added on the bank statement.

(h) The Evans Company discovered that Check No. 701, which was correctly written as $562 for the March rent, was recorded as $526 in the company's accounts.

Required: (1) Prepare a bank reconciliation on March 31, 2004.
(2) Record the appropriate adjustments in the company's accounts. Compute the ending balance in the Cash account.

9-28 You have been asked to help the Rancher Company prepare its bank reconciliation. You examine the company's accounting records and its bank statement at May 31, 2004, and find the following information:

(a) The Cash account has a balance of $7,753.24.

(b) The bank statement shows a bank balance of $3,783.04.

(c) The May 31 cash receipts of $4,926.18 were deposited in the bank at the end of that day but were not recorded by the bank until June 1.

(d) Checks issued and mailed in May but not included among the checks listed as paid on the bank statement were as follows:

| Check No. 949 | $518.65 |
| Check No. 957 | 699.95 |

(e) A bank service charge of $27 for May was deducted on the bank statement.

(f) A check received from a customer for $241 in payment of her account and deposited by the Rancher Company was returned marked "NSF" with the bank statement.

(g) Interest of $25.18 earned on the company's checking account was added on the bank statement.

(h) The Rancher Company discovered that Check No. 941, which was correctly written as $647.21 for the May utility bill, was recorded as $627.41 in the company's accounts.

Required: (1) Prepare a bank reconciliation on May 31, 2004.
(2) Record the appropriate adjustments in the company's accounts. Compute the ending balance in the Cash account.

9-29 The Huron Company keeps a petty cash fund of $80. On June 30 the fund contained cash of $36.87 and the following petty cash receipts:

Office supplies	$10.00
Postage	27.48
Miscellaneous	5.65

Required: (1) If the company's fiscal year ends June 30, should the petty cash fund be replenished on June 30? Why?
(2) How much cash is needed to replenish the petty cash fund?
(3) Prepare entries in the company's accounts to record the petty cash payments.

9-30 On December 31, 2004, the Bighorn Condominium Management Company had a balance of $70 in its petty cash fund, a reconciled balance of $1,283 in its checking account, and a $4,627 balance in its savings account.

Required: Show how the company would report its cash on its December 31, 2004 balance sheet.

9-31 The Snow-Be-Gone Company sells one type of snowblower and uses the perpetual inventory system. At the beginning of January, the company had a balance in its Cash account of $2,100 and an inventory of 8 units (snowblowers) costing $100 each. During January, it made the following purchases and sales of inventory:

Jan.	5	Purchases	4 units @ $102 per unit
	12	Sales	11 units @ $150 per unit
	18	Purchases	12 units @ $104 per unit
	25	Purchases	6 units @ $103 per unit
	29	Sales	13 units @ $150 per unit

All purchases and sales were for cash. The company uses "bar codes" to verify each sale. For the sales on January 12, 8 were units from the beginning inventory, and 3 were units purchased on January 5. For the sales on January 29, 9 were units purchased on January 18, and 4 were units purchased on January 25.

Required: (1) Record the beginning balances in the Cash and Inventory accounts. Using account columns, record the purchases and sales transactions during January and compute the ending balances of all the accounts you used.

(2) Assume that the company counted its inventory at the end of January and determined that it had 6 snowblowers on hand. Prove that the ending balance in the Inventory account that you computed in (1) is correct.

(3) Compute the company's gross profit.

9-32   The Kvam Lawn Mower Store sells one type of lawn mower at a price of $200 per unit. On June 1, it had an $800 accounts receivable balance and a $600 accounts payable balance, as well as an inventory of 10 mowers costing $120 each. During June, its purchases and sales of mowers were as follows:

	Purchases	**Sales**
June 8	7 mowers @ $125 each	
15		11 mowers
21	6 mowers @ $121 each	
26	4 mowers @ $124 each	
30		8 mowers

All purchases and sales were on credit. No payments or collections were made during June. The company has a perpetual inventory system and uses "bar codes" to verify each sale. For the June 15 sales, 8 were mowers from the beginning inventory, and 3 were mowers purchased on June 8. For the June 30 sales, 2 were mowers from the beginning inventory, 5 were mowers purchased on June 21, and 1 was a mower purchased on June 26.

Required: (1) Record the beginning balances in the Accounts Receivable, Inventory, and Accounts Payable accounts. Using account columns, record the purchases and sales transactions during June and compute the ending balances of all the accounts you used.

(2) Assume that the company counted its inventory at the close of business on June 30 and determined that it had 8 mowers in stock. Prove that the ending balance in the Inventory account that you computed in (1) is correct.

(3) Compute the company's gross profit percentage for June. How does this compare with its gross profit percentage of 40.8% for May? What might account for the difference?

9-33   The Bugs-Be-Gone Company sells two types of screen doors. Model A, which sells for $30, is the basic screen door, and Model B, which sells for $50, is the deluxe screen door that features removable glass panels so that it can be turned into a storm door during the winter. At the beginning of July, the company had a balance in its Cash account of $1,600 and an inventory consisting of 12 units of Model A costing $20 each and 15 units of Model B costing $35 each. During July, it made the following purchases and sales of inventory:

		Model A	**Model B**
July 6	Sales	8 units @ $30 each	10 units @ $50 each
13	Purchases	9 units @ $19 each	10 units @ $36 each
20	Sales	10 units @ $30 each	12 units @ $50 each
24	Purchases	7 units @ $21 each	6 units @ $37 each
29	Sales	7 units @ $30 each	3 units @ $50 each

All purchases and sales are for cash. The company has a perpetual inventory system, using "bar codes" to verify each sale. For the July 20 sales, 3 units of Model A were from the beginning inventory, and 7 were units purchased on July 13; 5 units of Model B were from the beginning inventory, and 7 were units purchased on July 13. For the July 29 sales, 1 unit of Model A was purchased on July 13, and 6 were units purchased on July 24; 1 unit of Model B was purchased on July 13, and 2 were units purchased on July 24.

On July 31, the company counted its inventory and determined that it had 3 units of Model A and 6 units of Model B on hand. However, 1 of the 3 units of Model A was run over by a customer's truck and had to be thrown away. This unit had been in the beginning inventory.

Required: (1) Record the beginning balances in the Cash and Inventory accounts. Using account columns (use one account column for inventory), record the purchases and sales transactions during July and compute the ending balances of all the accounts you used.
(2) Record the disposal of the damaged unit and prove the accuracy of the ending balance in the Inventory account.
(3) Compute the gross profit percentage. How was this affected by the damaged inventory?
(4) Do you think your work would have been easier if you had used two inventory accounts in (1)? How do you think a company with many items of inventory keeps track of these items under a perpetual inventory system?

Making Evaluations

9-34   Your younger brother, always bursting with curiosity, recently purchased a fishing rod from a catalog. While he was filling out the order form, he noticed the warning: "Don't Send Cash!" So, after pointing it out to you, he asked, "Does it seem odd to you that a company wouldn't appreciate receiving cash? You're taking accounting. Don't they teach you in there that companies need cash? Why would they say such a thing?"

Required: Tell your brother why you think the company puts this warning in its catalogs, give him some examples of what might happen if customers paid for their purchases with cash, and explain how checks and credit cards might prevent this from happening.

9-35   Sam Lewis has been operating a "full service" service station for several years. Although he occasionally has employed students part-time, he has collected the cash and checks for gas and repair work himself. He now has decided to open a second "full service" service station and put himself more in the role of a manager. He will hire employees to run the service stations and to pump gas and do repair work.

Required: How should Sam Lewis implement internal control procedures over cash receipts for the service stations?

9-36   Your dad's friend Frank was over for dinner the other night, and discussion turned to his business, Frank's Franks, which is responsible for street-corner vending of hotdogs, pretzels, beer, and soda. It's a small company, with an office downtown and four vending carts located in different areas of downtown. When you asked Frank what kind of internal controls his company has in place, Frank said, "We don't have a formal system of internal controls—don't need them. My employees are family members and friends, and I trust them completely! Now when the business grows, and I have to hire strangers, then I'll think about those controls. But now, the company's profitable, and I'm happy." After Frank left, you talked to your dad about what you had learned in accounting, and asked whether he thought Frank would appreciate hearing about it. Your dad assured you that Frank would be open to your suggestions.

Required: Write a letter to Frank explaining how you think his company would benefit from a system of internal controls, even though he trusts his employees. Also describe specific controls that Frank could use in his particular business.

9-37   Your friend Ruby Johnson works as a cashier in an upscale restaurant located in a business center that includes a bank. She works the late shift, and since the restaurant caters to the convention crowd, she generally doesn't leave work until 2:00 or 2:30 in the morning. One day, when you were having lunch with her, she began complaining about one aspect of her job: "My boss is a real stickler for procedures. Even though it's really late when the last customer leaves, and even though we are exhausted, we still have to follow *procedure*. Before the host and I can leave, we have to count the money in the register and match it against the register tape and match both amounts against the dollar total of the checks the customers paid. And, as if that's not enough, we have to make sure that every check number is accounted for. Every night the manager writes down the numbers of the checks each waitperson has been given to use that night for taking customer orders. At the end of the night, the waitpeople give the cashier all the checks they didn't use. If any money is missing, guess who takes the blame and has to make up the difference? Anyway, after we count the money, we have to put the money and the tape in a deposit bag, walk it across the parking lot to the bank, and deposit it in the bank's night-deposit box, *even though there is a safe right under the cash register!* Like we're not sitting ducks for anybody who wants to rob us. I don't understand why she would risk our lives like that. Furthermore, the boss unlocks the part of the register that contains a copy of the tape that we took to the bank, and uses that tape to enter the day's cash receipts amount into the accounting system. Like she really trusts me so much that she has to keep the tape copy under lock and key. What a jerk!" Now that you are taking accounting, you have a little better insight into why the boss is so interested in these procedures.

Required: Explain to Ruby what's going on before she does something rash, like quit her job.

9-38   The Anibonita Company is a retail store with three sales departments. It also has a small accounting department, a purchasing department, and a receiving department. All inventory is kept in the sales departments. When the inventory for a specific item is low, the manager of the sales department that sells the item notifies the purchasing department, which then orders the merchandise. All purchases are on credit. Anibonita pays the freight charges on all its purchases after being notified of the cost by the freight company. When the inventory is delivered, it is inspected and checked in by the receiving department and then sent to the sales department, where it is placed on the sales shelves. After notification that the ordered inventory has been received, the accounting department records the purchase. Upon receipt of the supplier's invoice or the freight bill, the accounting department verifies the invoice (or freight bill) against the purchase order and the receiving report before making payment.

Required: Briefly explain the internal controls that the Anibonita Company uses for its purchasing process. Include in your discussion what source documents it probably uses.

9-39   The JeBean Company makes only sales on credit. All JeBean's customers order through the mail. The company has a small accounting department, a credit department, an inventory department, and a shipping department. After approval of an order by the credit department, the merchandise is assembled in the inventory department and then sent to the shipping department. The shipping department packs the merchandise in cardboard boxes; then it is picked up by the freight company and shipped to the customer. The JeBean Company pays for freight charges on all items shipped to customers after being notified of the cost by the freight company. After verification of shipment, the accounting department mails an invoice to the customer and records the sale. On receipt of the customer's check, the accounting department records the collection.

Required: Briefly explain the internal controls that the JeBean Company uses for its sales process. Include in your discussion what source documents it probably uses.

9-40   Oliver Bauer, owner of Bauer's Retail Store, has been very careful to establish good internal control over inventory purchases for his store. The store has several employees, and since Ollie cannot devote as much time as he would like to running the store, he has entrusted a longtime employee with the task of purchasing inventory. This employee has worked for Ollie for 15 years and knows all of the store's suppliers. Whenever inventory must be purchased, the employee prepares a purchase order and mails it to the supplier. When a rush order is needed, the employee occasionally calls in the order and does not prepare a purchase order. This procedure is acceptable to the suppliers because they know the employee. When the merchandise is received from the supplier, this employee carefully checks in each item to verify the correct quantity and quality. This job is usually done at night after the store is closed, thus allowing the employee to help with sales to customers during regular working hours. After checking in the items, the employee initials the copy of the supplier invoice received with the merchandise, staples the copy to the purchase order (if there is one), records the purchase in the company's accounts, and prepares a check for payment. Oliver Bauer examines the source documents (purchase order and initialed invoice) at this point and signs the check, and the employee records the payment. Ollie has become concerned about the store's gross profit, which has been steadily decreasing even though he has heard customers complaining that the store's selling prices are too high. He has a discussion with the employee, who says, "I'm doing my best to hold down costs. I will continue to do my purchasing job as efficiently as possible—even though I am overworked. However, I think you should hire another salesperson and spend more on advertising. This will increase your sales and, in turn, your gross profit."

Required: Why do you think the gross profit of the store has gone down? Prepare for Oliver Bauer a report that summarizes any internal control weaknesses existing in the inventory purchasing procedure and explain what the result might be. Make suggestions for improving any weaknesses you uncover.

9-41   In the chapter, we mentioned that if Sweet Temptations came up short four boxes of candy, it should increase its Cost of Goods Sold account and decrease its Inventory account by the cost of those boxes. Suppose Anna wanted to keep a record of candy shortages in the accounting system.

Required: Design a way that Sweet Temptations' accounting system could be changed to accommodate Anna's request.

9-42   Suppose that one of your company's largest customers has written an NSF check for $9,734 and your boss has just found out about it. This morning he comes flying into your office (with smoke coming out of his ears) and demands to know how this NSF check will affect specific accounts in the company's financial statements. You examine the bank statement that came in the morning's mail and notice that, not only has the customer written an NSF check, but the bank has charged you a fee of $75 for processing this check.

Required: List the accounts that will be affected by this turn of events, and indicate by how much they will be affected. What do you think should happen next?

9-43   You are a consultant for several companies. The following are several independent situations you have discovered, each of which may or may not have one or more internal control weaknesses. The names of the companies have been changed to protect the innocent.
(a) In Company A, one employee is responsible for counting and recording all the receipts (remittances) received in the mail from customers paying their accounts. Customers usually pay by check, but they occasionally mail cash. Every day, after the mail is delivered, this employee opens the envelopes containing payments by customers. She carefully counts all remittances and places the checks and cash in a bag. She then lists the amount of each check or cash received and the customer's name on a sheet of paper. After totaling the cash and checks received, she records the receipts in the company's accounts, endorses the checks in the company's name, and deposits the checks and cash in the bank.
(b) Company B has purchased several programmable calculators for use by the office and sales employees. So that these hand calculators will be available to any employee who needs one, they are kept in an unlocked storage cabinet in the office. Anyone who takes and uses a calculator "signs out" the calculator by writing his or her name on a sheet of paper posted near the cabinet. When the calculator is returned, the employee crosses out his or her name on the sheet.
(c) Company C owns a van for deliveries of sales to customers. No mileage is kept of the deliveries, although all gas and oil receipts are carefully checked before being

paid. To advertise the store, Company C printed two signs with the store's name and hung one on each side of the van. These signs are easily removable so that the van can be periodically cleaned without damaging the signs. The company allows employees to borrow the van at night or on the weekends if they need the van for personal hauling. No mileage is kept of the personal hauling, but the employee who borrowed the van must fill the gas tank before returning the van.

(d) Employee Y is in charge of employee records for Company D. Whenever a new employee is hired, the new employee's name, address, salary, and other relevant information are properly recorded. Every payday, all employees are paid by check. At this time Employee Y makes out each employee's check, signs it, and gives it to each employee. After distributing the paychecks, Employee Y makes an entry in the company's accounts, increasing Salaries Expense and decreasing Cash for the total amount of the salary checks.

(e) To reduce paperwork, Company E places orders for purchases of inventory from suppliers by phone. No purchase order is prepared. When the goods arrive at the company, they are immediately brought to the sales floor. An employee then authorizes payment based on the supplier's invoice, writes and signs a check, and mails payment to the supplier. Another employee uses the paid invoice to record the purchase and payment in the company's accounts.

(f) All sales made by Company F, whether they are for cash or on account, are "rung up" on a single cash register. Employee X is responsible for collecting the cash receipts from sales and the customer charge slips at the end of each day. The employee carefully counts the cash, preparing a "cash receipts" slip for the total. Employee X sums the amount on the cash receipts slip and the customer charge slips, and compares this total with the total sales on the cash register tape to verify the total sales for the day. The cash register tape is then discarded, and the cash is deposited in the bank. The cash receipts slip and the customer charge slips are turned over to a different employee, who records the cash and credit sales in the company's accounts.

Required: (1) List the internal control weakness or weaknesses you find in each of the preceding independent situations. If no weakness can be found, explain why the internal control is good.

(2) In each situation in which there is an internal control weakness, describe how you would remedy the situation to improve the internal control.

9-44 Yesterday, you received the letter shown on the following page for your advice column in the local paper:

DR. DECISIVE

Dear Dr. Decisive:

Well, this takes the cake! I thought my boss was a little on the shady side, and now I'm pretty convinced, but some of my friends think I'm wrong. What do you think? Here's some background. My company uses the specific identification method to assign costs to inventory and cost of goods sold. Well, this year our inventory consisted of two batches of goods. We paid $6.00 per unit for each inventory item in the old batch and $6.75 for each item in the batch we purchased this year. As it turned out, most of the inventory items we sold this year came out of the new batch (the $6.75 ones). The effect was that our cost of goods sold for the year is higher than it would have been if we had sold the old batch of items before we sold items from the new batch. (Are you following me?) So my attitude is, "Well, que sera sera." Well, that's not my boss's attitude. This morning he came into my office and actually asked me to "recost" the inventory and cost of

goods sold assuming that we sold the items in the old batch first and
then sold items from the new batch. But we didn't!! Of course, his
method would make the cost of goods sold that we report in our income
statement lower and our net income higher. So the company would look
better. But something about this really galls me. My friends say: "So
what? What difference does it make?" Help! You can call me

"Ethical Ethyl" (or not)

Required: Meet with your Dr. Decisive team and write a response to "Ethical Ethyl."

INTRODUCTION TO CORPORATIONS: STRUCTURES, REPORTS, AND ANALYSES

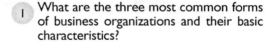

"WHEN SOMETHING GOES WRONG IN A LARGE CORPORATION, YOU BLAME THE BUREAUCRACY. WHEN SOMETHING GOES WRONG IN YOUR SMALL COMPANY, THERE IS NO ONE TO BLAME BUT YOURSELF."

—UNKNOWN

1. What are the three most common forms of business organizations and their basic characteristics?

2. What are the qualities that make accounting information about large corporations useful for decision making?

3. What do users need to know about the stockholders' equity section of a corporation's balance sheet and about its income statement?

4. How does a corporation provide information to external users?

5. How do users perform intracompany and intercompany analyses?

6. What is percentage analysis, and what are its three types?

I n its 2001 annual report, **Johnson & Johnson Corporation** indicates that it has 101,800 employees and 194 operating companies in 54 countries (a large company!). It also reports that it is organized on the principle of "decentralized" management, and that its Executive Committee is the principal management group responsible for the operations and allocation of the company's resources in its consumer, pharmaceutical, and medical devices & diagnostics businesses. Besides its Chief Executive Officer, Johnson & Johnson has 10 vice-presidents responsible for such areas as advertising, human resources, finance, administration, legal, government policy, information, public affairs, corporate development, and science and technology.

http://www.jnj.com

Starting with this chapter, we begin to study how accounting information helps managers, investors, and creditors of larger companies make business decisions. There are many similarities in how small and large companies operate. Both use cost-volume-profit (C-V-P) analysis, budgeting, and a GAAP-based accounting system. But, in part because large companies are more complex, there are many differences in the ways these companies make decisions and use accounting information. Large companies have a more formal and well-defined organizational and decision-making structure. In a small company, the owner-manager decides when to order additional inventory, to which customers to extend credit, or how to get the resources needed to expand. A large company usually has entire departments that deal with each of these issues. Large companies also usually have a complex ownership structure.

Do you think Target Corporation has an entire department responsible for ordering inventory, or do you think a store manager has that responsibility?

As we mentioned in Chapter 1, there are three main forms of business organizations: sole proprietorships, partnerships, and corporations. Corporations generally are larger than sole proprietorships or partnerships. In fact, although fewer than 20 percent of U.S. companies are corporations, they make nearly 90 percent of all U.S. sales.

Most large retail or manufacturing companies are corporations. As a result of incorporating, companies must follow additional laws and regulations. And although companies of all sizes face complicated business decisions, such as determining whether to rent or buy equipment, larger companies are more likely to do so.

In this chapter, we introduce several aspects of the environment in which larger corporations operate. First, we explain the legal forms and characteristics of different business organizations. Second, we discuss the structure of large corporations. Third, we discuss the qualities that make accounting information about large corporations useful for decision making. Fourth, we focus on the unique aspects of financial reporting for large corporations. Finally, we discuss what financial information about corporations external decision makers use and how they use it.

FORMS OF BUSINESS ORGANIZATIONS

1 What are the three most common forms of business organizations and their basic characteristics?

Think about the following names of companies:

ATI Technologies, Incorporated

Target Corporation

Ernst & Young LLP

Equus Capital Partners, LP

Delta Electronics, Incorporated

General Motors Corporation

Underwriters Laboratories Incorporated

http://www.atitech.com
http://www.target.com
http://www.ey.com
http://www.equuscap.com
http://www.deltaww.com
http://www.gm.com
http://www.ul.com

Did you notice several words the names have in common? Five of the companies include either "corporation" or "incorporated" in their name. Two companies include the letters "LP" or "LLP" in their title. Why do companies include these designations in their names? These designations reveal a company's form of organization. The word *corporation,* or *incorporated*, indicates that the company is a separate legal entity known as a corporation. The "P" in LP and LLP stands for *partnership*. (We will discuss the L and LL later.)

Choosing a company's legal form is an important decision for the company's owners to make. For example, as a company owner, this decision determines how laws and regulations affect your personal responsibility to pay the company's debts. When choosing among legal forms, you need to know the characteristics and advantages and disadvantages of each. Once you select a legal form and start operating your company, laws and regulations specific to your type of company will affect some of your business decisions. In this section we discuss the three most common forms of business organizations: sole proprietorships, partnerships, and corporations. In addition, we explain how some aspects of these forms have been combined to create three other forms: limited partnerships, limited liability partnerships, and Subchapter S corporations.

Sole Proprietorships

A **sole proprietorship** is a company owned by one person who is the sole investor of capital into the company. Because Anna Cox is the only investor in Sweet Temptations, this company is an example of a sole proprietorship. In general, sole proprietorships are small companies that focus either on selling merchandise or on performing a service. Many of the small shops you see downtown are sole proprietorships.

Usually, the owner of a sole proprietorship also manages the company. The owner makes the company's important decisions, such as when to purchase equipment, how much debt to incur, and to which customers to extend credit. In the United States, tax laws and regulations require each owner of a sole proprietorship to report and pay taxes on his or her company's taxable income. The company's taxable income is included in the owner's individual income tax return; there is no separate income tax return for a sole proprietorship. So, the owner adds the income from the sole proprietorship to his or her other sources of income, such as wages earned from other jobs and interest received from bank deposits. In the case of Sweet Temptations, Anna Cox includes with her personal income tax return a schedule that reports Sweet Temptations' taxable income. She includes this amount in her total personal taxable income. Anna calculates her personal income tax liability based on all her sources of income. In addition to income taxes, individuals who operate a sole proprietorship must pay self-employment taxes. These taxes are similar to social security taxes, and the owner also calculates and reports them on his or her individual tax return.

U.S. laws state that an owner of a sole proprietorship must assume personal responsibility for the debts incurred by the company. This requirement is referred to as **unlimited liability**. Unlimited liability may be a problem for the owner of a sole proprietorship because if the company cannot pay its debts, the company's creditors may force the owner to use his or her personal assets to pay them. So, if the sole proprietorship becomes insolvent, the owner may lose *more than* the amount of capital he or she invested in the company. Thus, unlimited liability adds additional financial risk for the owner of a sole proprietorship.

The life of a sole proprietorship is linked directly to its individual owner. Basically, a sole proprietorship ceases to exist when the owner decides to stop operating as a sole proprietor. If the owner of a sole proprietorship decides to sell the company, the owner's sole proprietorship dissolves, and the new owner or owners must choose the new company's form of business organization. Because of these characteristics, a sole proprietorship is said to have a **limited life**.

Partnerships

 Have you ever shared the purchase and use of an item with someone? Maybe you share a computer or an apartment. How do you decide how much money each contributes? How do you split the costs of software, rent, or insurance?

By definition, a sole proprietorship is owned by only one person. What if two or three people come up with a great business idea and want to start a company? What if the owner of a sole proprietorship wants someone else to invest in her company? One option is for the individuals to operate their company as a partnership. A **partnership** is a company owned by two or more individuals who each invest capital into the company.

Individuals must make many decisions before starting a partnership. These decisions include the following:

1. The dollar amount each partner will invest
2. The percentage of the partnership each individual will own
3. How to allocate and distribute partnership income to each partner
4. How business decisions will be made
5. The steps to be taken if a partner withdraws from the partnership or if a new partner is added

To limit disagreements, partners should always sign a contract, called a **partnership agreement**, before their company begins operations. This is a good idea even if partners are best friends or close relatives. This agreement specifies the terms of the formation, operation, and termination of the partnership. It defines the nature of the business, the types and number of partners, the capital contributions required of each partner, the duties of each partner, the conditions for admission or withdrawal of a partner, the method of allocating income to each partner, and the distribution of assets when the partnership is terminated.[1]

What should be included in the partnership agreement for this law firm?

Characteristics of Partnerships

 What concerns would you have about joining a partnership? Why?

Partnerships have many characteristics that are similar to those of sole proprietorships. Each partner is required by tax laws and regulations to report his or her share of the partnership's income on his or her individual income tax return. Laws and regulations regarding unlimited liability also apply to partnerships. In addition, a partnership has a **limited life**. It terminates whenever the partners change (i.e., when a partner leaves the partnership or when a new partner is added).

Of course, there is a basic difference between partnerships and sole proprietorships in that a partnership requires two or more owners. Several partnership characteristics relate to the co-ownership feature. To understand these characteristics, assume that Anna Cox invites her friend, Sanjeev Patal, to form Sweet Temptations as a partnership. If Sanjeev is like most other people, the first thing he would think is "What would I be getting myself into?" Because of a partnership's legal and business characteristics, he may be getting into more than he thinks. One important characteristic to know is that all the partners jointly own all the assets owned by a partnership; this is called **joint ownership**. Therefore, if Sanjeev contributes his property to the partnership, it no longer belongs to him alone.

Before entering a partnership, you should also know that each partner is an agent of the partnership. An **agent** is a person who has the authority to act for another. Thus a partner has the power to enter into and bind the partnership—and, therefore, all the partners—to any contract within the scope of the business. For example, either Anna or S̶a̶n̶j̶e̶e̶v̶ can bind the partnership to contracts for purchasing inventory, hiring employees, ̶r̶e̶n̶t̶i̶n̶g̶ a building, purchasing fixtures, or borrowing money. All of these activities are ̶w̶i̶t̶h̶i̶n̶ ̶t̶h̶e̶ ̶n̶ormal scope of a retail candy business.

̶I̶f̶ ̶a̶ ̶p̶a̶r̶t̶n̶e̶r̶s̶h̶i̶p̶ ̶d̶o̶e̶s̶ not have a formal agreement about how to operate, its partners resolve any disputes by referring to the ̶U̶n̶i̶f̶o̶r̶m̶ ̶P̶a̶r̶t̶n̶e̶r̶s̶h̶i̶p̶ ̶A̶c̶t̶. The Uniform Partnership Act, a set of laws adopted by most states, governs the formation, operation, ̶a̶n̶d̶ ̶t̶e̶r̶m̶i̶n̶a̶t̶i̶o̶n̶ ̶o̶f̶ ̶a̶ ̶p̶a̶r̶t̶ner̶ship in the absence of a partnership agreement.

The fact that each partner can obligate the partnership to honor contracts affects the unlimited liability requirements. **Unlimited liability** for a partnership means that each partner is liable for *all* the debts of the partnership. A creditor's claim is on the partnership, but if there are not enough assets to pay the debt, *each* partner's personal assets may be used to pay the debt. The only personal assets that are excluded are a partner's assets protected by bankruptcy laws, such as a personal residence. If one of the partners uses personal assets to pay the debts of the partnership, that partner has a right to claim a share of the payment from the other partners.

 Given the partnership characteristics we just discussed, if you were about to form a partnership, what specific items would you want to include in your partnership agreement?

Partnership Equity

Accounting for the owners' equity of a partnership differs from accounting for the owner's equity of a sole proprietorship (and a corporation). Company transactions that do not affect owners' equity are recorded in the same way regardless of the organizational form. But because a partnership's ownership is divided among the partners, its accounting system has a *Capital* account for each partner in which it records the partner's investments, withdrawals, and share of the partnership's net income.

A partnership's net income is computed in the same way as is the net income for a sole proprietorship. However, because there is more than one owner in a partnership, the net income must be allocated to each partner. Before their company begins operations, the partners need to decide how to split the partnership's net income among themselves and list this allocation in the partnership agreement. Two factors that usually affect the distribution of income among partners are (1) the dollar amount of capital contributed by each partner and (2) the dollar value of the time each partner spends working for the partnership. These factors are important because the portion of net income allocated to each partner represents the return on his or her investment of capital or time. A partnership includes a schedule at the bottom of its income statement that shows how, and how much, net income is allocated to each partner.

Corporations

Recall that Unlimited Decadence is a corporation that manufactures candy bars and sells them to companies like Sweet Temptations. Although a corporation is made up of individual owners, the law treats it as a separate "being." A **corporation** is a separate legal entity that is independent of its owners and is run by a board of directors. Hence, it has a *continuous* life beyond that of any particular owner. Therefore, it has a number of advantages. Because of the legal separation of the owners and the company, ownership in a corporation may be easily passed from one individual to another. Briefly, here's how it works. In exchange for contributing capital to the corporation, owners of a corporation receive shares of the corporation's *capital stock.* Hence, they are called **stockholders** (or *shareholders*). These shares of stock are the "ownership units" of the corporation and are *transferable.* That is, the current stockholders can transfer or sell their shares to new owners. As we discuss later, the capital stock of many corporations sells on organized stock markets such as the **New York Stock Exchange,** the **American Stock Exchange,** the **Tokyo Stock Exchange,** the **London Stock Exchange,** and the **NASDAQ Stock Market, Inc.** So stockholders of these corporations can sell their shares to new owners more easily.

http://www.nyse.com
http://www.amex.com
http://www.tse.or.jp
http://www.londonstock
   exchange.com
http://www.nasdaq.com

Because a corporation is a separate legal entity, a stockholder has no personal liability for the corporation's debts. Therefore each stockholder's liability is limited to his or her investment. Corporations tend to be larger than sole proprietorships and partnerships, so to operate, they need more capital invested by owners. Since transferring ownership is easy and since stockholders have *limited liability,* corporations can usually attract a large number of *diverse investors* and the large amounts of capital needed to operate. Corporations also can attract *top-quality managers* to operate the different departments, so stockholders are not involved in the corporations' operating decisions.

There also are several disadvantages of a corporation. As a separate legal entity, a corporation must pay federal and state income taxes on its taxable income. It reports this income on an income tax return for corporations. The maximum federal income tax rate for corporations is currently 35 percent, but since many of them also pay state income taxes, it is not unusual for the combined income taxes to be more than 40 percent of a corporation's taxable income. If some, or all, of the after-tax income of the corporation (the other 60 percent of the corporation's taxable income) is distributed to stockholders as dividends, the stockholders again may be taxed on this personal income. This is referred to as **double taxation**.

 Why do you think this is called double taxation? Is the stockholder taxed twice? Why or why not?

As we discussed earlier in this chapter, for a sole proprietorship or a partnership, the owners may have to use personal assets to pay the company's debts. However, since the owners (stockholders) of a corporation have limited liability, a corporation (particularly a smaller one) may find it more difficult to borrow money. Since the creditors can't go to the owners for payment, they may think there is more risk of not being paid.

Corporations also are subject to *more government regulation*. For instance, the federal and state governments have laws to protect creditors and owners. For example, the laws of the state in which it is incorporated usually limit the payment of dividends by a corporation. Since creditors cannot go to the owners of a corporation for payment of its debts, limiting the corporation's dividend payments is a way of protecting creditors—the corporation may have more resources with which to pay its debts. In addition, if a corporation's capital stock is traded in the stock market, the corporation must file specified reports with the Securities and Exchange Commission.

However, the advantages of a corporation usually exceed the disadvantages when a business grows to a reasonable size. Exhibit 10-1 summarizes the characteristics of each type of business organization.

EXHIBIT 10-1	GENERAL CHARACTERISTICS OF EACH FORM OF BUSINESS ORGANIZATION		
Characteristics	**Sole Proprietorships**	**Partnerships**	**Corporations**
Number of owner(s)	Single owner	Two or more owners (partners)	Usually many owners (stockholders)
Size of businesses	Small	Most are small; some professional partnerships (e.g., law firms) have several hundred partners.	Many are very large; some may have stock traded on an exchange.
Examples of businesses that typically have this legal form	Small retail shops; local service or repair shops; single practitioners such as CPAs, lawyers, doctors	Law firms; CPA firms; real estate agencies; family-owned businesses	Manufacturing companies; multinational companies; retail store chains; fast-food chains
Who makes business decisions	Owner	Depends on partnership agreement. Small partnerships will have all partners involved in business decisions; large partnerships will have managing partners. Partners are agents.	Decided by board of directors. Large corporations are managed by business professionals who often own little or no stock.
Liability of owner(s)	Unlimited	Unlimited	Limited
Life of organization	Limited	Limited	Continuous

Other Legal Forms of Business Organizations

Remember the company names we listed at the start of this section? Two companies had the letters LP or LLP in their names (Equus Capital Partners, LP and Ernst & Young LLP). The initials "LP" stand for *limited partnership,* and the initials "LLP" stand for *limited liability partnership.* Both types of partnerships were created by law to reduce the unlimited liability of partners.

Why should the government pass laws limiting a partner's liability? In the case of limited partnerships, the laws were enacted to protect partners who are not active in the management of a large partnership, so that they won't lose more money than they invested. In a limited partnership, only certain partners (usually the partners who originally set up the partnership) have unlimited liability. These partners are called *general partners.* Other investors (called *limited partners*) in a limited partnership do not have unlimited liability. Instead of having all their personal assets at risk, limited partners may lose only the amount of money they invested in the limited partnership.

In recent years, laws have been passed that allow large public accounting firms and other types of partnerships to become limited liability partnerships. The CPA profession actively lobbied the U.S. Congress and state legislatures to enact legislation allowing limited liability partnerships. These types of partnerships limit, to a prescribed amount, *each* partner's risk of losing his or her personal assets as a result of ownership in an LLP.

Another form of business organization, a Subchapter S corporation, is available to small businesses. A Subchapter S corporation maintains the attractive features of corporate ownership. The distinctive feature of these corporations is that the owners—not the corporation—pay income taxes on corporate income. Thus, owners of a Subchapter S corporation are not subject to double taxation.

In the sections that follow, we will make general statements about the characteristics of corporations. Because corporations are subject to state laws, which may differ, these general statements may not always be entirely accurate. However, the general overview will help you understand what you need to know for later discussions.

STARTING A CORPORATION

To operate as a corporation in the United States, a company must be incorporated in one of the states. **Incorporation** is the process of filing the required documents and obtaining permission from a state to operate as a corporation. The state-approved documents are called **articles of incorporation**.

After the state approves the incorporation, the individuals who filed for incorporation meet to complete several important tasks. First, they distribute among themselves the first issuance of capital stock. (As holders of the company's stock, these individuals have the right to vote on major corporate policies and decisions.) Then they (the stockholders) decide on, among other things, (1) a set of rules (bylaws) to regulate the corporation's operations, (2) a board of directors to plan and oversee the corporation's long-range objectives, and (3) a team of people to serve in top management positions.

http://www.3m.com
http://www.exxon.com

Not all corporations are huge companies like **3M** or **Exxon**. Some corporations are small companies that are owned by one person or a few people. The owners of small companies set them up as corporations because the owners want the legal benefits of incorporation, primarily limited liability. If a corporation is owned by a small number of investors, it is called a **closely held corporation**. The stock of a closely held corporation may not be purchased by the general public, and the board of directors places restrictions on the selling of stock by current stockholders. Generally, closely held corporations are

http://www.hallmark.com
http://www.mars.com

small, although some, such as **Hallmark Corporation** (the company that mak greeting cards) and **Mars Inc.** (the company that manufactures M&Ms, the Sr bar, Whiskas pet food, and Uncle Ben's rice), are very large multinational c

Publicly held corporations sell their stock to the general public, and are no restrictions on the selling of stock by current stockholders. After a c

A typical trading day on the New York Stock Exchange.

its stock to the public in what is called an **initial public offering** (or IPO), the stock begins to trade in a secondary equity market. The **secondary equity market** is where investors buy the stock of corporations from other investors rather than from the corporations. This means that the corporations have already issued this stock. In these secondary markets, the stock is traded between new and current stockholders, so the corporation is not involved in the trade.

Today, the secondary equity market is well established and plays an important role in the global economy. The **New York, American, Tokyo,** and **London Stock Exchanges,** as well as **NASDAQ,** are just a few of the national organizations in the secondary equity market, and they trade the stocks of thousands of corporations. For example, on an average day, the New York Stock Exchange (NYSE) trades over 1,360 million shares of stock.[2] By contacting a stockbroker, you can purchase stock in companies such as **Tootsie Roll Industries** or **Matsushita Electric.**

http://www.nyse.com
http://www.amex.com
http://www.tse.or.jp
http://www.londonstock
  exchange.com
http://www.nasdaq.com
http://www.tootsie.com
http://www.mei.co.jp

 Assume you plan to invest $1,000. Think of three publicly held corporations. Which would you invest in? Why? What factors would influence your decision?

THE ORGANIZATIONAL STRUCTURE OF LARGE CORPORATIONS

 Do you think the size of a company affects how it interacts with its customers? employees? suppliers? If so, why?

After the incorporation process is completed, a corporation begins operations. Recall that companies are created to achieve two specific goals: to earn a satisfactory profit by providing services or products to customers and to remain solvent. To accomplish these goals, each company must be organized into a structure that makes clear the jobs and working relationships of the employees of that company. This structure shows who is responsible for each task, whether it is an individual or a work team, and who reports to whom.

---

[2]*NYSE Quick Reference Sheet* (http://www.nyse.com, April, 2002).

In a small company, the structure may be simple. The owner makes all of the important business decisions, and all of the company's employees report directly to the owner. This is consistent with our discussion in Chapter 9 of working capital controls. However, as a company grows (e.g., hires more employees, opens additional stores, uses more suppliers), the owner will have more difficulty managing all of the company's activities. Also, the owner may not have the desire, or ability, to manage a larger organization. Eventually, the owner delegates some of these responsibilities to other managers. At this point, the small company starts to become a large company and may incorporate.

The organizational structure of a large corporation can be divided logically in a number of ways, such as by type of customer, by type of product, by geographical location, or by function. Whatever the organization, however, almost all large companies have functional areas involving the management of resources and activities related to marketing, production, human resources, finance, and distribution. All large companies also have an information technology component, but its personnel generally work in each functional area. In most companies, the **marketing** function is responsible for managing activities and resources to identify consumer needs, analyze consumer behavior, evaluate customer satisfaction, and promote the company's products. The **production** function is responsible for managing people and equipment to convert materials, components, and parts into products that the company will sell to customers. The **human resources** function is responsible for managing the company's employee-related activities, such as recruiting, hiring, training, and compensating employees, as well as providing a safe workplace. The **finance** function manages the company's capital requirements for both the short and the long term. This involves locating sources of capital and investing excess cash inside or outside of the United States. This function is also responsible for managing accounting activities. The **distribution** function is responsible for managing physical distribution systems to move products through the company and to customers. Exhibit 10-2 summarizes the management responsibilities within each function. The management of a corporation uses accounting information to identify, estimate, control, and report the costs of the marketing, production, human resources, finance, and distribution functions.

ENTERPRISE RESOURCE PLANNING SYSTEMS

In Chapter 1, we discussed an *integrated accounting system* in which accounting information about a company's activities is identified, measured, recorded, and accumulated so that it can be communicated in an accounting report. We pointed out that the managers of a company use the information in the accounting reports from the integrated accounting system to help them make decisions. Many corporations are now "reengineering" their operations by installing **enterprise resource planning (ERP) systems**, of which the integrated accounting system is one part. Unlimited Decadance uses an ERP system. The goal of an ERP system is to help all the functional areas of a company run smoothly in an integrated manner. An ERP system involves computer software that is "multi-functional." That is, the software records and stores many different types of data (e.g., units, quanti-

EXHIBIT 10-2	MANAGEMENT FUNCTIONAL AREAS			
Marketing	**Production**	**Human Resources**	**Finance**	**Distribution**
Identifying consumer needs	Converting materials into products	Recruiting, hiring, training, terminating employees	Locating cash resources	Moving products from company to customer
Analyzing consumer behavior	Overseeing purchasing	Compensating employees	Investing excess cash	Moving products within company
Evaluating customer satisfaction	Controlling manufacturing	Providing safe working conditions	Accounting	Protecting inventory
Promoting products				Processing orders

ties, times, prices, names, pay rates, and addresses, to name a few) that can be used to create a **data warehouse**. Each data entry is "coded"; that is, codes (letters and/or numbers) are assigned to similar entries. For instance, in the integrated accounting component of Unlimited Decadence's ERP system, a $1,000 cash sale and a $400 credit sale might be coded within its Sales Revenues account column as CA +1,000 and CR +400, respectively. Other information that might be coded with these sales include product numbers identifying the products sold and customer numbers identifying the customers who purchased the items. The software is "integrated" so that a manager who is in the process of making a decision can perform *data mining* to extract useful information from the data warehouse. For instance, a marketing manager might "mine" the data warehouse to identify when, what, and to whom credit sales were made, including the characteristics of the products sold (which can be mined using the product numbers) and the types of customers who purchased them (which can be mined using the customer numbers). Companies like **SAP, Oracle,** and **PeopleSoft** are developing and improving ERP systems to help the managers in a corporation's marketing, production, human resources, finance, and distribution functions use this computer technology to improve their decisions. For instance, Exhibit 10-3 shows a diagram of the Operations Management framework of an ERP system. The operating activities of a company are depicted at the top of the diagram and include purchasing, manufacturing, warehousing, promotion, and sales activities. The boxes at the bottom of the diagram illustrate how different "resources" (including information) of a corporation's operations "interact" with each other. For instance, the middle box shows the capital resources component. The corporation's finance function must manage the acquisition of capital resources to help obtain materials resources and human resources. The materials resources component (the left box) is managed by the corporation's distribution function. The distribution function, for instance, must manage the materials resources within the corporation for manufacturing its products, storing the products in the warehouse, and delivering the products to its customers. The human resources component (the right box) is managed by the corporation's human resources function. The human resources component hires employees so that, for instance, the production function can manage the purchase of materials and labor to manufacture the company's products, and the marketing function can manage the activities needed to promote the products. Together, the various functional areas interact with each other to allow the corporation to operate as efficiently as possible. We will discuss how corporations are reengineering various aspects of their operations relating to accounting in later chapters.

http://www.sap.com
http://www.oracle.com
http://www.peoplesoft.com

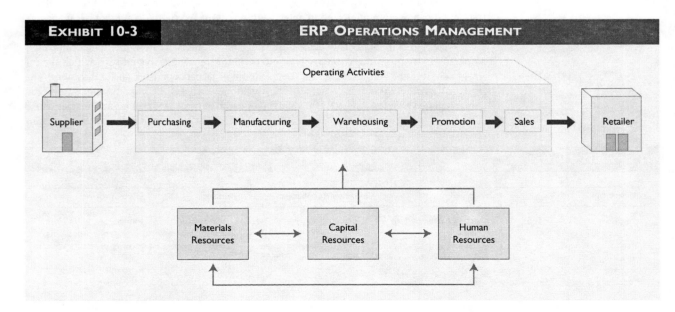

EXHIBIT 10-3 — **ERP OPERATIONS MANAGEMENT**

E-BUSINESS

Many large corporations, and some smaller ones, engage in e-business. Companies involved in **e-business** use the Internet to conduct transactions electronically. Most of these transactions involve online selling or purchasing. You are probably most familiar with *e-catalog stores,* where a customer can purchase an item from a company online through its Web site. For instance, **LandsEnd.com**, **Amazon.com**, and **Wilsonsports.com** sell clothing, books, and sporting goods, respectively, over the Internet. These and many other e-business companies also promote their products and services on the Internet, as you have noticed if you use a commercial Internet online service like **AOL.com**.

But e-business is not limited to retail companies selling online to consumers. Many e-business retailers, manufacturers, and suppliers establish "linked" online purchasing/selling relationships with each other. This allows each of the linked companies to have access to the other companies' inventory systems. So, for instance, when a retailer runs low on a particular inventory item, its system is programmed to automatically place an order with the manufacturer, which then produces sufficient quantities of the item in a timely manner for the retailer. Also, the manufacturer's system is programmed to place orders with its suppliers to provide the materials to the manufacturer for producing the inventory items.

Whether an e-business company sells only to consumers or is linked in its purchasing/selling function to other companies, the company's ERP system must be able to "interface" with its e-business system. This interface enables the company's managers to effectively and efficiently run its operations. We will discuss various e-business activities in later chapters.

http://www.landsend.com
http://www.amazon.com
http://www.wilsonsports.com

http://www.aol.com

QUALITIES OF USEFUL ACCOUNTING INFORMATION

Given the size, complexity, and wide dispersion of ownership of large corporations, summarizing and reporting useful accounting information to internal and external users is a challenging task. In later chapters, we will discuss how managers develop and use tools such as cost-volume-profit (C-V-P) analysis and master budgeting for internal decision making in large corporations. In the remainder of this chapter, we focus on financial reporting for external decision making. Two of the major advantages of the corporate form of organization—the ease of transferring ownership and the ability to attract large amounts of capital—have a significant effect on financial reporting.

As a small sole proprietorship, Sweet Temptations uses its financial reporting to provide information to Anna Cox and to the bank that makes loans to Sweet Temptations. A large corporation has many external users who also are interested in its financial information. They include individual investors, stockbrokers, and financial analysts who offer investment assistance. They also include consultants, bankers, suppliers, employees, labor unions, and local, state, and federal governments. In this section we discuss many of the qualities that accounting information should have in order to be useful to these external decision makers.

 Suppose you are a banker trying to decide whether to grant a loan to a company. When the company provides its financial information to you for your decision, what qualities would you want that information to have?

Conceptual Framework and Decision Usefulness

As an aid in establishing GAAP for financial reporting, the Financial Accounting Standards Board (FASB) has developed a **conceptual framework**. This framework is a set of concepts that provides a logical structure for financial accounting and reporting. Under this conceptual framework, the general objective of financial reporting is to provide useful information for external users in their decision making. Therefore, in the United States, *improving external decision making* is the primary purpose of financial accounting information. Other countries may have other objectives for financial accounting information, such as satisfying government requirements for computing income taxes, demonstrating compliance with the government's economic plan, or monitoring social responsibility activities.

To be useful for external decision making, financial accounting information must be relevant and reliable. Closely related to relevance and reliability are materiality and validity. We show these qualities of useful information in Exhibit 10-4. Although these concepts apply to financial reporting, they are equally applicable to management accounting information used for internal decision making.

2 What are the qualities that make accounting information about large corporations useful for decision making?

Relevant Accounting Information

Accounting information is **relevant** when it has the capacity to influence a user's decision. For example, suppose Second National Bank is considering whether to loan Unlimited Decadence money to be used for the purchase of new production equipment. The amount of cash in Unlimited Decadence's checking and savings accounts and the cash it expects to collect from sales and to pay its suppliers are relevant for the banker's decision.

 If you were buying a new car, what information about the car would be relevant to you?

Reliable Accounting Information

Even if information is relevant, users must have confidence that the information they are using for decision making is reliable. Accounting information is **reliable** when it is capable of being verified. Reliability does not always mean certainty. For example, Unlimited Decadence may be able to reliably *estimate* how many of its credit sales will not be collectible if there has been a uniform pattern of uncollectibility in the past. Source documents such as invoices, cash receipts, and canceled checks play an important part in verifying the reliability of accounting information.

 If you have a checking account, how do you verify the ending balance on your monthly bank statement? Do you consider the information reliable? Why or why not?

Materiality

Materiality is like relevance because both concepts relate to influencing a user of accounting information. Accounting information is **material** when the monetary amount is large enough to make a difference in a user's decision. Only material accounting information

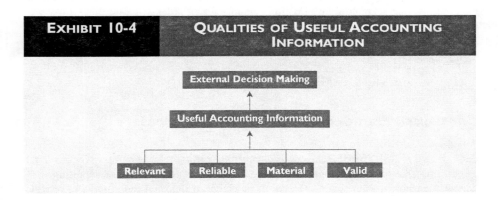

EXHIBIT 10-4   QUALITIES OF USEFUL ACCOUNTING INFORMATION

External Decision Making

Useful Accounting Information

Relevant   Reliable   Material   Valid

What dollar amount do you think would be material for Tiger Woods?

should be accumulated and communicated to users. Materiality is relative, however. A $1,000 purchase may be material to you but may be only a "drop in the bucket" to someone like Britney Spears or Tiger Woods. Similarly, what is material accounting information for Sweet Temptations may not be material for a company like Unlimited Decadence because of their different sizes.

Many users consider that an item *is not* material if its dollar amount is less than 5% and *is* material if the dollar amount is more than 10%. But the question here is, 5% or 10% of what? If an item is related to the income statement, then a user could use net income, gross profit, or some other summary amount (or even a single item) as the basis for comparison. So, for instance, if an income statement item is 12% of a company's net income, a user would generally consider the item to be material relative to net income. If another item is 3% of sales, the user would generally consider the item to be immaterial relative to sales. If an item is related to the balance sheet, then a user could use total assets, total liabilities, total owners' equity, or a summary amount such as current assets or working capital (or even a single item) as a basis for comparison. Furthermore, a user must decide whether an amount between 5% and 10% is, or is not, material. So you should understand that materiality is a concept that involves significant judgment and critical thinking.

How much would you consider as a material difference in prices in purchasing a new mountain bike? a new car?

Valid Accounting Information

Validity is closely related to reliability. Accounting information is **valid** when it shows a realistic picture of what it is meant to represent. To be valid, accounting information must realistically portray the results of a company's activities and financial position. Valid accounting information is like a good videotape or snapshot from a camera that does not distort the real picture (unlike your driver's license photo).

Keep these qualities in mind as you consider how to accumulate and use financial and management accounting information. We now turn to financial reporting for large corporations.

UNIQUE FEATURES OF A CORPORATION'S FINANCIAL STATEMENTS

3 What do users need to know about the stockholders' equity section of a corporation's balance sheet and about its income statement?

Financial statements for corporations follow the same structure that we discussed in earlier chapters and illustrated with Sweet Temptations. A corporation's balance sheet shows the dollar amounts of its assets, liabilities, and owners' equity on a specific date. Its income statement reports revenues, expenses, and net income for a specific period of time. The cash flow statement shows the corporation's cash receipts and payments from operating, investing, and financing activities for a specific period. Like Sweet Temptations, corporations prepare these financial statements according to GAAP.

However, the size and complexity of large corporations requires additional information. Some of this information relates to two aspects of a corporation's financial statements: (1) the equity section of its balance sheet, and (2) the presentation of earnings on its income statement.

Equity in a Corporation

The laws of each state apply to companies incorporated in that state. Various state laws require special accounting procedures for the owners' equity of a corporation. States passed these laws to help protect the absentee owners of a corporation (those not directly involved in the management of the company) as well as its creditors.

Corporate Capital Structure

The owners' equity of a corporation is called **stockholders' equity** because the owners of a corporation are called *stockholders*. Usually there are many stockholders of a corporation, and frequent changes in ownership can occur; as a result, maintaining separate capital accounts for each owner would be impractical. Instead, the stockholders' equity on a corporation's balance sheet is usually separated into two sections: contributed capital and retained earnings.[3] This division of stockholders' equity is required by state laws.

The **contributed capital** section shows the total investments made by stockholders in the corporation. As we will discuss in the next section, this section usually consists of two parts—capital stock and additional paid-in capital. The **retained earnings** reports the corporation's total lifetime net income that has been reinvested in the corporation and not distributed to stockholders as dividends.

Because stockholders' equity on a corporation's balance sheet consists of two sections, a corporation expands its accounting system to record information for these sections. So, under the accounting equation, it expands the stockholders' equity component to include (1) the permanent balance sheet sections of Contributed Capital and Retained Earnings and (2) the temporary revenue and expense accounts that comprise its current-year net income, as follows:

Assets	=	Liabilities	+	Stockholders' Equity	
				Permanent Sections	**Net Income**
				Contributed + Retained	Revenues − Expenses
				Capital Earnings	

In this accounting system, a corporation records transactions in account columns in the same manner as a sole proprietorship (as we discussed earlier in the book), except for contributed capital and retained earnings transactions (as we will discuss later in this chapter and the book).

Capital Stock and Legal Capital

Capital stock refers to the ownership units in the corporation. There are two types of capital stock: common stock and preferred stock. If a corporation issues only one type of stock, this is called **common stock**. If a corporation also issues another type of stock, that stock is called **preferred stock**. The differences between common stock and preferred stock involve stockholders' rights, and we will discuss these differences in Chapter 24. Here we focus on common stock.

A corporation may issue common stock for cash. It also may trade stock for an asset such as land or equipment. When an investor buys stock from the corporation, the corporation records the issuance (selling) price using both a Common Stock account and an Additional Paid-in Capital account. To protect creditors, state laws require corporations to keep in the company a minimum amount of the capital contributed by the owners. This is called *legal capital*. Usually, **legal capital** is a monetary amount per share of common stock, called the **par value**. The par value is stated in the articles of incorporation and is printed on each stock certificate. The par value of a share of common stock is often set very low—perhaps $10, $2, or even less per share—and has *no* relationship to the value of the stock. The total legal capital of a corporation is determined by multiplying the par value per share by the number of shares issued. Generally, states require companies to keep track of legal capital, so each time a corporation issues common stock, *it records the par value (legal capital) in the Common Stock account.*

Additional Paid-In Capital

The total dollar amount the corporation receives from selling its stock is called the **market value** of the stock. Corporations normally sell common stock at a market value much

---

[3]A corporation may also have a stockholders' equity section called "accumulated other comprehensive income." We will discuss this section in Chapter 23.

higher than the par value. So, legal capital is usually only a small part of the total selling price. When a corporation sells common stock, in addition to recording the par value in a Common Stock account it also records the excess of the market value over the par value. The excess value it receives is called *additional paid-in capital*. **Additional paid-in capital** is the difference between the selling price and the par value in each stock transaction and is recorded in an Additional Paid-in Capital account.

Now that you have this background, we can explain how the accounting system keeps track of one kind of common stock transaction—common stock issued for cash. We will explain other kinds of common stock transactions in Chapter 24.

Common Stock Sold for Cash

To understand the issuance of common stock, assume that Unlimited Decadence Corporation sells 30,000 shares of its $3 par value common stock for $16 per share. It records the transaction as follows:

> **FINANCIAL STATEMENT EFFECTS**
>
> Increases current assets and total assets, increases contributed capital and total stockholders' equity on *balance sheet*. Increase cash flows from financing activities on *cash flow statement*.

Assets	=	Liabilities	+	Stockholders' Equity	
				Contributed Capital	
				Common Stock	Additional Paid-in Capital
Cash					
+$480,000				+$90,000	+$390,000

As a result of this transaction, Unlimited Decadence increases assets (cash) by $480,000 (30,000 × $16), the total amount of capital invested in the corporation. Because Unlimited Decadence received the $480,000 from investors, it also increases stockholders' equity by $480,000 in the following manner. To adhere to state laws, Unlimited Decadence increases its common stock account by the $90,000 (30,000 shares × $3) par value (legal capital), and increases additional paid-in capital by the $390,000 difference between the total selling price and the total par value ($480,000 − $90,000 = $390,000).

Stockholders' Equity Section of Balance Sheet

As we said earlier, the stockholders' equity section of a corporation's balance sheet has two parts: (1) contributed capital and (2) retained earnings. Contributed capital includes common stock (the legal value) and additional paid-in capital, and shows the total capital invested in the corporation by its owners.

Exhibit 10-5 shows the stockholders' equity of Unlimited Decadence's balance sheet at December 31, 2004. As of this date, the stockholders of Unlimited Decadence have invested $8,140,000 in the corporation in exchange for 1,200,000 shares of common stock. Unlimited Decadence's common stock totals $3,600,000 ($3 par value × 1,200,000 shares), and additional paid-in capital totals $4,540,000 ($8,140,000 − $3,600,000). Because these numbers are large, Unlimited Decadence shows the amounts on its balance sheet in thousands of dollars. This is a common practice of many large corporations. Many corporations are so large that they show their balance sheet amounts in *millions* of dollars!

EXHIBIT 10-5   UNLIMITED DECADENCE CORPORATION

STOCKHOLDERS' EQUITY SECTION OF BALANCE SHEET
December 31, 2004
(in thousands of dollars)

Stockholders' Equity	
Contributed capital	
Common stock, 1,200,000 shares issued ($3 par value)	$ 3,600
Additional paid-in capital	4,540
Contributed capital	$ 8,140
Retained earnings	9,060
Total stockholders' equity	$17,200

The **retained earnings** reported in a corporation's stockholders' equity is the amount of its lifetime net income that has been reinvested in the corporation to date.[4] As we show in Exhibit 10-5, at December 31, 2004, the retained earnings for Unlimited Decadence is $9,060,000. The total stockholders' equity of Unlimited Decadence is $17,200,000, the sum of the $8,140,000 contributed capital and the $9,060,000 retained earnings.

Corporate Earnings

 When investors are evaluating a corporation, should they be concerned with how the corporation earned its net income? Do you think every dollar of income a corporation earns is equally important to investors? Why or why not?

The income statement of a corporation may contain several sections. Each section helps financial statement users better understand how a corporation earned its net income. Generally, a corporation will show separately (1) the earnings that resulted from its "continuing" operations, (2) the earnings that resulted from other, "nonrecurring" activities, and (3) the earnings per share of common stock. The nonrecurring earnings might include the income or loss from "discontinued operations" (such as **PepsiCo, Inc.**'s earnings from **Pizza Hut, Taco Bell,** and **KFC** in the year that it sold them) and any income or loss from "extraordinary" events (such as a tornado). We will discuss these nonrecurring items in Chapter 24.[5] Here, we discuss the two most common sections—income from continuing operations and earnings per share. Unlimited Decadence's income statement for the year ended December 31, 2004 (shown in Exhibit 10-6)[6] illustrates how the company reported these two sections (it did not have any nonrecurring earnings).

http://www.pepsico.com
http://www.pizzahut.com
http://www.tacobell.com
http://www.kfc.com

EXHIBIT 10-6	UNLIMITED DECADENCE CORPORATION

INCOME STATEMENT
For Year Ended December 31, 2004
(in thousands of dollars)

Sales (net)		$72,800
Cost of goods sold		(46,500)
Gross profit		$26,300
Operating expenses		
Selling expenses	$13,840	
General and administrative expenses	7,670	
Total operating expenses		(21,510)
Operating income		$ 4,790
Other items		
Interest revenue	$ 320	
Interest expense	(210)	
Nonoperating income		110
Pretax income from continuing operations		$ 4,900
Income tax expense		(1,960)
Net income		$ 2,940
Earnings per share		$ 2.45

---

[4]In its closing entries (which we discussed in Chapter 6), a corporation closes its revenue and expense accounts to its Retained Earnings account.

[5]A corporation may also have "other comprehensive income," which we will also discuss in both Chapter 23 and Chapter 24.

[6]For simplicity, we assume here that Unlimited Decadence sells only one type of candy bar. We will relax this assumption in Chapter 11.

Income from Continuing Operations

Income from continuing operations reports a corporation's revenues and expenses that resulted from its ongoing operations. For Unlimited Decadence, this section reports the revenues and expenses from the manufacture and sale of candy bars.

This section includes **operating income**, which is determined by subtracting cost of goods sold from net sales to obtain gross profit, and then by deducting the selling expenses and the general and administrative expenses. As we show in Exhibit 10-6, the operating income of Unlimited Decadence for 2004 is $4,790,000. (Remember, the numbers shown are rounded to the nearest thousand dollars.)

Income from continuing operations also includes nonoperating income, in a section called **other items**. These other (nonoperating) items include revenues and expenses that frequently occur in a corporation but do not relate specifically to its primary operating activities. Interest expense, interest revenue, and gains (or losses) are common examples of other items. The amounts of the other items are summed to determine the nonoperating income (or loss). Under other items, Unlimited Decadence reports $320,000 of interest revenue and $210,000 of interest expense, so that its nonoperating income is $110,000.

Since we have not previously discussed gains and losses, we briefly discuss them here. **Gains (losses)** are increases (decreases) in a corporation's income (and therefore its stockholders' equity) that result from transactions unrelated to providing goods and services. The rules for recording gains (losses) in a corporation's accounting system are the same as the rules for recording revenues (expenses). When a corporation has a gain or loss, it expands the net income section of its owners' equity in its accounting system as follows:

Assets	=	Liabilities	+	Stockholders' Equity		
					Net Income	
				Revenues (Gains)	**−**	**Expenses (Losses)**

For instance, suppose a corporation owns land that cost $8,000. If it sells the land for $10,000, it records a $2,000 gain ($10,000 selling price − $8,000 cost) as follows:

Assets		=	Liabilities	+	Stockholders' Equity		
					Net Income		
					Revenues (Gains)	**−**	**Expenses (Losses)**
Cash	Land				Gain on Sale of Land		
+$10,000	−$8,000				+$2,000		

The corporation would report the $2,000 gain in the other items section of its income statement. We will discuss gains and losses more later in the book.

A corporation's operating income is added to the nonoperating income to determine its pretax income from continuing operations. For Unlimited Decadence, the $4,790,000 operating income is added to the $110,000 nonoperating income to determine its $4,900,000 pretax income from continuing operations.

As we discussed earlier, corporations must pay income taxes on their earnings. **Income tax expense** is listed separately within this section. Unlimited Decadence is subject to a 40 percent income tax rate on its pretax (taxable) income, so its income tax expense is $1,960,000 ($4,900,000 × 0.40). This amount is subtracted from the pretax income from continuing operations to get the company's $2,940,000 income from continuing operations.

If Unlimited Decadence had nonrecurring earnings, it would report these items (and the related income tax expense) below income from continuing operations. Because Unlimited Decadence had no nonrecurring earnings during 2004, the income from continuing operations is called *net income.*

Earnings Per Share

A corporation's net income is earned for all the corporation's stockholders. Because the common stock for most large corporations is owned by many stockholders, it is useful to report a corporation's net income on a per-share basis. **Earnings per share (EPS)** is the amount of net income earned for each share of common stock. In its simplest form, earnings per share is computed by dividing the corporation's net income for the accounting period by the average number of shares of common stock owned by all of the corporation's stockholders during the period.

All corporations must report earnings per share on their income statements, and it is the last item they show. On its income statement in Exhibit 10-6, Unlimited Decadence shows earnings per share of $2.45. We can prove that this amount is correct by dividing the company's net income by the number of shares of common stock[7] reported in the stockholders' equity section of its balance sheet:

$$\text{Earnings Per Share} = \frac{\text{Net Income}}{\text{Average Number of Common Shares}} = \frac{\$2,940,000}{1,200,000} = \underline{\underline{\$2.45}}$$

If a corporation has several sections on its income statement or if the number of shares that stockholders own changes during the year, the calculation of earnings per share is more complicated. We will discuss this calculation in Chapter 24.

FINANCIAL INFORMATION USED IN DECISION MAKING

External users analyze the financial statements of a company to determine how well the company is achieving its two primary goals—remaining solvent and earning a satisfactory profit. In the case of a small company, owners and creditors can also discuss the company's financial performance with its managers. However, in a large corporation, this is usually not possible.

If investors and creditors in a large corporation are given few or no opportunities to ask its managers about its operations or plans, how do they get the information they need to make business decisions? Individuals investing in large corporations must rely on publicly available information when making their investment decisions. **Publicly available information** is any information released to the public; it may come directly from the corporation or from secondary sources.

Information Reported by Corporations

Much of the information that investors and creditors use to evaluate a corporation comes directly from the corporation. Corporations use four methods to supply information to external users: (1) annual reports, (2) Securities and Exchange Commission (SEC) reports, (3) interim financial statements, and (4) media releases.

4 How does a corporation provide information to external users?

Annual Reports

Corporations publish their annual financial statements as part of their **annual report**. In addition to the current year's financial statements, most corporations include the financial statements of the previous two years. These are called **comparative financial statements** and are included to help external users in their analyses. A corporation's annual report is always published in hard copy; it may also be available on the corporation's Web site.

In addition to the financial statements, a corporate annual report also includes notes to the financial statements, an audit report, financial highlights or a summary, and management's discussion and analysis of the corporation's performance. In a study commissioned

Internal & External directors

---

[7]The number of shares of common stock owned by all of Unlimited Decadence's stockholders was 1,200,000 for the entire year, so the average number is also 1,200,000.

http://www.potlatchcorp.com

http://www.coors.com

by **Potlatch Corp.**, a San Francisco-based paper manufacturer, a majority of portfolio managers and securities analysts ranked annual reports as the most important documents that a company produces, including documents on computer disks and on-line services.

Notes to the financial statements inform external users of the company's accounting policies and of important financial information that is not reported in the financial statements. For example, a note may inform users of a lawsuit against the company. **Adolph Coors Company**'s 2001 notes mentioned a lawsuit that the city of Denver has brought against it as a "potential responsible party" in regard to pollution-control issues. Although, in 2001, the lawsuit had not resulted in an additional liability for Coors, information about the lawsuit was relevant to users of its financial statements. Investors and creditors should analyze the notes to a company's financial statements before making their investment and credit decisions.

Because financial statements are so important, the Securities and Exchange Commission requires publicly held corporations to issue audited financial statements. Banks also may require a small company to provide its audited financial statements when applying for a loan. **Auditing** involves the examination of a company's accounting records and financial statements by an independent certified public accountant (CPA). This examination enables the CPA to attest to the fairness of the accounting information in the financial statements.

During an audit, an auditor must communicate with a company's managers to gain a better understanding of the company's accounting system and transactions. This relationship creates the possibility that the auditor will not maintain independence from the company's managers, which would reduce the reliability of the company's financial statements. To help facilitate auditor independence, most companies have audit committees. An **audit committee** is a part of a company's board of directors, and the committee's members usually are "outside directors" (not officers or employees of the company). The primary re-

A wide variety of users are interested in the information contained in a company's annual report.

sponsibility of the audit committee is to oversee the financial reporting process of the company and the involvement of both the company's managers and its auditor in that process. As part of this oversight, the company's audit committee generally selects the auditor for the company, and then acts as a "liaison" between the auditor and the company's managers. Although the specific duties of an audit committee vary from company to company, generally the audit committee of a company oversees the company's internal control structure, helps select the company's accounting policies, reviews the company's financial statements, and oversees the audit. A company will often mention in its annual report that it has an audit committee, which enhances the credibility of its financial statements.

A corporation's annual report includes an **audit report**, such as the one we show in Exhibit 10-7 for the 2001 financial statements of **Rocky Mountain Chocolate Factory, Inc.** Notice that the audit report consists of three paragraphs. The first paragraph states that an audit was performed for specific years and that the company's management is responsible for preparing the financial statements. In the second paragraph, the audit report briefly describes how an audit is conducted and explains that the auditor is *reasonably* sure (it would cost too much to gather enough information to be *absolutely* sure) that the financial statements do not contain material mistakes.

http://www.rmcfusa.com

 What do you think makes a mistake "material"?

The third paragraph, called the *opinion paragraph,* expresses the auditor's opinion about the fairness of the company's financial statements and states whether or not the audit firm believes that the statements were prepared according to GAAP.

EXHIBIT 10-7	ROCKY MOUNTAIN CHOCOLATE FACTORY, INC., AUDIT REPORT

Report of Independent Certified Public Accountants

**Board of Directors and Stockholders
Rocky Mountain Chocolate Factory, Inc.**

We have audited the accompanying balance sheets of Rocky Mountain Chocolate Factory, Inc. as of February 28, 2002 and 2001, and the related statements of income, stockholders' equity, and cash flows for each of the three years in the period ended February 28, 2002. These financial statements are the responsibility of the Company's management. Our responsibility is to express an opinion on these financial statements based on our audits.

We conducted our audits in accordance with auditing standards generally accepted in the United States of America. Those standards require that we plan and perform the audit to obtain reasonable assurance about whether the financial statements are free of material misstatement. An audit includes examining, on a test basis, evidence supporting the amounts and disclosures in the financial statements. An audit also includes assessing the accounting principles used and significant estimates made by management, as well as evaluating the overall financial statement presentation. We believe our audits provide a reasonable basis for our opinion.

In our opinion, the financial statements referred to above present fairly, in all material respects, the financial position of Rocky Mountain Chocolate Factory, Inc., as of February 28, 2002 and 2001, and the results of its operations and its cash flows for each of the three years in the period ended February 28, 2002, in conformity with accounting principles generally accepted in the United States of America.

GRANT THORNTON LLP

Dallas, Texas
April 12, 2002

http://www.grantthornton.com

An audit report is important to external users because it provides a measure of assurance that they can rely on the information presented in the financial statements. In Exhibit 10-7, the CPA firm **Grant Thornton LLP** is telling investors that it believes that Rocky Mountain's financial statements are fair in that they comply with all applicable accounting standards. This type of audit report is referred to as an *unqualified* or *clean* opinion. Around 90 percent of all audit reports are unqualified. If the company's audited financial statements did not comply with GAAP, the auditor would disclose this in the audit report to warn users.

Companies also include 5-, 10-, or 15-year summaries of key data from their financial statements in their annual report. These are titled **Financial Highlights** or **Financial Summaries**. Some companies merely list revenues, net income, total assets, and other key figures for recent years. Others use such items as graphs, pie charts, and ratio comparisons. As we show in Exhibit 10-8, **Bristol-Myers Squibb Company** used bar graphs in its 2001 annual report to illustrate how its net sales from continuing operations and net earnings from continuing operations have increased over time.

http://www.bms.com

Have you ever handed in a report and wanted the opportunity to explain what you think the report means? When a corporation releases its financial statements to the public, the corporation's managers want to provide their own analysis of its performance and financial condition. As part of the annual report, managers will include sections titled **Management's Discussion and Analysis** (MD&A) and **Letter to Shareholders**. In these sections, managers comment on how well (or poorly) the corporation performed over the past year, specifically in regard to its liquidity, capital, and results of operations. Usually these sections include a discussion about industry and market trends. Managers also discuss their plans for improvement. External users find this information useful because it explains how the corporation's current and planned operations are viewed "through the eyes of management."

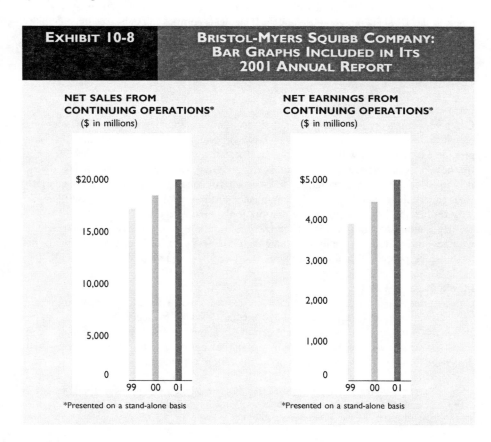

EXHIBIT 10-8   **BRISTOL-MYERS SQUIBB COMPANY: BAR GRAPHS INCLUDED IN ITS 2001 ANNUAL REPORT**

NET SALES FROM CONTINUING OPERATIONS*
($ in millions)

NET EARNINGS FROM CONTINUING OPERATIONS*
($ in millions)

*Presented on a stand-alone basis

*Presented on a stand-alone basis

In **The Campbell's Soup Company** annual report for 2001, Mr. Douglas Conant, President and CEO, used a straightforward approach to comment on the company's performance. His letter to shareholders stated: "In fiscal 2001, we took the first necessary steps aimed at restoring robust growth and capturing global opportunity. . . . Our plan is about re-energizing a great company, its brands and its people." His letter then detailed Campbell's strategy to improve its performance. Investors, creditors, and other interested parties certainly considered the president's comments as they made business decisions regarding Campbell.

http://www.campbellsoup.com

 If you were a Campbell Soup Company stockholder, what do you think your response would have been to Mr. Conant's comment? Why do you think you would have reacted that way?

Securities and Exchange Commission (SEC) Reports

In addition to distributing annual reports to the public, corporations file reports with the SEC that are also available to the public. A company offering stock for public sale must file a registration statement with the SEC and must provide potential investors with a "prospectus" containing most of the same information given the SEC. The **prospectus** typically contains the corporation's financial reports and other information, such as a description of the stock to be sold, the offering price, and how the cash received from the sale will be used.

Corporations that have their stock traded on any of the national stock exchanges, such as the New York Stock Exchange, must also file annual 10-K reports with the SEC. A **10-K report** includes the corporation's annual report and other information such as officers' names, salaries, and stock ownership. Corporations file these forms electronically with the SEC. For any company with revenues greater than $1.2 billion, the company's "chief executive" and "chief financial officer" both must "certify" that the company's annual report in the 10-K is both complete and accurate. This certification enhances the credibility of the company's financial statements. Companies' 10-K reports are located in the SEC's Electronic Data Gathering, Analysis, and Retrieval System (commonly known as *EDGAR*). When making business decisions, investors use information in these documents and other SEC filings.

http://www.sec.gov/info/edgar
.shtml

Interim Financial Statements

Companies normally use one year as their accounting period, but many companies also prepare interim financial statements. In fact, corporations registered with the SEC *must* provide external users with interim financial statements (called a *10-Q report)*. **Interim financial statements** are financial statements prepared for a period of less than one year. It is most common for corporations to issue interim financial statements on a quarterly basis (every three months). These are published at the end of each quarter, and the quarterly results are also included in the corporation's annual report.

By issuing interim financial statements, corporations provide investors and creditors with new information to use to reevaluate their business decisions (such as whether or not to sell stock purchased in an earlier period) on a more frequent basis.

Media Releases

Corporations also release important financial information in a more timely manner through press conferences and interviews with the media. Research by accounting professors has shown that corporate announcements of earnings, changes in dividend policies, major contracts with the government, and other important business activities affect investors' business decisions and result in changes in a corporation's stock price.

 Suppose Unlimited Decadence announced that last quarter's earnings were double its earnings in the same quarter the previous year. How do you think this announcement would affect Unlimited Decadence's stock price? Why?

Secondary Sources of Information about Corporations

http://www.dnb.com

External users do not depend solely on themselves to gather and evaluate information about a corporation's financial position or performance. Data regarding corporations' performances are available through many information services, both in "hard copy" and "on-line." Furthermore, information about standard industries is regularly published. These industries are identified by their North American Industry Classification System (NAICS) code assigned by the U.S. government. Firms such as **Dun & Bradstreet** annually publish statistics about the average financial performance of corporations in each NAICS code. Increasingly, computer technology provides vast amounts of data regarding market trends, industry trends, and specific corporate information.

http://www.agedwards.com
http://www.morganstanley.com

In addition, the financial services industry provides investment advice to investors and creditors. Financial services companies use corporations' publicly available information to evaluate the financial performance of these corporations. Stockbrokerage firms, such as **A. G. Edwards** and **Morgan Stanley**, closely follow market and industry trends and the performance of individual corporations in order to give advice to their clients. These firms provide services to thousands of investors by making purchases and sales of stock for them.

ANALYSIS OF FINANCIAL INFORMATION

5 How do users perform intracompany and inter-company analyses?

How do individuals and financial services firms evaluate a corporation's operating performance and financial condition? Basically, there are two approaches. One approach is to compare the corporation's current operating results and financial position with its past performance or with its expected performance. This approach is called **intracompany analysis**. One reason that external users perform intracompany analysis is to help determine *trends* in a corporation's financial performance. In other words, investors and creditors use trends to evaluate whether the corporation's performance is stable, improving, or declining.

 What trends in a corporation's performance do you think are most important? Why?

External users will probably want answers to the following questions about Unlimited Decadence: (1) Did Unlimited Decadence's sales increase over sales in the prior year? If so, what was the percentage growth? (2) Did Unlimited Decadence's net income change from that of the prior year? If so, what was the percentage growth or decline? (3) Is Unlimited Decadence's current ratio continuing to improve over the ratios of prior years? (4) Are Unlimited Decadence's sales growing at a higher rate than its operating expenses?

External users also use intracompany analysis to investigate whether a corporation is meeting *its* performance expectations. Investors and creditors learn about corporate performance expectations from annual reports, media releases, and financial analysts who study corporations' performances. With this data, they can answer important questions about Unlimited Decadence: (1) Did Unlimited Decadence meet the sales projections it disclosed in its prior year's annual report? (2) Did Unlimited Decadence's net income reach the level that financial analysts had predicted? (3) Did Unlimited Decadence make the capital expenditures it announced at earlier meetings? (4) What is the relationship between Unlimited Decadence's earnings and its stock price?

A second approach to analyzing a corporation's financial performance is known as intercompany analysis. **Intercompany analysis** involves comparing a company's operating results and financial position with that of competing companies, industry averages, or averages in related industries. For example, investors and creditors might want to compare Unlimited Decadence's operating results and financial position with those of **Hershey Corporation** or with the averages of all companies in the confectionary industry. They may perform intercompany analyses for a single period or for several past periods.

http://www.hersheys.com

Companies are of many sizes, and their sizes change from year to year. Because of these size differences, intracompany and intercompany analyses are easier to conduct and to interpret if percentage analysis is used. **Percentage analysis** involves converting financial statement information from dollars to percentages. For example, suppose that Unlimited Decadence's sales increased by $2,800,000—from $70,000,000 in 2003 to $72,800,000 in 2004. At the same time, a competing company's sales also increased by $2,800,000—from $35,000,000 to $37,800,000. Since both companies' sales increased by the same dollar amount, it might be tempting to think they performed equally well. But instead of evaluating the yearly change in each company's sales only as a dollar increase of $2,800,000, think of the change as a 4% ($2,800,000 ÷ $70,000,000) increase in Unlimited Decadence's sales and an 8% ($2,800,000 ÷ $35,000,000) increase in the competitor's sales. Thinking of the change in each company's sales as a *percentage change* leads to a different conclusion. Percentage analyses are especially useful when making intercompany comparisons.

There are three basic types of percentage analyses: ratio analysis, horizontal analysis, and vertical analysis. Recall that **ratio analysis** involves dividing an item on the financial statements by another related item (for example, net income divided by average owner's equity). We used ratio analyses in Chapters 6 through 8 to perform intracompany analysis for Sweet Temptations, and to perform intercompany analysis of JCPenney and Sears.

Horizontal analysis shows the changes in a company's operating results over time in percentages as well as in dollar amounts. Unlimited Decadence's yearly sales change that we talked about earlier is an example of horizontal analysis. **Vertical analysis** shows each item in a financial statement of a given period or date both as a percentage of another item on that statement (for example, every item on the income statement stated as a percentage of sales, or every asset stated as a percentage of total assets) and as a dollar amount. Investors and creditors use each of these types of analyses to help evaluate a corporation's performance.

6 What is percentage analysis, and what are its three types?

BUSINESS ISSUES AND VALUES

Should society expect corporations to do more than comply with laws and regulations, try to remain solvent, and earn a satisfactory profit for the owners? If so, what else should society expect? If not, what impact can this have on society?

Large corporations have the same basic goals as small companies—to remain solvent and to earn a satisfactory profit for the owners. Our society has placed limits on what companies, both large and small, can do to meet these goals. These limits may be formal laws or regulations, such as those controlling misleading advertising, product safety, and management fraud. Other limits are not laws or regulations but are norms about what society expects from the business community. For example, society expects companies to treat employees, investors, and creditors fairly, support the local economy, and "give back" some of their profits by supporting local and national charities.

There is a continuing debate over how large corporations should balance these sometimes conflicting expectations. How much severance pay should corporations offer employees who are terminated? How much profit should corporations give back to the local community?

For large corporations, these questions are difficult, in part because managers usually do not own the corporation. For example, the question of supporting charities is difficult to answer because the managers of a corporation are spending the stockholders' (not their own) resources on charities that it decides to support. How would you like someone else deciding where your donations to charity should go? On the other hand, through a corporation's identification with the charity, this support may help it increase customer loyalty and improve overall financial performance. Throughout the book we will discuss how accounting plays a role in shaping these types of corporate decisions.

SUMMARY

At the beginning of the chapter we asked you several questions. During the chapter, we asked you to STOP and answer some other questions to build your knowledge about specific issues. Be sure you answered these additional questions. Below are the questions from the beginning of the chapter, with a brief summary of the key points relating to the answers. Use your creative and critical thinking skills to expand on these key points to develop more complete answers to the questions and to determine what other questions you have that might lead you to learn more about the issues.

1 What are the three most common forms of business organizations and their basic characteristics?

The three most common forms of business organizations are sole proprietorships, partnerships, and corporations. A sole proprietorship is usually small, has a limited life, and is owned by one person who has unlimited liability. A partnership is owned by two or more people (called *partners*) who have unlimited liability; it can be large or small, is usually governed by a partnership agreement, and has a limited life. A corporation is a legal entity incorporated in one of the states. It is usually large and usually has many owners (called *stockholders*) who have limited liability. A corporation may have many managers, is subject to income taxes and other government regulations, and has an unlimited life.

2 What are the qualities that make accounting information about large corporations useful for decision making?

To be useful for decision making, accounting information must be relevant, reliable, material, and valid. Relevant information has the capacity to affect a user's decision. Reliable information is capable of being verified. Information is material when the dollar amount is large enough to make a difference in a decision. Information is valid when it shows a realistic picture of what it is meant to represent.

3 What do users need to know about the stockholders' equity section of a corporation's balance sheet and about its income statement?

The stockholders' equity section of a corporation's balance sheet is divided into two sections: contributed capital and retained earnings. Contributed capital includes the par value (legal capital) of the common stock as well as additional paid-in capital. Retained earnings reports the total lifetime net income that has been reinvested into the corporation and not distributed to stockholders as dividends. A corporation's income statement differs from the income statement of other types of companies because the former includes income from both continuing operations and nonrecurring earnings. Since corporations are subject to income taxes, a corporation's income statement also includes income tax expense. Finally, a corporation reports earnings per share on its income statement.

4 How does a corporation provide information to external users?

A corporation provides information to external users through its annual report, Securities and Exchange Commission (SEC) reports, interim financial statements, and media releases. A corporation's annual report includes its financial statements, related notes, audit report, financial highlights, management's discussion and analysis, and other information. SEC reports include a registration statement and an annual 10-K report. A corporation usually issues interim financial statements quarterly. Media releases provide information related to important business activities.

5 How do users perform intracompany and intercompany analyses?

Users perform intracompany analysis by comparing a corporation's current operations and financial position with its past results and expected results. They perform intercompany analysis by comparing a corporation's performance with the performance of competing companies, and with industry averages or averages in related industries.

6 **What is percentage analysis, and what are its three types?**

Percentage analysis involves converting financial statement information from dollars to percentages. There are three types of percentage analysis: ratio analysis, horizontal analysis, and vertical analysis. Ratio analysis involves dividing a financial statement item by another related item. Horizonal analysis shows the changes in a company's operating results over time as percentages. Vertical analysis shows the items on a financial statement of a given period or date as percentages of another item on that statement.

KEY TERMS

additional paid-in capital *(p. 316)*
agent *(p. 305)*
annual report *(p. 319)*
articles of incorporation *(p. 308)*
auditing *(p. 320)*
audit committee *(p. 320)*
audit report *(p. 321)*
capital stock *(p. 315)*
closely held corporation *(p. 308)*
common stock *(p. 315)*
comparative financial statements
   (p. 319)
conceptual framework *(p. 312)*
contributed capital *(p. 315)*
corporation *(p. 306)*
data warehouse *(p. 311)*
distribution *(p. 310)*
double taxation *(p. 307)*
e-business *(p. 312)*
earnings per share (EPS) *(p. 319)*
enterprise resource planning (ERP)
   systems *(p. 310)*
finance *(p. 310)*
Financial Highlights *(p. 322)*
Financial Summaries *(p. 322)*
gains *(p. 318)*
horizontal analysis *(p. 325)*
human resources *(p. 310)*
income from continuing operations
   (p. 318)
income tax expense *(p. 318)*
incorporation *(p. 308)*
initial public offering *(p. 309)*
intercompany analysis *(p. 324)*
intracompany analysis *(p. 324)*

interim financial statements *(p. 323)*
joint ownership *(p. 305)*
legal capital *(p. 315)*
Letter to Shareholders *(p. 322)*
limited life *(pp. 304, 305)*
losses *(p. 318)*
Management's Discussion and Analysis
   (p. 322)
marketing *(p. 310)*
market value *(p. 315)*
material *(p. 313)*
notes to the financial statements *(p. 320)*
operating income *(p. 318)*
other items *(p. 318)*
par value *(p. 315)*
partnership *(p. 305)*
partnership agreement *(p. 305)*
percentage analysis *(p. 325)*
preferred stock *(p. 315)*
production *(p. 310)*
prospectus *(p. 323)*
publicly available information *(p. 319)*
publicly held corporation *(p. 308)*
ratio analysis *(p. 325)*
relevant *(p. 313)*
reliable *(p. 313)*
retained earnings *(pp. 315, 317)*
secondary equity market *(p. 309)*
sole proprietorship *(p. 304)*
stockholders *(p. 306)*
stockholders' equity *(p. 315)*
10-K report *(p. 323)*
unlimited liability *(p. 304)*
valid *(p. 314)*
vertical analysis *(p. 325)*

SUMMARY SURFING

Here is an opportunity to gather information on the Internet about real-world issues related to the topics in this chapter. Go to http://www.cunningham.swlearning.com and click on the Interactive Study Center. Click on this chapter number then click on Summary Surfing and answer the following questions.

- Click on **TI** (Texas Instruments). Find the company's balance sheets. What types of stock does TI have? What is the par value per share of common stock, and how many shares were issued at the end of the most current year? What were the amounts of its retained earnings and its total stockholders' equity on December 31 of the most current year?

- Click on **Intel (Intel Corporation)**. Find the company's income statements. What was Intel's net income for the most current year, and how does this compare with its income in the previous year? What was Intel's earnings per share for the most current year, and how does this compare with its earnings per share in the previous year? Find the Management's discussion and analysis of financial condition and results of operations. By what percent did Intel's net revenues increase or decrease from the previous year to the most current year, and what were the reasons for this change?

INTEGRATED BUSINESS AND ACCOUNTING SITUATIONS

Answer the Following Questions in Your Own Words.

Testing Your Knowledge

10-1   What is a sole proprietorship? What is meant by the terms *unlimited liability* and *limited life* as they apply to a sole proprietorship?

10-2   What is a partnership? What is meant by the terms *limited life, joint ownership, agent,* and *unlimited liability* as they apply to a partnership?

10-3   What is included in a partnership agreement?

10-4   What is a corporation? What is meant by the terms *capital stock, limited liability,* and *double taxation* as they apply to a corporation?

10-5   What is the secondary equity market? Name three organizations that are part of the secondary equity market.

10-6   Name the typical functional areas of a large corporation. How is accounting information used to manage these functions?

10-7   What is an enterprise resource planning (ERP) system and how does it relate to a data warehouse and data mining?

10-8   What is the FASB's conceptual framework, and what is the primary purpose of financial accounting information?

10-9   What is relevant accounting information, and how does it relate to materiality?

10-10   What is reliable accounting information, and how does it relate to validity?

10-11   What is the owners' equity of a corporation called, and into what two sections is it separated on a corporation's balance sheet?

10-12   What is included in contributed capital and retained earnings on a corporation's balance sheet?

10-13   What amounts are included in a corporation's common stock and additional paid-in capital accounts?

10-14   What is reported in the income from continuing operations section of a corporation's income statement?

10-15   What is the last item shown on a corporation's income statement, and how is it computed?

10-16   What is included in a corporation's annual report?

10-17   What is included in the three paragraphs of an audit report? Why is an audit report important to external users?

10-18 What is included in the management's discussion and analysis (MD&A) section of a corporation's annual report?

10-19 What is the difference between intracompany analysis and intercompany analysis?

10-20 What is the difference between horizontal analysis and vertical analysis?

Applying Your Knowledge

10-21 Companies have different forms of organization.

Required: Discuss the differences among sole proprietorships, partnerships, and corporations regarding ownership, decision making, income taxes, the responsibility of owners for the company's debts, and the life of the company.

10-22 Suppose you were starting a new business with a friend. Your friend and you have agreed that he will invest most of the capital and that you will do most of the work. You will invest some cash and a two-year-old truck, which will be used as a delivery vehicle in the business.

Required: Briefly discuss what information you would include in the partnership agreement for the new company. Be as specific as possible.

10-23 At the end of 2004, before allocating net income, the Simon and Art partnership had total owners' equity of $100,000, consisting of Simon, capital: $60,000, and Art, capital: $40,000. During 2004 the partnership earned net income of $35,000. The partnership agreement specifies that net income is to be allocated according to three factors as follows: (a) first, each partner is to be allocated a share of net income equal to 10% of her capital amount, (b) second, Simon is to be allocated a salary of $8,000, and Art is to be allocated a salary of $12,000 as a share of net income, and (c) the remaining net income is to be allocated 60% to Simon and 40% to Art.

Required: (1) Prepare a schedule that allocates the net income to Simon and Art according to the partnership agreement. (*Hint:* The salaries paid to the partners are used only to allocate the net income; they are not included as salaries expense on the income statement.)
(2) Explain why you think factors (a) and (b) for allocating net income were included in the partnership agreement.

10-24 During 2004, the Fame and Fortune partnership had sales revenue of $200,000, cost of goods sold of $120,000, and operating expenses of $25,000. At the end of 2004, before allocating net income, the A. Fame, Capital account had a balance of $140,000, and the B. Fortune, Capital account had a balance of $70,000. In reviewing the partnership agreement, you find that annual net income is to be allocated to each partner based on three factors. (a) First, A. Fame is to be allocated a salary of $5,000, and B. Fortune is to be allocated a salary of $20,000 as a share of net income. (b) Second, each partner is to be allocated a share of net income equal to 10% of his capital account balance. (c) Third, the remaining net income is to be allocated 2/3 to A. Fame and 1/3 to B. Fortune.

Required: (1) Prepare a 2004 income statement for the Fame and Fortune partnership. At the bottom of the income statement, include a schedule that allocates the net income to each partner based on the factors in the partnership agreement. (*Hint:* The salaries paid to the partners are used only to allocate the net income; they are not included as salaries expense on the income statement.)
(2) Explain why you think factors (a) and (b) for allocating net income were included in the partnership agreement.

10-25 One way to logically divide the organization structure of a large corporation is by function.

Required: Identify the typical functional areas of a corporation, and briefly discuss what activities are performed in each area. Briefly explain how managers would use accounting information in the operations of these functional areas.

10-26 A friend of yours has recently completed a course in bookkeeping at his high school. He has been browsing through this chapter of your book and noticed the term "conceptual framework." He says, "We never had a conceptual framework in our bookkeeping class. What is this framework anyhow? Please tell me about the qualities of useful accounting information, and define each one."

Required: Prepare a written response to your friend's question.

10-27 Ryland Carpet Corporation sells 10,000 shares of its common stock for $10 per share.

Required: (1) Using account columns, show how Ryland Carpet Corporation would record this transaction under each of the following independent assumptions:
(a) The stock has a par value of $2 per share.
(b) The stock has a par value of $5 per share.
(c) The stock has a par value of $7 per share.
(2) If you were a stockholder of Ryland Carpet Corporation, which par value would you prefer? Why?

10-28 Tiger Corporation previously had issued 10,000 shares of its $2 par value common stock for $15 per share. On December 28, 2004, it sells another 5,000 shares to investors for $20 per share.

Required: (1) Using account columns, (a) enter the balances in the applicable accounts for the common stock that had previously been issued and (b) record the sale of the 5,000 shares of common stock on December 28, 2004.
(2) Prepare the contributed capital section of Tiger Corporation's December 31, 2004 balance sheet.

10-29 On December 29, 2004, Lion Corporation sells 4,000 shares of its common stock with a $5 par value to investors for $25 per share. This is the only sale of common stock during 2004. Before 2004, the corporation had issued 12,000 shares of this common stock for $21 per share. At the end of 2004, the corporation had retained earnings of $124,000.

Required: (1) Using account columns, (a) enter the balances in the Common Stock and Additional Paid-in Capital accounts at the beginning of 2004 and (b) record the sale of the 4,000 shares of common stock on December 29, 2004.
(2) Prepare the stockholders' equity section of Lion Corporation's December 31, 2004 balance sheet.

10-30 During all of 2004, stockholders of the Planet Pluto Corporation owned 15,000 shares of its $3 par value common stock. They had purchased this stock from the corporation for $29 per share. At the end of 2004, the Planet Pluto Corporation had an ending balance of $247,000 in its retained earnings account.

Required: Prepare the stockholders' equity section of the Planet Pluto Corporation's December 31, 2004 balance sheet.

10-31 The stockholders of Riglets Corporation owned 10,000 shares of its $5 par value common stock during all of this year. The corporation is subject to a 40% income tax rate. The corporation's balance sheet information at the end of this year and its income statement information for this year are as follows:

Common stock, $5 par value	$ (a)
Gross profit	113,000
Pretax income from continuing operations	(b)
Operating expenses	33,000
Total contributed capital	228,000
Income tax expense	(c)
Retained earnings	181,000
Net income	48,000
Additional paid-in capital	(d)
Earnings per share	(e)
Cost of goods sold	(f)
Sales (net)	240,000
Total stockholders' equity	(g)

Required: Fill in the blanks lettered (a) through (g). All the necessary information is listed. (Hint: It is not necessary to calculate your answers in alphabetical order.)

10-32 Braiden Corporation is subject to a 40% income tax rate. During all of this year, stockholders owned 20,000 shares of its $2 par value common stock. The corporation's income statement information for this year and its balance sheet information at the end of this year are as follows:

Sales (net)	$311,000
Net income	(a)
Total stockholders' equity	500,000
Operating expenses	(b)
Additional paid-in capital	198,000
Income tax expense	(c)
Common stock, $2 par value	(d)
Pretax income from continuing operations	(e)
Earnings per share	3.15
Total contributed capital	(f)
Gross profit	147,000
Retained earnings	(g)
Cost of goods sold	164,000

Required: Fill in the blanks lettered (a) through (g). All the necessary information is listed. (Hint: It is not necessary to calculate your answers in alphabetical order.)

10-33 Ringland Glass Corporation showed the following balances in its income statement accounts at the end of 2004:

Interest expense	$ 1,400
Sales (net)	585,000
Selling expenses	54,200
Interest revenue	3,200
Cost of goods sold	350,700
General expenses	31,900

The corporation pays income taxes at a rate of 40%. It had average stockholders' equity during 2004 of $500,000, and stockholders owned 30,000 shares of its common stock during all of 2004.

Required: (1) Prepare a 2004 income statement for Ringland Glass Corporation.
(2) Compute the return on owners' (stockholders') equity of Ringland Glass Corporation for 2004. How does this compare with the industry average of 14.8%?

10-34 The stockholders of Buffalo Chips Corporation owned 40,000 shares of its common stock during all of 2004. The corporation pays income taxes at a rate of 40% and had the following balances in its income statement accounts at the end of 2004:

Administrative expenses	$ 67,400
Cost of goods sold	302,000
Interest expense	3,500
Interest revenue	1,700
Sales (net)	563,800
Selling expenses	40,600

Required: (1) Prepare a 2004 income statement for Buffalo Chips Corporation.
(2) Compute the profit margin of Buffalo Chips Corporation for 2004. How does this compare with the industry average of 17.4%?

10-35 On July 1, 2004, Kelly Corporation sold for $15,000 an acre of land that it was not using in its operations. Kelly had purchased the land four years ago for $9,000. In other transactions during 2004, Kelly incurred $1,500 interest expense.

Required: (1) Using account columns, record the sale of the land by Kelly.
(2) Prepare the Other Items section of Kelly's 2004 income statement.

10-36 Stockholders of the Tomar Export Corporation owned 5,000 shares of common stock during all of this year. The corporation listed the following items in its financial statements on December 31 of this year:

Net income	$ 12,000
Current assets	15,000
Average stockholders' equity	70,000
Cost of goods sold	72,000
Total liabilities	26,000
Net sales	100,000
Current liabilities	6,000
Average inventory	9,000
Total assets	100,000

Required: Using the preceding information, compute the following ratios of the Tomar Export Corporation for this year: (1) earnings per share, (2) gross profit percentage, (3) profit margin, (4) return on owners' (stockholders') equity, (5) current ratio, (6) inventory turnover, and (7) debt ratio.

10-37 Taboue Cutlery Corporation showed the following income statement information for the years 2004 and 2005:

TABOUE CUTLERY CORPORATION
Comparative Income Statements
For Years Ended December 31

			Year-to-Year Increase (Decrease)	
	2004	2005	Amount	Percent
Sales (net)	$60,000	$65,000	$ (a)	(b) %
Cost of goods sold	(33,600)	(c)	(d)	(e)
Gross profit	$26,400	$27,950	$ (f)	(g)
Operating expenses	(h)	(19,050)	400	(i)
Pretax operating income	$ (j)	$ 8,900	$1,150	(k)
Income tax expense	(3,100)	(3,560)	(l)	(m)
Net Income	$ 4,650	$ (n)	$ (o)	(p)
Number of common shares issued	(q)	2,700	(r)	12.6
Earnings per share	$ 1.94	$ 1.98	$ (s)	(t)

Required: (1) Determine the appropriate percentages and amounts for the blanks lettered (a) through (t). Round to the nearest tenth of a percent.
(2) Did you just do horizontal or vertical analysis? Briefly comment on what your analysis reveals.

10-38 The Clovland Corporation presents the following comparative income statements for 2004 and 2005:

CLOVLAND CORPORATION
Comparative Income Statements
For Years Ended December 31

	2004	2005
Sales (net)	$90,000	$108,000
Cost of goods sold	(45,000)	(60,000)
Gross profit	$45,000	$ 48,000
Operating expenses	(20,000)	(22,000)
Pretax operating income	$25,000	$ 26,000
Income tax expense	(10,000)	(10,400)
Net Income	$15,000	$ 15,600
Number of common shares	6,800	7,000
Earnings per share	$ 2.21	$ 2.23

Required: (1) Based on the preceding information, prepare a horizontal analysis for the years 2004 and 2005. (*Hint:* To the right of the income statements, add an *Amount* column and a *Percent* column as in 10-37.)

(2) Calculate the corporation's profit margin for each year. What is this ratio generally used for, and what does it indicate for the Clovland Corporation?

10-39 The Anton Electronics Corporation presents the following income statement for 2004:

ANTON ELECTRONICS CORPORATION
Income Statement
For Year Ended December 31, 2004

Sales (net)	$140,000
Cost of goods sold	(81,340)
Gross profit	$ 58,660
Operating expenses	(28,560)
Pretax operating income	$ 30,100
Income tax expense	(12,100)
Net Income	$ 18,000
Earnings per share	$ 3.20

In addition, the average inventory for 2004 was $10,000.

Required: (1) Based on the preceding information, prepare a vertical analysis of the income statement for 2004. (*Hint:* To the right of the income statement, add a *Percent* column, and assign 100% to net sales.)

(2) Compute the corporation's inventory turnover for 2004, and briefly explain what this ratio tells you about a company.

Making Evaluations

10-40 The Toys "R" Us 2001 annual report contains the following information in the section called "Report of Management": ". . . Management has established a system of internal controls to provide reasonable assurance that assets are maintained and accounted for in accordance with its policies and that transactions are recorded accurately on the company's books and records. . . . The company has distributed to key employees its policies for conducting business affairs in a lawful and ethical manner. . . . The financial statements of the company have been audited by Ernst & Young LLP, independent auditors, in accordance with auditing standards generally accepted in the United States, including a review of financial reporting matters and internal controls to the extent necessary to express an opinion on the . . . financial statements."

http://www.toysrus.com

Required: Suppose you were using these financial statements to make a decision about investing in Toys "R" Us. Would this information help you make your decision? If so, how would it help? If not, why do you think Toys "R" Us includes the information with its financial statements?

10-41 You have a close friend from high school who is attending college in another state. You see each other at holidays, and in between visits you correspond on a regular basis by e-mail. Here is the latest e-mail you received from your friend:

From: HT246@UA.EDU <friend>

To: Ace@AAU.EDU <you>

Subject: HELP!!!

Hey, how's it going? Sorry I don't have time to chat. I need some help. I'm taking a personal finance course. In this course, I estimate how much my annual income will be after I graduate, establish a personal monthly budget, and decide how to invest my monthly savings. As part of this, I must select a real-life corporation to invest in.

I estimated my annual income to be $75,000 (hey, you know me—I think positive), so I have a lot of savings to invest. Here's where you come in. I am having trouble getting started with the part of the assignment where I pick a real-life corporation to invest in. Luckily, dear friend, I remembered that you are taking this accounting course.

Remember how I helped you out when you were taking Poli Sci? Now it's return-the-favor time. Tell me what you know about (1) the kinds of information I need to decide whether a corporation is a good investment, (2) where I can get the information, and (3) how to analyze the information once I get it.

By the way, the assignment is due day after tomorrow, so don't send back 20 pages of techno-jargon I can't even understand. Just send what you think is most important, and tell me in words I can understand. Thanks a bunch.

Required: Using the material presented in this chapter, write an e-mail message back to your friend. Make sure you respond specifically to the three items he needs help with.

http://www.dell.com

10-42 Below are condensed income statements of **Dell Computer Corporation** for fiscal years 2001 and 2000.

	Fiscal Year Ended	
(in millions, except earnings per share)	February 2, 2001	January 28, 2000
Net revenue	$31,888	$25,265
Cost of revenue	25,445	20,047
Gross margin	$ 6,443	$ 5,218
Operating expenses:		
Selling, general and administrative	3,193	2,387
Research, development and engineering	482	374
Special charges	105	194
Total operating expenses	$ 3,780	$ 2,955
Operating income	$ 2,663	$ 2,263
Investment and other income, net	531	188
Income before income taxes	$ 3,194	$ 2,451
Provision for income taxes (and other)	1,017	785
Net Income	$ 2,177	$ 1,666
Earnings per common share:	$ 0.84	$ 0.66

Required: (1) Based on the preceding information, prepare a horizontal analysis for the fiscal years 2001 and 2000. (*Hint:* To the right of the income statements, add an *Amount* column and a *Percent* column as in 10-37.)

(2) For each year, compute (a) the gross profit percentage and (b) the profit margin.

(3) Briefly explain what your results from (1) and (2) tell you about Dell's change in its profitability, and why this change took place.

http://www.dell.com

10-43 Below are condensed comparative balance sheets of **Dell Computer Corporation** for the fiscal years ended February 2, 2001 and January 28, 2000.

(in millions)	February 2, 2001	January 28, 2000
Assets		
Current assets:		
Cash and cash equivalents	$ 4,910	$ 3,809
Short-term investments	528	323
Accounts receivable, net	2,895	2,608
Inventories	400	391
Other	758	550
Total current assets	$ 9,491	$ 7,681
Property, plant, and equipment, net	996	765
Investments	2,418	2,721
Other noncurrent assets	530	304
Total assets	$13,435	$11,471

Liabilities and Stockholders' Equity

Current liabilities:

Accounts payable	$ 4,286	$ 3,538
Accured and other	2,257	1,654
Total current liabilities	$ 6,543	$ 5,192
Long-term debt	509	508
Other	761	463
Commitments and contingent liabilities (Note 7)	—	—
Total liabilities	$ 7,813	$ 6,163
Total stockholders' equity	5,622	5,308
Total liabilities and stockholders' equity	$13,435	$11,471

Required: (1) Based on the preceding information, prepare a vertical analysis of each balance sheet. (*Hint:* To the right of each column, add a *Percent* column and assign 100% to total assets, as well as to total liabilities and stockholders' equity.)

(2) For each year, compute the (a) current ratio and (b) debt ratio.

(3) Briefly explain what your results from (1) and (2) tell you about Dell's liquidity and financial flexibility.

10-44 Refer to the income statements and balance sheets for the **Dell Computer Corporation** in 10-42 and 10-43. In addition to this information, you determine that at the end of fiscal year 1999, Dell had accounts receivable (net) of $2,094 million, inventories of $273 million, total assets of $6,877 million, and total stockholders' equity of $2,321 million.

http://www.dell.com

Required: (1) For each fiscal year (2001 and 2000), compute the (a) accounts receivable turnover (for simplicity, assume all Dell's net revenue is from credit sales), (b) inventory turnover, (c) return on owners' (stockholders') equity, and (d) return on total assets (assume that Dell has interest expense of $47 million in fiscal year 2001 and $34 million in fiscal year 2000).

(2) Briefly explain what your results from (1) tell you about the change in (a) Dell's operating capability and (b) Dell's profitability.

(3) Explain whether you are using intercompany analysis or intracompany analysis.

10-45 In this chapter, we described how investors and creditors obtain information about publicly held corporations, and we began our coverage of the stockholders' equity section of a corporation's balance sheet and a corporation's income statement. Read the following item that appeared on May 16 of a recent year, on the first page of the Money section of *USA Today*:

> TOY SALES: The world's largest toy retailer, hurt by weak video game sales, said [yesterday] that its first-quarter net income slipped 51%. **Toys "R" Us** net income fell to $18.4 million, or 7 cents a share, from $37.58 million, or 13 cents a share, a year earlier. Revenue was up slightly, at $1.49 billion, compared with $1.46 billion in the year-ago quarter. Revenue for stores open at least a year fell 10%. Toys "R" Us stock dropped $1\frac{1}{8}$ to $26 in heavy trading. The company has 618 toy stores in the USA, 300 international toy stores and 206 Kids "R" Us children's clothing stores.

http://www.toysrus.com

Required: Respond to the following questions:

(1) In what month does the annual accounting period for Toys "R" Us end? How did you figure that out? Why would Toys "R" Us end its accounting period then?

(2) How many shares of stock do you think are owned by Toys "R" Us investors at the end of the first quarter of the year? How did you calculate that number?

(3) What was the market value of Toys "R" Us stock on the morning of May 15? What about on the morning of May 16? Explain why you think this change occurred.

(4) Assume you had purchased 500 shares of Toys "R" Us stock on May 15 of the previous year for $24 a share. How would you evaluate your investment's performance? As an investor, what specific information in the newspaper item would concern you the most? Why?

10-46 Almost everyone has eaten at **McDonald's**. Big Macs and Quarter Pounders are part of many people's weekly diet. At last count, McDonald's serves over 46 million people

http://www.mcdonalds.com

per day. McDonald's Corporation is known not only for its fast food but also for its corporate social responsibility efforts. We have listed several social responsibility activities that McDonald's Corporation decided to undertake:

(a) McDonald's decided to sponsor Ronald McDonald Houses, where families with hospitalized children can stay near the hospital for free.

(b) McDonald's decided to become a major contributor to Jerry Lewis's fund-raising campaign to fight muscular dystrophy.

(c) McDonald's decided to employ physically and/or mentally challenged adults who find it more difficult to get jobs.

(d) Although McDonald's believed that styrofoam containers helped to maintain the quality of its food, it decided to reduce its use of styrofoam containers in response to news reports about the dangers of fluorocarbons and in response to pressures applied by nonprofit environmental groups.

In each of these cases, McDonald's senior executives probably consulted managers from some or all of the functional areas (marketing, production, human resources, finance, distribution) before making their decision.

Required: Answer the following questions. Think through your answers carefully so that you can defend the conclusions you reach.

(1) For each of the four activities listed above, list the questions you would ask to determine whether McDonald's decision to undertake this activity helps or hinders the corporation's ability to meet its basic objectives of remaining solvent and earning a satisfactory profit.

(2) For each of the four activities, analyze what functional areas will be affected and how they will be affected. Then, assuming you are the finance manager, use your analysis to write McDonald's senior executives a memo explaining the impact of activities (a) and (d) on your functional area.

(3) Now, think back through your answers to (1) and (2) with one thing in mind— information from other functional areas. More specifically, brainstorm about information that the other functional areas could provide that would improve your ability to answer questions (1) and (2). Use the knowledge you have gained about costs, profits, budgeting, and reporting to help formulate your answer.

10-47 In this chapter, we said that individuals investing in large corporations must rely on publicly available information when making their investment decisions. But some individuals may have more than just publicly available information. Consider the following situation.

Suppose you and your mom were playing golf with two of her business "cronies." While waiting to tee off at the eighth hole, one of them, the chief financial officer of her company, mentioned ("off the record") that her company had just finished the development of a new product that would "blow the socks off" the health care industry. Although she was not at liberty to discuss the product (it won't be released until early next year), she speculated that once the product was released, the company stock price would "skyrocket." She also mentioned that she intended to purchase a large number of shares of the company's stock (which is being traded on the New York Stock Exchange) later in the day, and to sell them after the product was released.

WHOA!!! This could be an opportunity for you to invest the $2,000 your wealthy Aunt Bertha gave you for your birthday. Also, your boss, coworkers, and friends might like to know about this opportunity.

Required: (1) What are the relevant facts for your decision?

(2) Are there any ethical issues involved in your decision?

(3) Who has a stake in your decision (who stands to gain and who stands to lose), and how might your decision affect these stakeholders?

(4) What do you think you should do? Why?

(5) Do you see any ethical issues in the chief financial officer's discussion and decision?

(6) Who are the stakeholders in her discussion and decision?

(7) Suppose that when you got home after the golf game, your mom asked you what you thought of the conversation at the eighth hole. How would you answer her?

(8) What do you think should happen to "level the playing field"?

10-48 Yesterday, you received the following letter for your advice column at the local paper:

DR. DECISIVE

Dear Dr. Decisive:

I am having a crisis with my professional life right now and could use an objective opinion. Please help! I have been working for a partnership for ten years (let's just call it "Seedman and Seedman") and have built up a strong relationship with the company's customers. I am currently earning a salary of $42,000. But last week I received from one of Seedman and Seedman's competitors a job offer (unsolicited) that included a salary of $50,000. The offer is tempting, but I like my current job. So, I met with the Seedmans to see if they could increase my salary to match the offer. Well, I wasn't prepared for their response. They offered me a share of the partnership!! But here's where it gets complicated. In order to join the partnership, I would contribute $25,000 cash but would be paid no salary until the end of the year. Then, after the net income of the partnership was determined for each year, it would be allocated as follows. First, I would be allocated a salary of $38,000, Sadie Seedman would be allocated a salary of $42,000, and Johnny Seedman would be allocated a salary of $38,000. The remaining net income (after we withdrew our salaries) would be assigned to our individual capital accounts. Sadie's share would be 50%, Johnny's 30%, and mine 20%. (If I become a partner, we could call ourselves Seedman, Seedman, and Seedling—I'd be the seedling.) If I take the other job, I plan to invest $25,000 in the stock of a high-tech company in my city.

What do you see as the advantages and disadvantages of each of my alternatives?

"Seedling"

Required: Get together with your Dr. Decisive consulting team and write a response to "Seedling."

CHAPTER 11

DEVELOPING A BUSINESS PLAN FOR A MANUFACTURING COMPANY: COST-VOLUME-PROFIT PLANNING AND ANALYSIS

"WE SHOULD ALL BE CONCERNED ABOUT THE FUTURE BECAUSE WE WILL HAVE TO SPEND THE REST OF OUR LIVES THERE."

—CHARLES KETTERING

1. How does the fact that a manufacturing company makes the products it sells affect its business plan?

2. How does a manufacturing company determine the cost of the goods that it manufactures?

3. Why are standard costs useful in controlling a company's operations?

4. How do manufacturing costs affect cost-volume-profit analysis?

5. What is the effect of multiple products on cost-volume-profit analysis?

6. How does a manufacturing company use cost-volume-profit analysis for its planning?

ave you ever read the ingredients in a candy bar and then wondered how
these ingredients were combined to form the candy bar? It is the manufacturing
process that precisely mixes these ingredients and performs the other activities
necessary to form and package a tasty candy bar. This process involves a combination
of materials (ingredients), employees who work in the factory, and manufacturing activities. To succeed, a manufacturing company must plan all these aspects of its manufacturing processes.

Earlier in this book, we discussed many aspects of planning that apply to entrepreneurial service and retail companies. Now we will expand our discussion to include
additional aspects of planning that apply to larger companies and specifically to manufacturing companies. Recall from our early discussion in Chapter 1 that the fact that manufacturing companies *make* the products they sell to their customers distinguishes them
from service and retail companies. As you will see, although the planning processes of
all types of companies are basically the same, this characteristic of a manufacturing company influences its planning processes and its business plan.

A MANUFACTURING COMPANY'S BUSINESS PLAN

Like the entrepreneurs we discussed in Chapter 3, a manufacturing company's managers
use its business plan to help them plan their company's activities, visualize the results of
implementing their plans, carry out the company's plans, and later evaluate how well the
company performed. A company may decide to share its business plan with potential investors and creditors. If so, they can use the company's business plan to help them learn
as much as they can about the company, its plans, and how it operates so that they can
make decisions about whether to invest in or lend money to the company.

A manufacturing company's business plan has much in common with the business
plans of retail and service companies and contains the same components that we discussed
in Chapter 3. However, since a manufacturing company makes the products that it sells,
its business plan contains an additional component useful to internal and external decision makers—the production plan. The production plan has considerable impact on the
other sections of the business plan, particularly the financial plan, which we will discuss
later in this chapter.

1 How does the fact that a
manufacturing company
makes the products it sells
affect its business plan?

© LISA KRANTZ/SYRACUSE NEWSPAPER/THE IMAGE WORKS

How does the production and
sale of these candies fit into
this corporation's business plan?

The Production Plan

The production plan included in a manufacturing company's business plan describes how the company plans to efficiently produce its goods while maintaining a desired level of product quality. It also describes the company's plans for achieving specific levels of productivity through the use of materials, labor, equipment, and facilities. For example, **Techknits, Inc.**, a sweater manufacturer in New York, purchased computerized looms to improve its production. So Techknits would have described in its production plan how it planned to complete orders for sweaters faster and more efficiently than it did previously. For example, the company planned to use these looms 24 hours a day. Whereas previously one person was needed per manual loom, after the purchase of the computerized loom, one person would be able to run four computerized looms.[1] Techknits also would have explained that with this process, it would be able to turn out 60,000 sweaters a week. A company's production plan also describes the raw materials that make up the company's products, the company's production processes, and the finished products.

The Raw Materials

Raw materials are the materials, ingredients, and parts that make up a company's products. They also include materials that the company needs for production but that do not become a part of the products, such as production supplies and grease for lubricating machine parts. For example, some of the raw materials that Unlimited Decadence uses in manufacturing one of its candy bars include cocoa nibs, cocoa liqueur, sugar and other sweeteners, cocoa butter, milk products, emulsifiers, and paper (for packaging the candy bars). When Unlimited Decadence lists its raw materials in this section, it specifies any criteria that these materials must meet (such as standards or grades of cocoa). It also indicates which of them are perishable and how quickly they perish.

A company also lists its raw materials suppliers in this section. This list includes such information as the company's major suppliers and alternative suppliers, as well as a comparison of these suppliers' characteristics such as quality, delivery time and method, dependability, cost, and payment methods and schedules.

Another aspect of raw materials that a company describes in this section is the way it handles the raw materials once it receives them from its suppliers. This description includes the company's delivery inspection procedures, warehousing, and security for the raw materials. Since Unlimited Decadence's candy bar ingredients are perishable, Unlimited Decadence follows special procedures to ensure that fresh, quality raw materials go into its candy bars. In this section of its production plan, Unlimited Decadence describes how it inspects incoming raw materials for grade (percent defective) and for freedom from bugs and other contamination. Unlimited Decadence also describes unique aspects of its raw materials storage, including regulated temperature, humidity, and exposure to air.

 Picture the movement of ingredients from the warehouse of the supplier to that of Unlimited Decadence. If you were a manager at Unlimited Decadence, at what point or points in this movement would you want the ingredients to be inspected for grade, freedom from bugs, and other contamination? What is your rationale?

The Production Processes

This section typically includes a description of the employees, facilities, and equipment necessary for the manufacture of a company's products. Here, the company describes the sequence of production steps necessary to manufacture its products and how the raw ma-

---

[1]John S. DeMott, "Small Factories' Big Lessons," *Nation's Business,* April 1995, 29, 30.

terials flow through this sequence. For example, Exhibit 11-1 lists the production steps necessary to manufacture Unlimited Decadence's Darkly Decadent candy bar. A company also describes the employees needed for each of the steps, including what their skills must be, the availability of employees with these skills, how much time they will be spending on each product, and how much this time will cost the company.

Fairchild of California, a sofa manufacturer near Los Angeles, has employed 100 highly skilled immigrant workers to produce sofas at wages ranging from $9 to $13 per hour (depending on how many sofas each employee worked on in an hour).[2] What information about these employees do you think Fairchild would have included in this section of its business plan?

http://www.fairchildfurniture .com

A company also lists in this section all the equipment and facilities it uses or plans to use in the production process, as well as the costs associated with the use of the equipment and facilities. These associated costs include mortgage, rent, utilities, maintenance, and insurance payments. The diagram in Exhibit 11-2 illustrates the equipment that Unlimited Decadence uses to manufacture the Darkly Decadent candy bar and other candy bars. Notice that it uses an electronic control panel to regulate and automate the production process. For example, the control panel dictates how much of each ingredient is fed into the mixer from the hoppers. Since the proportion of ingredients is different for each type of candy bar, Unlimited Decadence programs different formulas into the control panel; the formulas determine the mix of ingredients for each type of candy bar. The control panel also monitors such production variables as temperature and candy density, and it adjusts the production process when these variables deviate from the acceptable range of values.

Spangler Candy Co. in Bryan, Ohio, a manufacturer of candy canes, would describe in its production processes section its partially automated factory, which doubles Spangler's output by automatically wrapping the canes in a thin plastic film, packing them in boxes of twelve (in tiny cradles to keep them from breaking), and bundling the boxes into cases. Before automation, Spangler hand-packaged the candy canes. It also would describe the costs associated with this automation, including the cost of the 215,000-square-foot warehouse where it stores the millions and millions of candy canes from February, when it begins candy cane production, until they are sold during the holiday season.[3] Can you think of any other costs that might be associated with this automated factory?

http://www.spanglercandy.com

Another aspect of production that a company addresses in this section is the regulations with which it must comply and the permits and licenses that it must maintain. For example, manufacturers in California, like those in other states, must follow national rules

EXHIBIT 11-1	PRODUCTION STEPS NECESSARY TO MANUFACTURE THE DARKLY DECADENT CANDY BAR

(1) Preparing ingredients—pulverizing cocoa nibs, grinding sugar
(2) Mixing ingredients—producing chocolate paste of a rough texture and plastic consistency
(3) Refining chocolate paste—smoothing texture of paste
(4) Conching chocolate paste—dispersing sugar and milk solids in liquid fat
(5) Tempering chocolate paste—stabilizing chocolate, causing good color and texture
(6) Molding candy bars—shaping candy bars
(7) Wrapping candy bars—packaging candy bars

[2]Ibid.
[3]Ibid.

EXHIBIT 11-2   **EQUIPMENT NECESSARY TO MANUFACTURE THE DARKLY DECADENT CANDY BAR**

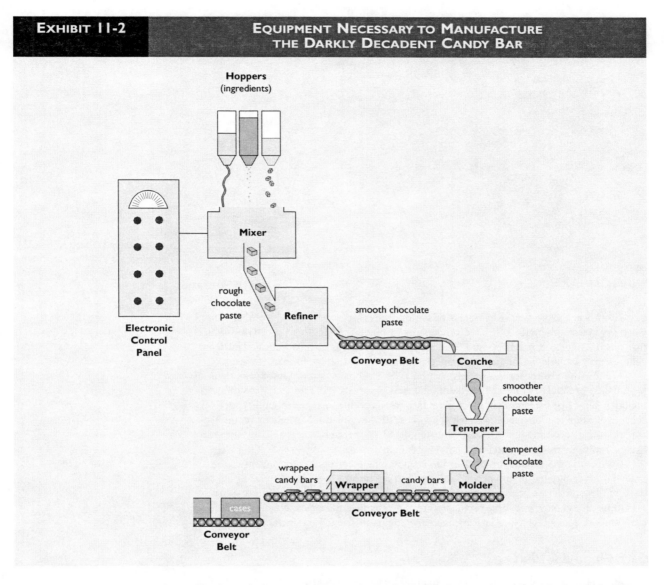

for clean air. In most states, regulators dictate how companies must meet these rules. However, in California, regulators allow companies to determine how they will comply with these rules. These companies include their clean air plans in this section of their business plans.[4] Recall from our discussion of regulations in Chapter 1 that these regulations, permits, and licenses also include such items as operating permits, certification and inspection licenses, state and local building codes, and recycling systems.

The Finished Products
Virtually all manufacturing companies have some form of product inspection that occurs during the production process to ensure a quality finished product. In this section of the production plan, a company describes where in the production process this quality inspection occurs, the inspection criteria that it uses, and how it handles defective products. Since Unlimited Decadence manufactures a food product, it also discusses sanitation and

---

[4]Laura M. Litvan, "A Breath of Fresh Air," *Nation's Business,* March 1995, 53.

As a potential investor or creditor, would you have an interest in this company's clean air plan?

pest control in this section. Regardless of the rigorous inspection procedures that a company may have, it should still have procedures in place for handling defective products that reach its customers and should describe these procedures in this section. These procedures may include warranties, guarantees, and repair and replacement policies.

A company also describes in this section how it protects, stores, and keeps track of its finished products before selling them, and how it transports them to its customers (it includes names of carriers and alternative carriers, as well as carrier reliability, delivery schedules, shipping fees, and payment schedules). If the company's products require special shipping accommodations, such as a regulated temperature, the company describes these conditions in this section of the business plan.

By including a production plan in its business plan, a manufacturing company gives potential investors and creditors vital information about how well it can execute its production plans, meet its sales orders, stay in business, pay back its loans, and provide a return to its investors. These same production plan details give managers a plan of action and a benchmark against which to later measure the company's actual production performance.

The Financial Plan

As we mentioned earlier, the production plan influences a company's financial plan. This influence not only results from the unique production function of a manufacturing company, but is also a direct reflection of how accounting in a manufacturing company differs from that in a retail or service company. As we will discuss later in this chapter and in the next chapter, these differences add some new dimensions to both the cost-volume-profit (C-V-P) analysis and the budgeting of a manufacturing company. Before we look at these dimensions, however, we will first explore some of the ways that accounting differs between a manufacturing company and its retail and service counterparts.

MANUFACTURING COSTS

A major difference between the accounting of a manufacturing company and that of a retail company is the way in which a manufacturing company accounts for inventories and cost of goods sold. As you know, a retail company has one type of inventory—goods available (ready) for sale. When it sells these goods, the retail company moves the cost of these goods from its inventory account into its cost of goods sold account. Since a

2 How does a manufacturing company determine the cost of the goods that it manufactures?

manufacturing company *makes* the goods that it sells, it has *three* types of inventories: (1) the raw materials it uses either directly or indirectly in manufacturing its products, called **raw materials inventory**, (2) the products that it has started manufacturing but that are not yet complete, called **goods-in-process inventory** (also called *work-in-process inventory*), and (3) finished products that are ready to be sold, called **finished goods inventory**. Raw materials inventory and goods-in-process inventory are unique to manufacturing companies. However, a manufacturing company's finished goods inventory is the equivalent of the inventory of a retail company. Both of these inventories contain goods that are ready to be sold.

Since a manufacturing company makes the products it sells rather than purchasing them in a form ready for sale, determining the cost of the three inventories is more complex than is determining the cost of the one inventory of a retail company. For example, the cost of a retail company's inventory is usually the sum of the inventory's invoice price and shipping costs. On the other hand, the cost of a manufacturing company's finished goods inventory is the sum of all of the costs of manufacturing that inventory.

In a simple manufacturing process, three elements, or production inputs, contribute to the cost of manufacturing a product: the cost of direct materials, the cost of direct labor, and the cost of factory overhead. The sum of the costs of these three elements eventually becomes the cost of the manufactured products. Then, as each product is sold, its cost (composed of the costs of direct materials, direct labor, and factory overhead) becomes part of the total cost of goods sold. Exhibit 11-3 shows the relationships among these cost elements, the manufacturing process, the three inventories, and the cost of goods sold for Unlimited Decadence. We will discuss these elements and relationships next.

Direct Materials

Direct materials are the raw materials that physically become part of a manufactured product. In other words, direct materials are the raw materials and parts from which the product is made. Think again about that list of ingredients on the wrapper of a candy bar. These ingredients are the direct materials from which that candy bar was made. Now consider the Girl Scout cookies made at the Little Brownie Bakery in Louisville, Kentucky.[5] The direct materials the company uses each week to make the cookies include:

- 21 truckloads of flour (875,000 pounds)
- 3.5 truckloads of sugar (650,000 pounds)
- 7 truckloads of shortening (300,000 pounds)
- 115,000 pounds of peanut butter
- 45,000 pounds of cocoa
- 72,000 pounds of toasted coconut
- 475,000 pounds of chocolate coating

Direct materials include materials the company acquires from natural sources, such as the honey that Unlimited Decadence includes in some of its candy bars. Direct materials also include processed or manufactured products that the company purchases from other companies, such as the milk products that Unlimited Decadence purchases from Mo-o-oving Milk Products and the corn syrups that it purchases from Corn Syrups Are Us. The direct materials of many manufacturing companies also include parts or components that they purchase from other companies. For example, the microchips inside your computer were probably manufactured by a different company from the one that manufactured your computer. Since direct materials become a part of the finished product, a company includes their costs in the cost of the finished product.

---

[5]Melinda Hemmelgarn, "Trip to factory reveals Girl Scout cookie facts." *Columbia Daily Tribune,* January 19, 2000, 2C.

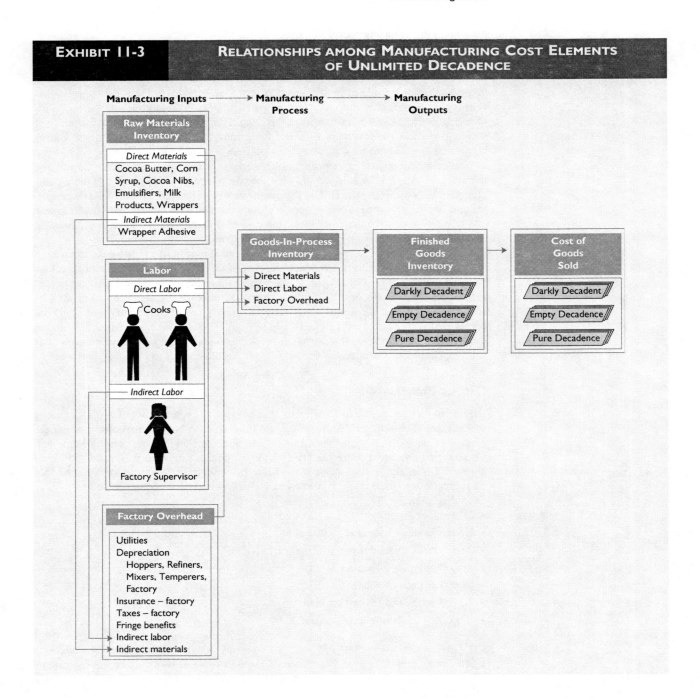

EXHIBIT 11-3 — **RELATIONSHIPS AMONG MANUFACTURING COST ELEMENTS OF UNLIMITED DECADENCE**

Direct Labor

Unless a factory is fully automated, factory employees help convert or assemble the direct materials into a finished product. **Direct labor** is the labor of the employees who work with the direct materials to convert or assemble them into the finished product. For example, when Unlimited Decadence manufactures some of its candy bars, the labor of all the employees who operate the equipment that prepares the candy bar ingredients, mixes the ingredients, refines and tempers the chocolate paste, forms the candy bars, and packages the candy bars is direct labor. The Little Brownie Bakery employs 900 people

Suppose the conveyor belt is set at an unrealistically high speed. As a result, the employees in charge of packaging can't keep up and try to "hide" the unpackaged candy. Do you think this candy manufacturing process has any "hidden" costs?

to convert the direct materials into Girl Scout cookies.[6] The cost of the direct labor is the wages earned by these employees. Since direct labor is the labor necessary to convert or assemble direct materials into a finished product, manufacturing companies include the cost of direct labor in the cost of the finished product. Since labor in some countries is cheaper than that in the United States, some U.S. manufacturing companies ship direct materials to those countries to be assembled. By doing this, the companies save on direct labor costs.

 Do you think the cost of shipping the direct materials to other countries and back should be included in the cost of the finished product? Why or why not?

Factory Overhead

Factory overhead includes all items, other than direct materials and direct labor, that are necessary for the manufacture of the product. Factory overhead is often called *manufacturing overhead,* or simply *overhead.* Although factory overhead items are necessary for the manufacture of products, they usually cannot be traced directly to individual products. For example, Unlimited Decadence's factory overhead includes repair and maintenance of its factory equipment. It also includes depreciation of this equipment, utilities used in the manufacturing process, insurance and property taxes on the factory and factory equipment, depreciation of the factory, and other factory costs. Unlimited Decadence's factory overhead also includes raw materials and labor that are not traceable to individual products. Because they are not traceable to products, these raw materials and labor are called **indirect materials** and **indirect labor**.

For example, although adhesive is used for each candy bar wrapper, managers think of the adhesive as an indirect material and include the cost of the adhesive in factory overhead costs. They choose this treatment because the amount and cost of adhesive per candy bar is so small that tracing it to individual candy bars, or even to cases of candy bars, is difficult. Notice in Exhibit 11-3 that before Unlimited Decadence uses indirect materials in production, it includes them with the direct materials in the raw materials inventory. Indirect labor includes the salaries of employees like custodians and maintenance workers, as well as supervisors. Managers consider the factory supervisor's salary to be indirect labor and include this salary in factory overhead costs because her job activities are too broad to be able to assign portions of her salary to individual products. Since factory overhead is necessary for the manufacture of the product, manufacturing companies include the costs of factory overhead in the cost of the finished product.

---

[6]Ibid.

The Little Brownie Bakery has quality control employees who make sure that the Girl Scout cookies meet size, weight, percent coating, moisture, and baking specifications. The bakery also has five metal detectors through which the cookies must pass.[7] Would the quality control employees and the metal detectors be a part of the bakery's overhead? Why, or why not?

Notice that all of the overhead costs relate to what goes on in the factory. Factory overhead does not include selling costs, general and administrative costs, or other costs that we discussed earlier in the book. Although we discussed salaries, utilities, and depreciation expenses in previous chapters, these expenses did not occur in the factory and, therefore, did not relate to the manufacturing process. Since both manufacturing companies and retail companies have selling and administrative activities, the two types of companies treat these items in exactly the same way.

A manufacturing company includes manufacturing costs (direct materials, direct labor, and factory overhead) in the cost of its manufactured products and, therefore, in the cost of its inventories. Then, as the company sells its products, it transfers these costs from finished goods inventory to cost of goods sold. A retail company shows the cost of its products on hand (the invoice price and transportation costs) on its balance sheet as inventory and then as it sells these products, it transfers these costs from inventory to cost of goods sold. Exhibit 11-4 shows selling expenses, and general and administrative

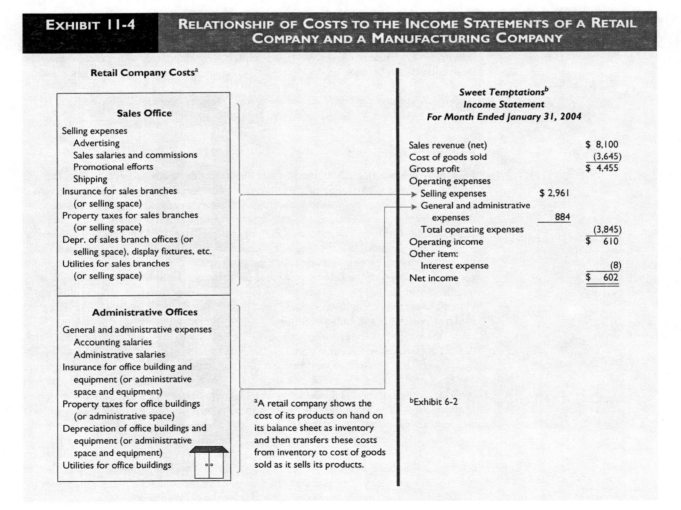

EXHIBIT 11-4   RELATIONSHIP OF COSTS TO THE INCOME STATEMENTS OF A RETAIL COMPANY AND A MANUFACTURING COMPANY

Retail Company Costs[a]

Sales Office

Selling expenses
  Advertising
  Sales salaries and commissions
  Promotional efforts
  Shipping
Insurance for sales branches
  (or selling space)
Property taxes for sales branches
  (or selling space)
Depr. of sales branch offices (or
  selling space), display fixtures, etc.
Utilities for sales branches
  (or selling space)

Administrative Offices

General and administrative expenses
  Accounting salaries
  Administrative salaries
Insurance for office building and
  equipment (or administrative
  space and equipment)
Property taxes for office buildings
  (or administrative space)
Depreciation of office buildings and
  equipment (or administrative
  space and equipment)
Utilities for office buildings

[a]A retail company shows the cost of its products on hand on its balance sheet as inventory and then transfers these costs from inventory to cost of goods sold as it sells its products.

Sweet Temptations[b]
Income Statement
For Month Ended January 31, 2004

Sales revenue (net)		$ 8,100
Cost of goods sold		(3,645)
Gross profit		$ 4,455
Operating expenses		
Selling expenses	$ 2,961	
General and administrative expenses	884	
Total operating expenses		(3,845)
Operating income		$ 610
Other item:		
Interest expense		(8)
Net income		$ 602

[b]Exhibit 6-2

expenses for a retail company, and how a retail company (Sweet Temptations) shows these expenses on its income statement. It also shows selling expenses, and general and administrative expenses, as well as the manufacturing costs for a manufacturing company, and how a manufacturing company (Unlimited Decadence) shows these items on its income statement. However, note that cost of goods sold includes only the cost of inventory sold, and not the total manufacturing cost for inventory produced.

 Since products have three types of inputs—direct materials, direct labor, and factory overhead—why do you think candy bar labels list only the candy bars' ingredients?

STANDARD COSTS

A knowledge of how many direct materials, how many hours of direct labor, and approximately how much overhead should go into the manufacture of each product gives managers information with which to plan the materials, labor, and overhead necessary for expected levels of production. It also gives managers a benchmark against which to measure the actual usage of each of these production inputs in the manufacture of the product. Similarly, a knowledge of what the *costs* of these production inputs should be gives managers information with which to plan production costs and cash flows, and also gives them

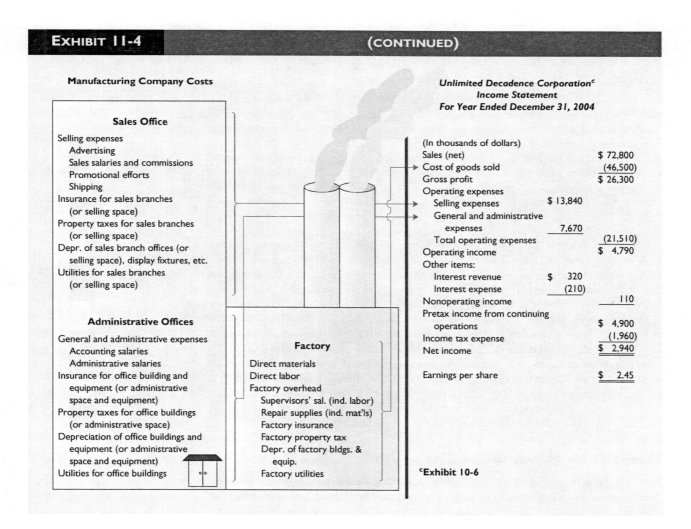

EXHIBIT 11-4 (CONTINUED)

Manufacturing Company Costs

Sales Office

Selling expenses
  Advertising
  Sales salaries and commissions
  Promotional efforts
  Shipping
Insurance for sales branches
  (or selling space)
Property taxes for sales branches
  (or selling space)
Depr. of sales branch offices (or
  selling space), display fixtures, etc.
Utilities for sales branches
  (or selling space)

Administrative Offices

General and administrative expenses
  Accounting salaries
  Administrative salaries
Insurance for office building and
  equipment (or administrative
  space and equipment)
Property taxes for office buildings
  (or administrative space)
Depreciation of office buildings and
  equipment (or administrative
  space and equipment)
Utilities for office buildings

Factory

Direct materials
Direct labor
Factory overhead
  Supervisors' sal. (ind. labor)
  Repair supplies (ind. mat'ls)
  Factory insurance
  Factory property tax
  Depr. of factory bldgs. &
    equip.
  Factory utilities

Unlimited Decadence Corporation[c]
Income Statement
For Year Ended December 31, 2004

(In thousands of dollars)

Sales (net)		$ 72,800
Cost of goods sold		(46,500)
Gross profit		$ 26,300
Operating expenses		
Selling expenses	$ 13,840	
General and administrative		
expenses	7,670	
Total operating expenses		(21,510)
Operating income		$ 4,790
Other items:		
Interest revenue	$ 320	
Interest expense	(210)	
Nonoperating income		110
Pretax income from continuing		
operations		$ 4,900
Income tax expense		(1,960)
Net income		$ 2,940
Earnings per share		$ 2.45

[c]Exhibit 10-6

a benchmark against which to measure the actual costs of manufacturing each product. The accounting system supplies managers with standard costs and actual costs of materials, labor, and overhead to aid them in planning and controlling the operations of the company.

What Are Standard Costs?

Standard costs are the costs that *should* be incurred in performing an activity or producing a product under a given set of planned operating conditions. Management accountants, engineers, and others involved in a manufacturing activity establish the standard costs, or predetermined costs, of that activity based on a careful study of the manufacturing process. Many factors can influence these costs, so in setting standards, the group must assume that a certain set of conditions exists. For example, Unlimited Decadence bases the standard direct labor costs in its factory on the assumption that the company has direct materials of the proper specification (for example, dark chocolate versus milk chocolate) and quality entering the process as needed, equipment adjusted properly for the particular candy bar being manufactured, cooks who have the proper training and experience and who earn the normal wage rate, and so forth. The standard costs are the costs that managers expect the company to incur when these conditions exist. Together, these costs become the standard cost for each manufactured product.

 If direct materials do not meet the company's specifications or quality requirements, how might using these materials affect direct labor costs? Why would maladjusted manufacturing equipment affect direct labor costs?

Uses of Standard Costs and Variances

One reason that standards are useful in planning is that they aid in the development of budgets, as you will see in the next chapter. Later in this chapter you will see how the standard costs for direct materials, direct labor, and factory overhead become part of a manufacturing company's variable costs.

3 Why are standard costs useful in controlling a company's operations?

    Standards also are a valuable source of information for decision making. If they reflect current operating conditions, standard costs provide a more reliable basis for estimating costs than do actual past costs, which may reflect abnormal conditions or past inefficiencies. It is also normally less time-consuming and costly to develop cost estimates from standard costs than to perform an analysis of actual past costs each time a decision is required.

    The most valuable use of standard costs, however, is in controlling company operations. Standard costs provide the benchmark against which managers compare actual costs to help them evaluate an activity. As we said earlier, the standard cost is the amount of cost that *should* be incurred if the planned conditions under which an activity is to be performed actually exist when the activity is performed. If the actual cost incurred differs from the standard cost, one or more of the planned conditions must not have existed. This difference between a standard cost and an actual cost is called a **variance**. Reporting a variance provides a *signal* that an operating problem (such as a machine being out of adjustment) is occurring and may require managers' attention. If actual costs do not differ from standard costs, managers assume that no operating problems are occurring and that no special attention is needed. In other words, timely *feedback* of variance information helps managers implement the *management by exception* principle that we discussed in Chapter 4. We will talk more about analyzing variances in Chapter 17.

Standards for Manufacturing Costs

You might be wondering how a company determines its standard costs. Managers establish standard costs for each manufactured product (or *output* of the manufacturing process). Remember, a company uses three inputs (direct materials, direct labor, and factory overhead) to manufacture a product, and the costs of these inputs become costs of its manufactured products. Therefore, to establish the standard cost of each product *output* of the manufacturing process, managers must first determine two standards for each *input* to the manufacturing process: a quantity standard and a price standard.

Quantity Standard

A **quantity standard** is the *amount* of an input that the company should use to produce a unit of product in its manufacturing process. Examples of quantity standards for direct materials and direct labor are the 10 pounds of cocoa beans and the 30 minutes of direct labor that Unlimited Decadence expects to use to produce one case of Darkly Decadent candy bars.

Price Standard

A **price standard** is the *cost* that the company should incur to acquire one unit of input for its manufacturing process. Examples of price standards are the expected cost per pound of the cocoa beans that Unlimited Decadence uses to produce a case of Darkly Decadent candy bars, and the expected cost per hour for the wages of the cooks who produce a case of Darkly Decadent candy bars. We will discuss quantity standards and price standards more in Chapter 17.

COST-VOLUME-PROFIT ANALYSIS

4 How do manufacturing costs affect cost-volume-profit analysis?

One of the most common forms of analysis in which managers use cost estimates based on standards is cost-volume-profit (C-V-P) analysis, or break-even analysis. Recall from our discussion in Chapter 3 about C-V-P analysis in retail companies that this is a tool that managers use to evaluate how changes in sales volume, selling prices of products, variable costs per unit, and total fixed costs affect a company's profit. Managers in manufacturing companies use the same type of analysis.

Behaviors of Manufacturing Costs

As in C-V-P analysis for a retail company, the first step in C-V-P analysis for a manufacturing company is to identify the behaviors of the company's costs. Remember, a manufacturing company has three additional costs over those of a retail company: the manufacturing costs of direct materials, direct labor, and factory overhead. Like the other costs, these can ultimately be classified as variable or fixed costs.

Variable Manufacturing Costs

We just discussed how managers develop standard quantities and costs for the inputs into the production process. Since the standard quantity of direct materials and direct labor are for one unit of product, the total quantity of direct materials and direct labor that should be used in manufacturing the company's products will increase as the level of production increases. For example, assume that the standard amount of cocoa beans used to produce a case (unit) of Darkly Decadent chocolate bars is 10 pounds. If Unlimited Decadence plans to produce 25 cases of Darkly Decadent candy bars, it expects to use 250 pounds of cocoa beans (10 pounds of cocoa beans per case × 25 cases). If it plans to produce 30 cases, it expects to use 300 pounds of cocoa beans (10 pounds of cocoa beans per case × 30 cases).

Since companies develop standard costs for each unit of input, the total cost of production input increases as the quantity of direct materials and direct labor increases. If Unlimited Decadence's standard cost per pound of cocoa beans is $0.45, the cost of the cocoa beans it expects to use in producing 25 cases of Darkly Decadent candy bars is $112.50 ($0.45 per pound × 10 pounds per case × 25 cases). If it plans to produce 30 cases of the candy bar, the cost of cocoa beans it expects to use is $135 ($0.45 per pound × 10 pounds per case × 30 cases).

A **variable manufacturing cost** is constant for each unit produced but varies in total in direct proportion to the volume produced. Since the costs of direct material and direct labor are constant for each unit produced and since the total costs of direct materials and direct labor increase as total production increases, we classify these costs as variable manufacturing costs.

We also classify some (but not all) factory overhead costs as variable costs. For example, as we mentioned earlier, Unlimited Decadence classifies the adhesive used to close the candy bar wrappers as factory overhead rather than direct material because tracing this adhesive to any particular product is difficult. Since the amount of adhesive used increases as the production of cases of candy bars increases, the total cost of the adhesive increases as the production of cases of candy bars increases. We classify the adhesive's cost, then, as a variable manufacturing cost.

Fixed Manufacturing Costs
We classify all factory overhead costs that are not affected in total by changes in the volume of production within a specific period as fixed costs. At Unlimited Decadence, for example, the factory supervisor's salary, the depreciation on factory machines, and the property tax on the factory are all fixed costs.

Mixed Manufacturing Costs
Mixed costs (sometimes called *semivariable costs*) are costs that behave as would the sum of a fixed cost and a variable cost. That is, mixed costs have a fixed cost component and a variable cost component.

For example, suppose that the local power company charges Unlimited Decadence a constant amount, say $0.10, for each kilowatt hour (kwh) of electricity it uses. If the amount of electricity that Unlimited Decadence uses for factory lighting remains constant each year regardless of the volume of production (say at 420,000 kwh per month), the power cost for this use is fixed at $42,000 per month (420,000 kwh $\times$ $0.10 per kwh). The amount of electricity required by the equipment used in production, however, is directly proportional to the number of cases of candy bars produced (the volume). Normally, the power cost for this use varies in proportion to the number of cases of candy bars produced. Suppose, for the sake of illustration, that we know it takes 0.5 kwh of electricity for each case of candy bars produced. The total power cost is a mixed cost because it equals the sum of a fixed component of $42,000 (from factory lighting) and a variable component (from equipment use) that increases at a rate of $0.05 per case of candy bars produced ($0.10 per kwh $\times$ 0.5 kwh per case of candy bars).

The general cost equation for the total amount of a mixed cost is as follows:

$$\text{Total mixed cost} = F + vX$$

where:
    F = The fixed component
    v = The rate at which the variable component increases per unit of volume
    X = The volume

This equation should look familiar to you. In Chapter 3 we used this same equation for total costs. What is the difference between total costs and a mixed cost?

The cost equation describing Unlimited Decadence's total power cost for one year is as follows:

$$\text{Total power cost} = \$42,000 + \$0.05X$$

where:
    X = The number of cases of candy bars produced per year

Thus, the total power cost is $242,000 at a production volume of 4,000,000 cases of candy bars [$42,000 + $0.05(4,000,000)]. At a production volume of 4,800,000 cases of candy bars, the total power cost would be $282,000 [$42,000 + $0.05 (4,800,000)].

Exhibit 11-5 shows a graph of the mixed cost we just described. Notice that a mixed cost increases in a straight line. In other words, the mixed cost increases at a constant rate equal to the rate of its variable component ($0.05 per case of candy bars produced) as volume increases. However, unlike a variable cost, it intersects the vertical (cost) axis above the origin at an amount equal to its fixed cost component ($42,000).

Managers often separate the fixed and variable components of mixed costs and treat them independently. The fixed components of mixed costs are grouped with (and treated like) other fixed costs, and the variable components are grouped with (and treated like) variable costs. This is how we will treat them in this chapter.

Estimating Costs

So far, we have treated variable, fixed, and mixed costs as if they are linear; if we graphed them, they would form a straight line. However, many factors other than volume can also affect costs. In reality, costs do not always behave linearly, although they *approximate* that behavior. For example, assume that Unlimited Decadence's electric bills for the factory and for candy bar production for the last six years were the following:

Year	Volume (cases)	Total Power Cost
1	2,800,000	$182,000
2	3,000,000	198,700
3	3,500,000	208,200
4	4,000,000	255,100
5	4,620,000	268,900
6	5,000,000	292,000

We can visualize the behavior of these costs by plotting them on a graph, as we show in Exhibit 11-6. This pattern of points on a graph is called a **scatter diagram**. Although the points on the scatter diagram don't form a straight line, they approximate one.

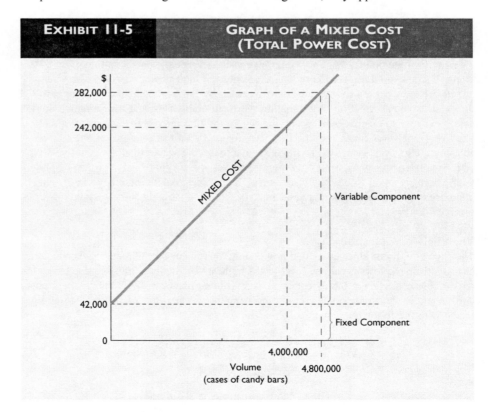

EXHIBIT 11-5	GRAPH OF A MIXED COST (TOTAL POWER COST)

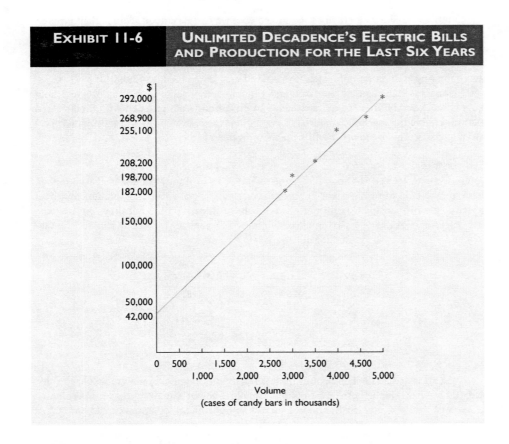

EXHIBIT 11-6   **UNLIMITED DECADENCE'S ELECTRIC BILLS AND PRODUCTION FOR THE LAST SIX YEARS**

If we visualize a straight line that runs through these points and that best represents the pattern of the points, this line would intersect the vertical axis above the origin of the graph, just as would a mixed cost. Therefore, for the total power cost, we can assume a mixed behavior pattern that would be represented by the equation for a mixed cost. We then can solve the equation by estimating the fixed component and the variable component. One way to do this is by observing where the line intersects the vertical axis (the fixed cost) and how much the line increases for every unit change in volume (the variable cost). However, visually drawing the line through the scatter diagram is not very precise; if you drew a line through the points and compared your line with that of one of your classmates, you would likely see two different lines. Another method that managers often use to give a quick, shorthand estimate of fixed and variable costs is called the *high-low method*.

The High-Low Method

The **high-low method** assumes that any change that occurs in total costs when volume changes from the lowest volume level to the highest volume level must be due to the total variable cost change that occurs with that volume increase. (Remember, fixed costs don't change with volume.) Therefore, the variable cost per unit of volume can be estimated with the following equation:

$$v = \frac{\text{Cost change from lowest to highest volume}}{\text{Volume change from lowest to highest volume}}$$

where:

v = The variable cost per unit of volume

Going back to Unlimited Decadence's power cost and volume data for the last six years, we can figure the volume and corresponding cost changes from the lowest to the highest volume levels as follows. Since it is a change in volume that causes the total variable cost to increase or decrease, we first identify the highest and the lowest volume levels. The difference between these volume levels is the volume change from the lowest to the highest volume. Next, we identify the total cost at each volume level. The difference between these costs is the cost change—that is, the increase in total variable costs—that occurred as a result of the change in volume.

	Volume (cases)	Total Power Cost
High volume	5,000,000	$292,000
Low volume	2,800,000	182,000
Change	2,200,000	$110,000

Now we can estimate the variable cost per unit:

$$v = \frac{\$110,000}{2,200,000} = \$0.05$$

Once we determine that the variable cost estimate is $0.05 per case of candy bars, we can estimate the total variable component of the mixed cost at either the high or the low volume level. For example, at the high volume level, the total variable cost is $250,000 (5,000,000 cases × $0.05). The total cost at that level of volume was $292,000. If we subtract the variable cost component from the total cost ($292,000 − $250,000), what we have left is the fixed cost component of $42,000. Thus, by using the high-low method, Unlimited Decadence can use the following equation to estimate the total annual power cost at any volume level:

$$\text{Total power cost} = \$42,000 + \$0.05X$$

where:

$$X = \text{Volume}$$

Since fixed costs don't change with changes in volume, the fixed cost component should be the same at the low volume level. Furthermore, the variable cost per case of candy bars should be the same at both volume levels. We can test this by substituting the low volume level into the above equation:

$$\text{Total power cost} = \$42,000 + \$0.05(2,800,000) = \$182,000$$

As expected, the $182,000 we computed equals the $182,000 total cost measured at the low volume level of 2,800,000 cases.

It is important to be careful when studying past cost data to establish future cost behavior patterns. In making the analysis, you should be able to answer *yes* to the following questions before using this method:

• Were conditions essentially the same in all the periods the data represent?

• Will the same conditions continue to exist in the future, so that the cost behavior patterns will remain the same?

• Does the computed cost pattern make sense?

It is important that similar conditions existed in the time periods from which the data were obtained because abnormal conditions would distort the data and, consequently, the estimates made from the data. Suppose, for example, that during one of the years from which we obtained Unlimited Decadence's power costs, Unlimited Decadence had to add supplemental heat from electric furnaces in the factory because of an unusually cold winter. If that year happened to be the high- or low-activity year, the high-low method would produce a distorted result because of the abnormally high power cost during that year.

We also must expect that the conditions on which we are basing our estimates will continue. Otherwise, we can't count on the cost behavior patterns to stay the same. If, for example, the power company adds a surcharge or if Unlimited Decadence increases its lighting requirements, these changed conditions will also change the cost equation.

Finally, the resulting cost pattern must make sense. (Is it reasonable?) For instance, using the high-low method to estimate the relationship between sales volume and depreciation of factory equipment would give misleading results; sales volume and the depreciation of factory equipment are not directly related.

The high-low method allows decision-makers to quickly estimate the fixed and variable components of mixed costs. However, it is not the most precise method. Statistical methods such as regression analysis, which you may learn about in another class, provide more accurate results but are not as convenient. Even with these methods, however, careless interpretation can lead to misleading results. Furthermore, even the most precise estimates apply only within a certain range of volumes.

Relevant Range

Each time a management accountant prepares an analysis that involves cost estimates for a manager's decision, the requirements of the decision determine a range of volumes over which the estimates must be especially accurate. For example, if Unlimited Decadence expects to produce between 3,500,000 and 6,000,000 cases of candy bars each year and wants to decide whether or not to change its manufacturing process, cost estimates for both the existing process and the alternative process over that specific range (3,500,000 to 6,000,000 cases) would be useful in making the decision. The company would not be helped by knowing which process is less expensive to operate when producing below 3,500,000 candy bars, nor would the company care about comparing manufacturing costs for volumes above 6,000,000 candy bars. Only the range of volumes from 3,500,000 to 6,000,000 is relevant (useful) to the decision.

The **relevant range** is the range of volumes over which cost estimates are needed for a particular use and over which observed cost behaviors are expected to remain stable. The relevant range concept is extremely important because by focusing on the range of volumes for which cost estimates should be accurate, a management accountant can make cost estimates that are useful for decision making. Decision makers will be able to ignore cost behavior patterns outside of the relevant range. Furthermore, if the management accountant states the relevant range whenever providing cost estimates, potential users of the estimates will be alerted to the range of volumes over which the estimates are reliable.

Some costs do not fit the fixed, variable, or mixed cost behavior patterns we described. They may vary, but not in a straight line over all possible volumes. For example, Unlimited Decadence may pay less for each tub of cocoa butter that it purchases above a certain volume level than it pays for each tub that it purchases below that volume level. Other costs may be fixed over a wide range of volumes but increase abruptly to a higher amount if the upper limit of that volume range is exceeded. For example, once Unlimited Decadence's production of candy bars reaches a certain level, the capacity of its mixers will be reached. To exceed that level of production, Unlimited Decadence will need to purchase an additional mixer. When this occurs, Unlimited Decadence's fixed depreciation cost for mixers will jump by the amount of the new mixer's depreciation. Exhibit 11-7 presents graphs of these two cost behavior patterns.

Despite these potential cost behavior patterns, the management accountant must determine only how the costs behave within the relevant range, and in most cases, these costs fit one of the three common behavior patterns. For example, the behavior patterns shown in the graphs in Exhibit 11-7 might be estimated within the relevant range as a variable cost and a fixed cost, as we show in Exhibit 11-8.

EXHIBIT 11-7	GRAPHS OF COSTS THAT ARE NOT FIXED, VARIABLE, OR MIXED

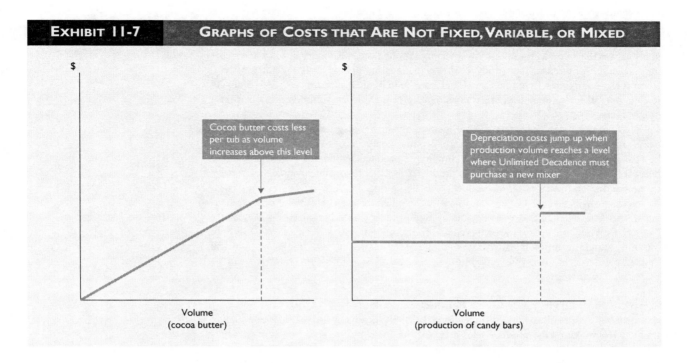

Cocoa butter costs less per tub as volume increases above this level

Depreciation costs jump up when production volume reaches a level where Unlimited Decadence must purchase a new mixer

$

Volume
(cocoa butter)

$

Volume
(production of candy bars)

C-V-P Computations Using the Profit Equation
(Multiple Products)

5 What is the effect of multiple products on cost-volume-profit analysis?

Remember from our discussion in Chapter 3 that C-V-P analysis is an examination of how profit is affected by changes in the sales volume, in the selling prices of products, and in the various costs of the company. Decision makers use C-V-P analysis to gain an

EXHIBIT 11-8	RELEVANT RANGE OF COSTS THAT ARE NOT FIXED, VARIABLE, OR MIXED

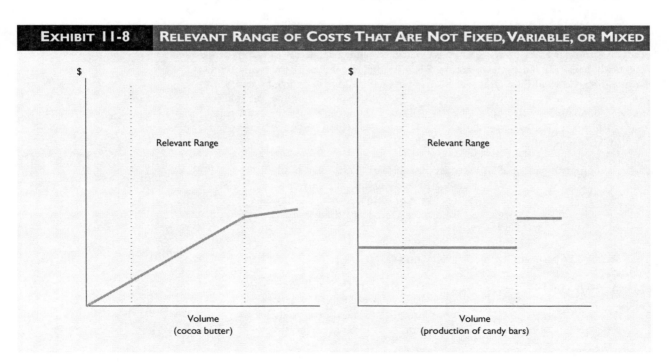

$

Relevant Range

$

Relevant Range

Volume
(cocoa butter)

Volume
(production of candy bars)

understanding of the profit impact of plans that they are making. This understanding can produce more informed decisions during the planning process.

When a manufacturing company produces and sells only one product, its C-V-P computations are the same as those that Sweet Temptations used in Chapter 3, with sales volume measured in units of that product. However, most companies sell several products. When these companies (manufacturing and retail companies) perform a C-V-P analysis of their overall operations, they must consider the relative sales volumes of their different products.

This doesn't mean that the single-product analysis we described in Chapter 3 is not useful, however. Managers can use a single-product analysis to study the relationship of the cost, volume, and profit of one of a company's products as long as they can separate the costs and the sales revenues of that product from the costs and revenues caused by the production and sales of other products.

Next, we will discuss C-V-P analysis in companies that sell more than one product. Our discussion assumes that you remember single-product C-V-P analysis from Chapter 3. A brief review, on your part, of the concepts and computations in Chapter 3 will help you understand the following discussion. To keep the computations simple, we will use an example involving two products, but *the procedure we are about to describe can be used with any number of products*.

Contribution Margin for a Given Product Sales Mix

Managers can apply the single-product form of analysis to any *group* of a company's products if they know the product sales mix and the costs and sales revenues of the individual products within that group of products. The **product sales mix** is the relative proportion of units of the different products sold. In C-V-P analysis when there is more than one product, the product sales mix is considered to be one "unit." For example, if Unlimited Decadence has typically sold five cases of Darkly Decadent candy bars for every one case of Pure Decadence candy bars sold, this combination of cases of candy bars is considered to be one "unit." (Visualize a basket containing five cases of Darkly Decadent candy bars and one case of Pure Decadence candy bars.) As long as Unlimited Decadence plans to continue to sell the Darkly Decadent and Pure Decadence candy bars in that proportion, we can say that it has the same product sales mix as it had in the past.

Suppose that Unlimited Decadence estimates that its fixed costs will be $24,800,000 in the coming year and that its product sales mix will be five cases of Darkly Decadent candy bars sold for every one case of Pure Decadence candy bars. Cases of Darkly Decadent candy bars will sell for $16, require $11 of variable cost, and earn a contribution margin of $5. Cases of Pure Decadence candy bars will sell for $20 each, require $14 of variable cost, and earn a contribution margin of $6. Thus, the sales revenue, variable cost, and contribution margin per "unit" (5 cases of Darkly Decadent candy bars and 1 case of Pure Decadence candy bars) will be $100, $69, and $31, respectively. The sales revenue per "unit" is computed as follows:

$$
\begin{aligned}
\text{Selling price per "unit"} = \ &(\$16 \text{ per case of Darkly Decadent candy bars} \times \\
&5 \text{ cases of Darkly Decadent candy bars}) + \\
&(\$20 \text{ per case of Pure Decadence candy bars} \times \\
&1 \text{ case of Pure Decadence candy bars}) \\
= \ &\underline{\underline{\$100}}
\end{aligned}
$$

The variable cost per "unit" is computed as follows:

$$
\begin{aligned}
\text{Variable cost per "unit"} = \ &(\$11 \text{ per case of Darkly Decadent candy bars} \times \\
&5 \text{ cases of Darkly Decadent candy bars}) + \\
&(\$14 \text{ per case of Pure Decadence candy bars} \times \\
&1 \text{ case of Pure Decadence candy bars}) \\
= \ &\underline{\underline{\$69}}
\end{aligned}
$$

Therefore, the contribution margin per unit is $31 ($100 selling price per "unit" − $69 variable cost per "unit"). We summarize this information in the following schedule:

	Darkly Decadent Candy Bars		Pure Decadence Candy Bars		Total per "Unit"
Selling price/case	$16		$20		
Variable cost/case	(11)		(14)		
Contribution margin/case	$ 5		$ 6		
Expected sales mix (cases)	5		1		
Sales revenue	$80	+	$20	=	$100
Less: variable cost	(55)	+	(14)	=	(69)
Contribution margin	$25	+	$ 6	=	$ 31

Finding the Break-Even Point

We can calculate the break-even point in units of product sales mix by beginning with the same equation that we used in calculating the break-even point for one product:

$$\frac{\text{Unit sales volume}}{\text{(to earn zero profit)}} = \frac{\text{Total fixed cost}}{\text{Contribution margin per unit}}$$

However, keep in mind that when Unlimited Decadence uses this equation, its contribution margin per unit represents a "unit" that now consists of five cases of Darkly Decadent candy bars and one case of Pure Decadence candy bars. Since fixed costs are $24,800,000, Unlimited Decadence can find the break-even point expressed as "units" of sales volume by dividing the $24,800,000 total fixed cost by the $31 contribution margin per "unit." We can write this in equation form as follows:

$$\frac{\text{Break-even point}}{\text{(in "units")}} = \frac{\text{Total fixed cost}}{\text{Contribution margin per "unit"}}$$

$$= \frac{\$24,800,000}{\$31}$$

$$= 800,000 \text{ "units"}$$

To break even, Unlimited Decadence would have to sell 800,000 "units," composed of 4,000,000 cases of Darkly Decadent candy bars (800,000 × 5 cases of Darkly Decadent candy bars) and 800,000 cases of Pure Decadence candy bars (800,000 × 1 case of Pure Decadence candy bars).

We can verify the break-even "unit" sales volume of 800,000 "units" that we just computed for Unlimited Decadence with the following profit computation:

Sales revenue:		
Cases of Darkly Decadent candy bars..................	$64,000,000	
(4,000,000 cases @ $16 per case)		
Cases of Pure Decadence candy bars....................	16,000,000	
(800,000 cases @ $20 per case)		
Total sales revenue		$80,000,000
Less variable costs:		
Cases of Darkly Decadent candy bars..................	$44,000,000	
(4,000,000 cases @ $11 per case)		
Cases of Pure Decadence candy bars....................	11,200,000	
(800,000 cases @ $14 per case)		
Total variable cost...		(55,200,000)
Total contribution margin		$24,800,000
Less total fixed cost..		(24,800,000)
Profit...		$ 0

Note in this computation that total sales revenue ($80,000,000) equals total cost ($55,200,000 variable + $24,800,000 fixed). Also note that total contribution margin equals total fixed cost ($24,800,000). In either case, profit equals $0.

Finding the Unit Sales Volume to Achieve a Target Pretax Profit

A company often states its profit goal at an amount that results in a satisfactory rate of return on owner's equity. Remember, the return on owner's equity measures how effectively managers have earned income on the amount the owners invested in the company.

In Chapter 7, in our discussion of entrepreneurial companies, we calculated the return on owner's equity as follows:

$$\text{Return on owner's equity} = \frac{\text{Net income}}{\text{Average owner's equity}}$$

The calculation is similar for a corporation, but with two differences. One difference is that the denominator is stockholders' equity. The other difference is caused by the fact that a corporation pays income taxes and deducts the amount of its income tax expense on its income statement to determine its net income. Since the return on stockholders' equity measures a company's efficiency in earning income on the stockholders' investment, and since income taxes are a tax *on* that income, both internal and external users are interested in the return *after* income taxes. The calculation then becomes as follows:

$$\text{Return on stockholders' equity} = \frac{\text{Income after taxes}}{\text{Average stockholders' equity}}$$

But since internal users usually can't control the income tax expense imposed by the U.S. government, they are interested in what pretax income (profit) the company must earn to achieve its targeted after-tax income. For example, assume Unlimited Decadence estimates that its average stockholders' equity for the coming year will be $18,000,000, and assume that its managers would like it to earn a return on stockholders' equity of 18%, the industry average for confectioners. To achieve this return on stockholders' equity, Unlimited Decadence will have to earn $3,240,000 income after taxes ($18,000,000 × 18%). Suppose, then, that you want to know how many units Unlimited Decadence will have to sell to earn this desired after-tax income. To use the C-V-P equations, you must first convert the desired after-tax income to pretax income, using the following equation:[8]

$$\text{Pretax income} = \frac{\text{After-tax income}}{(1 - \text{Tax rate})}$$

If we assume that Unlimited Decadence is subject to a 40 percent income tax rate, it must earn $5,400,000 [$3,240,000 ÷ (1 − 0.40)] pretax income to achieve a $3,240,000 after-tax income. The computation to determine the "unit" sales volume to achieve this target pretax income is as follows:

$$\frac{\text{"Unit" sales volume needed to}}{\text{earn a desired pretax income}} = \frac{\text{Total fixed cost + Desired pretax income}}{\text{Contribution margin per "unit"}}$$

$$\frac{\text{"Unit" sales volume needed to}}{\text{earn \$5,400,00 pretax income}} = \frac{\$24,800,000 \text{ fixed cost} + \$5,400,000 \text{ pretax income}}{\$31}$$

$$= \underline{974,194} \text{ "units" (rounded)}$$

---

[8] Recall that we use the following formula to compute after-tax income:

$$\text{After-tax Income} = \text{Pretax Income} - \text{Income Tax Expense}$$

Income tax expense is computed with the following formula:

$$\text{Income Tax Expense} = \text{Pretax Income} \times \text{Tax Rate}$$

We can substitute the second formula into the first formula, yielding the following revised formula for after-tax income:

$$\text{After-tax Income} = \text{Pretax Income} - (\text{Pretax Income} \times \text{Tax Rate})$$

Then, by eliminating pretax income from each of the two terms on the right side of the equation, we can restate the equation as follows:

$$\text{After-tax Income} = \text{Pretax Income} \times (1 - \text{Tax Rate})$$

Then we can find pretax income by isolating it on one side of the equation:

$$\text{Pretax Income} = \frac{\text{After-tax Income}}{(1 - \text{Tax Rate})}$$

Therefore, at a sales mix of five cases of Darkly Decadent candy bars and one case of Pure Decadence candy bars per "unit," Unlimited Decadence will have to sell 4,870,970 cases of Darkly Decadent candy bars (974,194 × 5 cases of Darkly Decadent candy bars) and 974,194 cases of Pure Decadence candy bars (974,194 × 1 case of Pure Decadence candy bars) to earn a pretax income of $5,400,000.

Finding the Dollar Sales Volume to Achieve a Target Pretax Profit

When a company sells more than one product, it is often easier to represent sales volume in dollars instead of units. Earlier, we said that Unlimited Decadence would break even if it sold 800,000 "units," composed of 4,000,000 cases of Darkly Decadent candy bars and 800,000 cases of Pure Decadence candy bars. Since Darkly Decadent candy bars sell for $16 per case and Pure Decadence candy bars sell for $20 per case, Unlimited Decadence would break even when its sales revenue was $80,000,000 [(4,000,000 cases of Darkly Decadent candy bars × $16) + (800,000 cases of Pure Decadence candy bars × $20)].

A more direct way to find the dollar sales volume to achieve a desired profit is to use the company's contribution margin percentage. The *contribution margin percentage* is the ratio of the contribution margin to sales revenue or, stated another way, the contribution margin as a percentage of sales revenue. Since the contribution margin of one "unit" of Unlimited Decadence's product sales mix is $31 and the sales revenue of this "unit" is $100, its contribution margin percentage is 31% ($31 ÷ $100). We could say, then, that the contribution margin is 31% of sales revenue for this product sales mix.

Since break-even occurs when

$$\text{Contribution margin} = \text{Fixed costs},$$

we could also say that it occurs when

$$\text{Sales revenue} \times \text{Contribution margin percentage} = \text{Fixed costs}.$$

Therefore, Unlimited Decadence could compute its break-even point (in sales dollars) as shown here:

$$\text{Break-even point (in dollars)} = \frac{\text{Total fixed cost}}{\text{Contribution margin percentage}}$$

$$= \frac{\$24,800,000}{31\%}$$

$$= \underline{\underline{\$80,000,000}}$$

Similarly, it could compute the total dollar sales volume needed to earn a desired amount of pretax income. For example, if managers want Unlimited Decadence to earn $5,400,000 pretax income, sales will have to be $97,419,355. The computation is as follows:

$$\begin{array}{l}\text{Dollar sales volume} \\ \text{needed to earn a} \\ \text{desired pretax income}\end{array} = \frac{\text{Total fixed cost} + \text{Desired pretax income}}{\text{Contribution margin percentage}}$$

$$\begin{array}{l}\text{Dollar sales volume} \\ \text{needed to earn} \\ \$5,400,000 \text{ desired} \\ \text{pretax income}\end{array} = \frac{\$24,800,000 \text{ Fixed cost} + \$5,400,000 \text{ Pretax income}}{31\%}$$

$$= \underline{\underline{\$97,419,355}}$$

 How can you verify the dollar sales volumes needed to break even and to earn a pretax profit of $5,400,000?

What If the Company Changed Its Product Sales Mix?

Remember that Unlimited Decadence's contribution margin percentage would not be 31% if its product sales mix did not remain five cases of Darkly Decadent candy bars and one case of Pure Decadence candy bars. Consider what will happen if sales of cases of Darkly Decadent candy bars change relative to sales of Pure Decadence candy bars so that Unlimited Decadence sells three cases of Darkly Decadent candy bars for every four cases of Pure Decadence candy bars. The new contribution margin percentage will be 30.47%, shown as follows:

	Darkly Decadent Candy Bars		Pure Decadence Candy Bars		Total per "Unit"
Selling price/case	$16		$20		
Variable cost/case	(11)		(14)		
Contribution margin/case	$ 5		$ 6		
Expected sales mix (cases)	3		4		
Sales revenue	$48	+	$80	=	$128
Less: variable costs	(33)	+	(56)	=	(89)
Contribution margin	$15	+	$24	=	$ 39
Contribution margin percentage ($39/$128)					30.47% (rounded)

With this new product sales mix, the dollar sales volume will have to be $99,113,883 [($24,800,000 fixed cost + $5,400,000 income before taxes) ÷ 30.47%] to earn a profit of $5,400,000 before income taxes. Furthermore, with the new product sales mix, total dollar sales volume will have to be $81,391,533 ($24,800,000 total fixed cost ÷ 30.47%) for Unlimited Decadence to break even.

Applications of C-V-P Analysis in a Manufacturing Environment

So far, we have described fixed and variable costs as they relate to changes in sales. However, costs do not always vary with sales. They may vary with production, number of hours worked by employees, or some other measure of operating activity. Manufacturing companies incur costs to produce the products as they manufacture them. For this reason, managers of manufacturing companies usually classify the behavior of production costs based on whether or not these costs vary with the level of *production*.

6 How does a manufacturing company use cost-volume-profit analysis for its planning?

As in a service or retail company, C-V-P analysis in a manufacturing company is useful in planning because it shows the potential impact of alternative plans on profit. The analysis can help managers make planning decisions and help investors and creditors evaluate the risk associated with their investment and credit decisions. In the following discussion, we illustrate how C-V-P analysis in a manufacturing company can show the potential profit impact of alternative plans.

Suppose that Unlimited Decadence is considering manufacturing the new fat-free, sugarless candy bar (to be called "Empty Decadence") that we discussed in Chapter 2. Marketing thinks it can sell 450,000 cases of these candy bars at $20 per case during the first year on the market. Purchasing has located suppliers for the sweetener and for the fat substitute. Together, all the direct materials for a case of Empty Decadence candy bars will cost $4. The production department expects to hire two new cooks to produce the Empty Decadence candy bars and has set a standard of 45 minutes of direct labor for every case of these candy bars. The human resources department expects the pay rate for these cooks to be $12 per hour. The production department estimates that variable overhead costs will increase by $0.50 per case of Empty Decadence candy bars. Marketing expects variable selling and administrative costs to increase by $1.00 because of the sale of this new product. There is currently space in the factory to manufacture the Empty Decadence candy bar, but Unlimited Decadence will need to purchase new equipment to process the sweetener and the fat substitute. The finance department has located financing for the new equipment.

The production department expects depreciation and insurance on this equipment, together with the additional advertising expense, to add $1,188,000 to Unlimited Decadence's fixed costs.

Unlimited Decadence must decide whether to manufacture the Empty Decadence candy bar. Recall from Chapter 2 that there are numerous questions that Unlimited Decadence's managers should answer before they make a decision. Furthermore, there may be more than two alternatives (manufacturing or not manufacturing) to solving this issue. However, assume that Unlimited Decadence's managers are very interested in adding this new product and are gathering product information that will help them make their decision. C-V-P analysis will contribute useful information for evaluating this decision.

Managers first may want to know whether the revenue from sales of the Empty Decadence candy bar will even cover the additional costs of producing it, or whether it will "break even." We can calculate the contribution margin of each case of Empty Decadence candy bars as follows:

Sales revenue per case		$20.00
Variable costs per case		
Direct materials	$4.00	
Direct labor ($12/hour × 3/4 hour)	9.00	
Additional variable overhead	0.50	
Additional variable selling and administrative costs	1.00	(14.50)
Contribution margin per case		$ 5.50

Now we can calculate the break-even point for the Empty Decadence candy bar:

$$\text{Break-even unit sales} = \frac{\text{Additional fixed costs}}{\text{Contribution margin per unit}}$$

$$= \frac{\$1,188,000}{\$5.50}$$

$$= \underline{216,000} \text{ cases}$$

Unlimited Decadence will need to sell 216,000 cases of Empty Decadence candy bars in order for this line of candy bars to break even. Each case sold above the break-even point will cause Unlimited Decadence's profit to increase by $5.50. So, at the predicted sales level of 450,000 cases, Unlimited Decadence should earn $1,287,000 additional profit [(450,000 cases − 216,000 cases) × $5.50] on sales of the Empty Decadence candy bar.

Based on this information alone, it seems like a good idea to manufacture this new candy bar. The manager of the purchasing department, though, thinks that there is a possibility that the costs of the sweetener and the fat substitute (both direct materials) may increase in the next few months. It would be useful to know how much these costs can increase before the Empty Decadence line of candy bars will "lose money," or not be able to earn enough revenue to cover its costs. We know that costs can increase until they equal revenues, or until Empty Decadence "breaks even" at predicted sales of 450,000 cases (assuming that the marketing department's estimates are correct and that the selling price of a case of Empty Decadence candy bars doesn't change). So, we use the same formula:

$$\text{Break-even unit sales} = \frac{\text{Additional fixed costs}}{\text{Contribution margin per unit}}$$

$$450,000 = \frac{\$1,188,000}{X}$$

$$X = \underline{\$2.64}$$

The contribution margin per case can decrease to $2.64 before the Empty Decadence product line will break even. If it decreases more than that, the Empty Decadence product line will lose money. In other words, the contribution margin can decrease by $2.86 (the old contribution margin of $5.50 minus the new contribution margin of $2.64). Therefore, if the only factors that change are the costs of the sweetener and the fat substitute, the direct materials can increase by $2.86 from $4.00 to $6.86 per case without the Empty Decadence product line generating a loss. We can verify this as follows:

Sales revenue per case..		$20.00
Variable costs per case		
Direct materials ($4.00 + $2.86).........................	$6.86	
Direct labor ($12/hour × 3/4 hour)...................	9.00	
Additional variable overhead	0.50	
Additional variable selling and		
administrative costs	1.00	(17.36)
Contribution margin per case..		$ 2.64

If Unlimited Decadence sells 450,000 cases, each will contribute $2.64, for a total contribution margin of $1,188,000, just enough to cover the additional fixed costs.

But what if marketing overestimates sales of the Empty Decadence candy bar? How much below its sales estimate can Unlimited Decadence's actual sales be before it loses money, if the costs of the sweetener and fat substitute do not change? The amount that sales (in units) can decrease without a loss, or the difference between the estimated sales volume and the break-even sales volume, is called the **margin of safety**. Managers use the margin of safety as a measure of the risk of a new plan. The higher the margin of safety is, the lower is the risk.

At the current cost estimate, break-even sales is 216,000 cases of Empty Decadence candy bars. That means that the margin of safety is 234,000 cases (450,000 cases − 216,000 cases). Sales could drop 234,000 cases below the estimate before Empty Decadence would experience a loss.

Notice that in making this calculation, we considered only the revenues and costs attributable to the Empty Decadence candy bar. Isolating these costs and revenues gives managers a clearer picture of the expected effect of adding the Empty Decadence candy bar. However, Unlimited Decadence must also consider the other effects of adding this product. For example, will the manufacture and sale of the Empty Decadence candy bar cause the sales of Unlimited Decadence's other candy bars to decrease? If so, it is possible that even though the sale of the Empty Decadence candy bar will probably generate a profit for the candy bar, *total income* for Unlimited Decadence may decrease. The decrease in the sales of other candy bars could be larger than the increase generated by the sale of the Empty Decadence candy bar. (On the other hand, maybe it would be better for Unlimited Decadence to introduce this new product before its competition does!) Wherever possible, when deciding whether to introduce a new product, managers should consider quantitative information about how the new product would affect the profits generated by the company's other products.

Summary of the C-V-P Analysis Computations for Multiple Products

Exhibit 11-9 summarizes the profit, break-even point, and sales volume computations used in our discussion of C-V-P analysis with multiple products. As in Chapter 3, we present these computations as equations that can be used to answer the basic questions that occur frequently in C-V-P analysis. Although it may be tempting to try to commit them all to memory, you should instead strive to understand how these equations relate to one another and how managers can use the answers found by applying these formulas in certain types of decision making.

BUSINESS ISSUES AND VALUES

To make smart decisions, managers must combine the results of C-V-P analysis with other factors. For example, many factors other than C-V-P analysis influence Unlimited

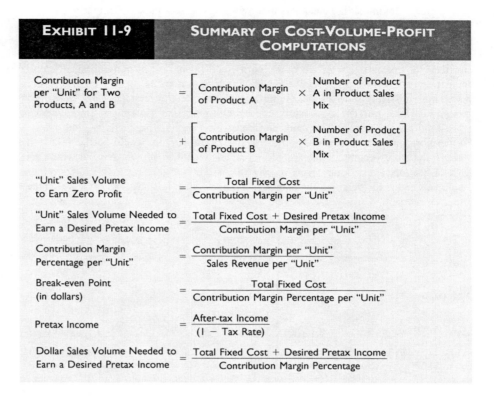

EXHIBIT 11-9	SUMMARY OF COST-VOLUME-PROFIT COMPUTATIONS

$$\text{Contribution Margin per "Unit" for Two Products, A and B} = \left[\text{Contribution Margin of Product A} \times \text{Number of Product A in Product Sales Mix}\right]$$

$$+ \left[\text{Contribution Margin of Product B} \times \text{Number of Product B in Product Sales Mix}\right]$$

$$\text{"Unit" Sales Volume to Earn Zero Profit} = \frac{\text{Total Fixed Cost}}{\text{Contribution Margin per "Unit"}}$$

$$\text{"Unit" Sales Volume Needed to Earn a Desired Pretax Income} = \frac{\text{Total Fixed Cost} + \text{Desired Pretax Income}}{\text{Contribution Margin per "Unit"}}$$

$$\text{Contribution Margin Percentage per "Unit"} = \frac{\text{Contribution Margin per "Unit"}}{\text{Sales Revenue per "Unit"}}$$

$$\text{Break-even Point (in dollars)} = \frac{\text{Total Fixed Cost}}{\text{Contribution Margin Percentage per "Unit"}}$$

$$\text{Pretax Income} = \frac{\text{After-tax Income}}{(1 - \text{Tax Rate})}$$

$$\text{Dollar Sales Volume Needed to Earn a Desired Pretax Income} = \frac{\text{Total Fixed Cost} + \text{Desired Pretax Income}}{\text{Contribution Margin Percentage}}$$

Decadence's decision about whether to manufacture and sell the Empty Decadence candy bar; one of these factors is the hiring of the two new cooks. How much responsibility does Unlimited Decadence have when it hires new employees? Perhaps the two cooks will have to quit their secure jobs in order to work for Unlimited Decadence. Given the fact that it may be difficult for them to return to their old jobs, how sure should Unlimited Decadence be that its new product will succeed before it hires new employees? What if the two cooks are currently unemployed? Should that make a difference in Unlimited Decadence's decision?

Another factor that influences this decision is the presence or absence of health risks associated with the sweetener and fat substitute used in the Empty Decadence candy bar. Unlimited Decadence prides itself on its concern for its customers and would like to maintain its reputation of producing quality candy bars. Therefore, even if C-V-P analysis indicates that the Empty Decadence candy bar should be produced, the presence of health risks (if any) in the sweetener or fat substitute may cause managers to delay production until the company can locate safer alternatives.

Another example of other factors that companies must consider is the raising of the minimum wage. For example, several years ago, Congress raised the minimum wage from $4.25 per hour to $5.15 per hour. While this change in policy positively affected millions of minimum-wage earners, it also affected the decisions of the companies that hire these workers because an increase in the minimum wage causes companies' variable costs to increase. For instance, the owners of fast-food franchises typically hire minimum-wage workers to staff their restaurants. The fast-food industry is a "low margin" industry, which means that the average contribution margin of restaurants in this industry is low relative to that of companies in other industries. An increase in the minimum wage causes the already low contribution margins of these companies to drop lower. So companies faced with raising the wages of their minimum-wage employees must make a difficult decision, with alternatives such as accepting lower profits, reducing staff, or raising prices. Using their critical thinking skills, managers and owners of these companies might ask the following questions when making this decision: Would accepting lower profits threaten the long-term solvency of the company? If the company raises prices, will competitors also raise prices, keeping competition about the same within the industry? Will raising prices hurt the industry as a whole? (In other words, will people eat at home more often?) If competitors don't raise prices, will the company lose customers?

A more difficult issue for these companies is the effect of reducing staff. In many of these restaurants, workers are already assigned multiple tasks. How many employees can a fast-food restaurant lay off without hurting quality, service, productivity, and morale? Furthermore, in many cases, the poorest, least-skilled workers would be the first to be laid off; the more versatile, multiskilled individuals would keep their jobs but would perhaps be assigned more duties. How much responsibility do the restaurants have to these employees and to the neighborhoods in which they conduct business and from which they hire their workers? Clearly, the answers to these questions are not as clear-cut as the results of C-V-P analysis alone.

 Can you think of any other factors that the managers of Unlimited Decadence should consider when deciding whether to produce Empty Decadence candy bars?

SUMMARY

At the beginning of the chapter we asked you several questions. During the chapter, we asked you to STOP and answer some additional questions to build your knowledge about specific issues. Be sure you answered these additional questions. Below are the questions from the beginning of the chapter, with a brief summary of the key points relating to the answers. Use your creative and critical thinking skills to expand on these key points to develop more complete answers to the questions and to determine what other questions you have that might lead you to learn more about the issues.

1   How does the fact that a manufacturing company makes the products it sells affect its business plan?

A manufacturing company's business plan has much in common with the plans of retail and service companies. However, because a manufacturing company makes the products it sells, its business plan also includes a production plan, which affects other parts of the business plan, particularly the financial plan. The production plan describes how a company intends to manufacture and maintain the quality of the products it will sell. The plan describes the raw materials that the company will use in manufacturing its product. A description of the production process specifies how these raw materials will be converted into finished products. The production plan also describes where the company intends to store the finished products before it sells them.

2   How does a manufacturing company determine the cost of the goods that it manufactures?

Managers of a manufacturing company determine the cost of the goods that it manufactures by adding together the direct materials costs, the direct labor costs, and the factory overhead costs that it incurs in manufacturing its products. To determine the direct materials costs, managers must know how much of each raw material went into the company's products and multiply the amount of raw materials by their costs. To determine the direct labor costs, managers must know how much of each type of direct labor was necessary to manufacture the company's products and multiply the number of hours of each type of direct labor by the appropriate wages per hour. To determine the factory overhead costs, managers must add together the costs of each type of factory overhead.

3   Why are standard costs useful in controlling a company's operations?

A manufacturing company uses standard costs as a way of controlling its production. By measuring actual costs against standard costs (what the costs actually were against what the costs should have been), the managers of a company can evaluate its activities. They can determine which activities caused costs to be too high or too low and decide whether to change those activities. A company develops standards for each manufactured product by determining price and quantity standards for each input to its manufacturing process.

4 How do manufacturing costs affect cost-volume-profit analysis?

Managers of manufacturing companies treat manufacturing costs like other costs when performing cost-volume-profit analysis. Like other costs, manufacturing costs can be classified as fixed or variable costs. So, managers add the variable manufacturing costs to the other variable costs and add the fixed manufacturing costs to the other fixed costs when performing cost-volume-profit analysis.

5 What is the effect of multiple products on cost-volume-profit analysis?

C-V-P analysis is similar for a company that sells only one product and for a company that sells multiple products. With multiple products, however, managers must know the product sales mix and the costs and sales revenues of the group of products being analyzed. The C-V-P calculation for a company with multiple products results in the number of "units" of product mix that the company must sell to break even or earn a desired profit. The managers then must convert the number of "units" of product mix to the number of units of individual products. Managers may find dollar sales volume to be more useful than unit sales volume in determining the level of sales necessary to achieve a desired profit.

6 How does a manufacturing company use cost-volume-profit analysis for its planning?

C-V-P analysis helps the managers of a manufacturing company make planning decisions by showing the potential impact of alternative plans on profit. These alternative plans might include changing suppliers of direct materials, hiring production workers, buying new equipment, adding or dropping a product line, raising or lowering product costs or selling prices, or changing the planned sales volume on some of its products (and therefore also changing the planned production of those products).

KEY TERMS

direct labor *(p. 345)*
direct materials *(p. 344)*
factory overhead *(p. 346)*
finished goods inventory *(p. 344)*
goods-in-process inventory *(p. 344)*
high-low method *(p. 353)*
indirect labor *(p. 346)*
indirect materials *(p. 346)*
margin of safety *(p. 363)*
mixed costs *(p. 351)*

price standard *(p. 350)*
product sales mix *(p. 357)*
quantity standard *(p. 350)*
raw materials *(p. 340)*
raw materials inventory *(p. 344)*
relevant range *(p. 355)*
scatter diagram *(p. 352)*
standard costs *(p. 349)*
variable manufacturing cost *(p. 350)*
variance *(p. 349)*

SUMMARY SURFING

Here is an opportunity to gather information on the Internet about real-world issues related to the topics in this chapter. Go to http://www.cunningham.swlearning.com and click on the Interactive Study Center. Click on this chapter number then click on Summary Surfing and answer the following questions.

• Click on **Planning Your Business**. Scroll down and click on *Business plan case studies.* Click on *Case study—manufacturer's business plan.* Scroll down and click on *sample language.* Click on

Manufacturer's business plan. What business plan components are represented? Identify Break-away Bicycle Company's direct materials costs and direct labor costs.

- Click on **Quicken.com**. Click on *Start a Business.* Under *Tools & Services,* click on *Write a business plan.* Scroll down and click on *Sample Business Plans.* Click on *Willamette Furniture.* Use Willamette's Sales Forecast to calculate Willamette's break-even point. Does it match Willamette's break-even analysis?

INTEGRATED BUSINESS AND ACCOUNTING SITUATIONS

Answer the Following Questions in Your Own Words.

Testing Your Knowledge

11-1 Describe the purpose of the production plan.

11-2 Describe how raw materials in a manufacturing company are different from the goods that a retail company sells.

11-3 What type of information about raw materials does a manufacturing company include in its production plan?

11-4 What type of information about labor does a manufacturing company include in its production plan?

11-5 What type of information about equipment and facilities does a manufacturing company include in its production plan?

11-6 What type of information about finished products does a manufacturing company include in its business plan?

11-7 Describe the three types of inventories in a manufacturing company.

11-8 Describe the three types of production input. Give an example of each type of input for a company that builds houses.

11-9 Describe the difference between direct and indirect materials.

11-10 Describe the difference between direct and indirect labor.

11-11 How do you know if a cost should be classified as a factory overhead cost?

11-12 Define *standard costs,* and describe how managers use them.

11-13 What is a variance? How do managers use variance information?

11-14 What is the difference between a price standard and a quantity standard?

11-15 Describe a mixed cost and how it behaves.

11-16 Describe three methods that can be used to estimate the components of a mixed cost. What are the advantages and disadvantages of each?

11-17 What is a relevant range? How is C-V-P analysis affected by this concept? Why would it be useful for users of cost estimates to know the relevant range?

11-18 What is the difference between C-V-P analysis with one product and C-V-P analysis with multiple products?

11-19 Explain what is meant by a company's product sales mix. Give an example.

11-20 What happens to the results of C-V-P analysis if a company's product sales mix changes? Why does this happen?

11-21 How is a contribution margin for a product sales mix different from a contribution margin for one product?

11-22 How does a desired return on stockholders' equity help managers determine the company's target profit?

11-23 How do income taxes affect C-V-P analysis?

Applying Your Knowledge

11-24 Visit your favorite grocery store and select a packaged product.

Required: What raw materials went into the production of that product?

11-25 Think about how decision makers use a business plan.

Required: Why do you think it is important to include information about a manufacturing company's suppliers in its production plan?

11-26 Suppose you work for a company called Split Decisions and the boss has selected you to develop the production steps necessary to manufacture a banana split.

Required: Describe the sequence of the production steps and how the raw materials flow through this sequence.

11-27 Western Brands produces western accessories. It manufactures one product, a bolo tie, from a thin 36-inch strip of leather, 1/3-inch wide, and four fancy brass rings. The leather strip is threaded through two of the rings, which act as a clasp. The remaining two rings are sewn onto the ends of the leather strip.

  Western Brands purchases leather strips in 72-inch lengths that are 2 inches wide for $4.80 each and cuts them to 1/3-inch widths and then 36-inch lengths. Brass rings cost $9.60 per dozen. It takes two people 5 minutes each to produce one bolo tie. Each person earns $6 per hour.

Required: (1) What is the cost, per tie, of the direct materials?
        (2) What is the cost, per tie, of the direct labor?
        (3) What are the total direct costs for 500 ties?

11-28 The utility costs and production levels for The Cat's Pajamas for the last four months were as follows:

Month	Production Levels (pairs of pajamas)	Cost
1	1,700	$1,280
2	2,600	1,600
3	3,200	1,970
4	2,900	1,680

Required: (1) Assuming the utility cost is a mixed cost, use the high-low method with the above data to determine the variable cost per pair of pajamas and the total fixed cost per month.
        (2) If 3,150 pairs of pajamas are manufactured, what should the utility cost be?

11-29 Western Brands has estimated its factory overhead costs for the next year at two production volumes, 100,000 and 200,000. These estimates are shown below.

	Factory Overhead Costs	
	100,000 ties	200,000 ties
Depreciation on factory equipment	$ 5,000	$ 5,000
Factory rent	30,000	30,000
Factory supervisor's salary	45,000	45,000
Maintenance of factory equipment	1,920	2,240
Factory utilities	7,000	14,000
Factory supplies	3,000	6,000

Required: Estimate total factory overhead at a production level of 140,000 ties.

11-30 A company with multiple products has a break-even point of 5,000 "units."

Required: Explain the preceding sentence.

11-31 Bathtub Rings Corporation manufactures shower curtains that sell for $5.00 each and cost $3.48 to produce. It also manufactures shower-curtain rings that sell for $1.60 per box and cost $1.22 per box to produce. Fixed costs total $30,000 per year. Bathtub Rings sells five shower curtains for every two boxes of shower-curtain rings that it sells.

Required: (1) How many of each product must Bathtub Rings sell to break even?
(2) How many of each product must Bathtub Rings sell to earn a pretax income of $49,800?
(3) How many of each product must Bathtub Rings sell to earn an after-tax income of $49,800 if the income tax rate is 30%?

11-32 Refer to 11-31.

Required: (1) Compute Bathtub Rings' contribution margin percentage per "unit."
(2) Compute Bathtub Rings' break-even point in sales dollars.
(3) What total pretax income (loss) would Bathtub Rings have if its total sales revenue amounted to $216,000?

11-33 Sí, Your Dinner! Mexican Food Company manufactures three strengths of salsa: Lightweight, Hot Stuff, and Burn Your Tongue. The selling prices and manufacturing costs of these salsas are as follows:

	Lightweight	Hot Stuff	Burn Your Tongue
Selling price per jar	$3.00	$4.00	$6.00
Manufacturing cost per jar	2.00	2.00	4.50

For every 9 jars sold, 3 are Lightweight, 2 are Hot Stuff, and 4 are Burn Your Tongue.

Required: (1) What is the contribution margin percentage per "unit"?
(2) By how much would profits change if a $1,000 advertising campaign increased the sales of Lightweight salsa by 900 jars but left sales of the other salsas unchanged?
(3) Sales are currently $9,430, and fixed costs total $2,125.
    (a) How many jars of Lightweight salsa were sold?
    (b) How many jars of Burn Your Tongue salsa were sold?
    (c) What is the current pretax income?
(4) If the sales mix changes to 3 jars of Lightweight, 3 of Hot Stuff, and 4 of Burn Your Tongue, what would be the new contribution margin percentage?

11-34 The Grandma Corporation manufactures two products—cookies and candy. Cookies have a contribution margin of $4 per box, and candy has a contribution margin of $5 per bag. Grandma's total fixed cost is currently $450,000. Grandma expects to sell two boxes of cookies for every three bags of candy sold. Boxes of cookies and bags of candy each sell for $10.

Required: (1) Compute the contribution margin percentage per "unit."
(2) At the current product mix, what total dollar sales volume is required for Grandma to earn a pretax income of $217,000?

11-35 Greco Manufacturing Corporation produces two products—olives and baklava. Olives require $3 of variable costs per jar and sell for $5 per jar. Baklava has variable costs of $5 per box and sells for $10. Greco's total fixed costs amount to $72,250. This year Greco sold 30,000 jars of olives and 5,000 boxes of baklava. Greco believes that consumer tastes will shift dramatically next year. Although it expects total dollar sales volume to be the same as this year's sales volume, the product mix will change, so that one-third of the units sold will be boxes of baklava.

Required: (1) Compute Greco's pretax income for *this* year.
(2) At what dollar sales volume would Greco have broken even for *this* year?
(3) Compute Greco's expected pretax income for *next* year assuming total dollar sales volume does not change.
(4) Why does Greco expect more pretax income next year than it earned this year when the total dollar sales volume is expected to be the same? Explain.

11-36 The Boston Company (a sole proprietorship) expects to operate at a loss next year on its two products, as shown here:

	Commons	Not So Commons	Total
Production and sales (units)	100,000	20,000	120,000
Sales revenue	$200,000	$100,000	$300,000
Variable costs	(140,000)	(40,000)	(180,000)
Contribution margin	$ 60,000	$ 60,000	$120,000
Less: Fixed costs			(172,240)
Loss			$ (52,240)

Boston has two plans that it believes will improve its profit (reduce its loss) next year:

Plan A—to spend $53,000 on advertising to increase the number of *Not So Commons* sold without affecting the number of *Commons* sold

Plan B—to reduce the selling price of *Not So Commons* from $5 per unit to $4 per unit; this plan should change the product mix so that one *Not So Common* is sold for every two *Commons*

Required: (1) If Boston follows neither of the two plans, so that its product mix is one *Not So Common* sold for each five *Commons* sold, what must total dollar sales volume be for the company to break even?

(2) If Boston follows Plan A, how many *Not So Commons* must it sell next year to break even? (The number of *Commons* sold will still be 100,000.)

(3) If Boston follows Plan B, what total dollar sales volume is required for it to break even?

(4) Compare your answers to (1) and (3). Explain the result obtained from this comparison.

11-37 Professional Robotics Corporation manufactures three different robots in its "domestic servants" line. Sales information for this line of product is given below:

	Butler	Maid	Bartender
Selling price per unit	$3,000	$2,500	$4,000
Variable cost per unit	750	750	1,300
Expected sales mix	1	3	4

Fixed costs are $2,500,000.

Required: (1) What is Professional Robotics' break-even point, in "units" of sales mix?

(2) To break even, how many Butlers, Maids, and Bartenders must Professional Robotics sell?

(3) Suppose Professional Robotics' sales during the year were $4,800,000. Assuming a normal sales mix, how many units of each product were sold? What was Professional Robotics' pretax income?

(4) Assuming a normal sales mix and an income tax rate of 40%, how many Butlers, Maids, and Bartenders must Professional Robotics sell to earn an after-tax profit of $198,000?

Making Evaluations

11-38 Suppose you and your brother Noah, a veterinary medicine major, want to open a pet store. After long deliberations and lively discussion about the name of the corporation, you agree to name it Noah's Bark. You arrange to obtain retail space, to purchase supplies and pets, and to advertise in the newspaper. Now you are almost ready to open for business. But first you want to analyze how your plans will affect profit.

You and Noah plan to sell Labrador retrievers for $350 each, and you estimate that total fixed costs for Noah's Bark will be $6,375 ($2,000 rent, $4,000 salaries, and $375 advertising). The cost to you of purchasing the Labrador retriever puppies is $120 each. On average, you spend $60 to feed each puppy before it is sold and $30 per puppy on miscellaneous items such as dipping and grooming supplies.

Required: (1) How many dogs must Noah's Bark sell to break even?

(2) How much pretax income will Noah's Bark earn if it sells 750 dogs?

(3) How many dogs must Noah's Bark sell to earn $14,000 pretax income?

(4) What must dollar sales be to break even?

(5) What must dollar sales be to earn $14,065 pretax income?

Suppose you and your brother believe that if the selling price does not change, Noah's Bark will be able to sell 200 Labrador retrievers next year. At that unit sales volume, profit is expected to be $21,625 ([$140 contribution margin per dog × 200 dogs] − $6,375 total fixed costs). However, you are considering three alternative plans (only one of which will be followed) that you believe may allow Noah's Bark to earn even more than $21,625. These plans are as follows:

(a) Raise the selling price of the dogs to $450 per dog. With this alternative, variable costs per dog and total fixed costs do not change.

(b) Purchase only those Labrador retrievers that are descendants of American Kennel Club (AKC) ribbon winners, thus increasing the variable cost to $240 (these dogs can be purchased for $150 each). You are considering this alternative because you think the perceived improvement in the purity of the breed will increase the sales volume of dogs. With this change, neither the selling price per dog nor the total fixed cost changes.

(c) Increase total fixed costs by spending $1,000 more on advertising. With this alternative, the selling price per dog and the variable costs per dog do not change.

Required: (6) Basing your decision on C-V-P results alone, which plan should you choose? Why?

(7) What other factors might you want to consider in making this choice?

11-39 Suppose that after Noah's Bark (in 11-38) has been in business for a year, you and your brother Noah decide to sell beagles in addition to Labrador retrievers. Each beagle sells for $150, requires $60 of variable cost, and earns a contribution margin of $90. Each Labrador retriever still sells for $350, requires $210 of variable cost, and earns a contribution margin of $140. The contribution margin earned by Noah's Bark in the second year of business from beagles, from Labrador retrievers, and in total is shown here:

	Labrador Retrievers	Beagles	Total
Sales (dogs)	600	200	800
Sales revenue	$210,000	$30,000	$240,000
Less variable costs	(126,000)	(12,000)	(138,000)
Contribution margin	$ 84,000	$18,000	$102,000

Assume that fixed costs for Noah's Bark are now $76,500 per year.

Required: (1) Compute the break-even point (in sales dollars).

(2) What must sales be (in dollars) for Noah's Bark to earn $30,000 pretax income?

(3) What is the break-even point in beagles and Labs?

(4) How many beagles and Labs must Noah's Bark sell to earn a pretax income of $30,000?

(5) How many beagles and Labs must it sell to earn a net income of $30,000? The income tax rate is 40%.

Suppose you and your brother are considering three independent alternative plans (only one of which will be followed) that you believe may raise the company's income. These plans are as follows:

(a) Raise the selling price of the Labs to $450 per dog. With this alternative, variable costs per dog and total fixed costs do not change.

(b) Purchase only those Labrador retrievers that are descendants of American Kennel Club (AKC) ribbon winners, thus increasing the variable cost to $240 (these dogs can be purchased for $150 each). You are considering this alternative because you think the perceived improvement in the purity of the breed will increase the sales volume of dogs. With this change, neither the selling price per unit nor the total fixed cost changes.

(c) Increase total fixed costs by spending $1,000 more on advertising. With this alternative, the selling price per unit and the variable costs per unit do not change.

Required: (6) Based on C-V-P analysis alone, which plan should you choose? Why?

(7) What other factors might you want to consider in making this choice?

11-40 Lucas Air Service, a sole proprietorship, provides charter flights on weekdays only. Earl Lucas, the owner, is thinking of offering flying lessons on weekends. He has always protected his weekends for family time and will give up his family weekends only if he can earn at least $12,000 per year extra. Estimates of demand for lessons suggest that, weather permitting, he could teach a full six-lesson day every Saturday and Sunday for the entire year.

Earl's large plane could be used for about half of the lessons. He would borrow a smaller plane, owned by his friend Pat, for the other half. Fuel and maintenance would run about $45 per flying lesson with the large plane. With the small plane, fuel would cost $25 per lesson, and Earl would pay Pat $75 per day when he used the small plane regardless of how many lessons he gave in it.

Earl plans to use the large plane for advanced students and to charge them $85 for a lesson. He would use the small plane for beginners and charge them $75 per lesson. Each weekend day that Earl opens for flight lessons he will incur $35 for the salary of his receptionist, utilities, and other expenses to keep his office open.

Required: (1) Prepare a schedule showing the daily profit if
(a) Earl gives from 1 to 6 advanced lessons only
(b) Earl gives from 1 to 6 beginner lessons only
(c) Earl offers both types of lessons on a given day for a total of 6 lessons (make the profit computation for only two situations: the situation where Earl gives 3 lessons of each type and the situation where Earl gives 4 advanced lessons and 2 beginner lessons).
(2) Is it more or less profitable for Earl to give 4 advanced lessons and 2 beginner lessons each day than to give 3 lessons of each kind each day?
(3) If Earl gives 3 flight lessons of each type each day, how much profit could he earn if he gives lessons for 90 days?
(4) How much profit could Earl earn if he gives 6 advanced lessons in the large plane for each of 45 days and 6 beginner lessons in the small plane for each of 45 days?
(5) How much profit could Earl earn if he gives 6 advanced lessons in the large plane for each of 60 days and 6 beginner lessons in the small plane for each of 30 days?
(6) Explain why the mix of two-thirds advanced and one-third beginner lessons seems better than the mix of half advanced and half beginner lessons in the case where only advanced or only beginner lessons are scheduled on a given day.

11-41 Golden Chocolate Inc. of Brooklyn, a manufacturer of chocolate bars, learned that a Connecticut child suffered a mild allergic reaction to the nuts in its candy bar. The wafer in the chocolate bar contains ground hazelnuts, but the candy bar's label didn't include nuts in its list of ingredients.[9] Suppose you have been hired to advise Golden Chocolates about how to respond to this situation and how to prevent related potential problems.

Required: Assume Candace Sugarbaker is the company president. Write her a memo making a recommendation or recommendations. For each recommendation, describe the effect you think it will have on Golden Chocolate's sales, fixed costs, variable costs, and break-even point. Include any other factors you think Golden's managers should consider when they make this decision.

11-42 The National Fishing Heritage Center in Grimsby, a port in northeastern England, is luring a record number of visitors (up 27 percent from last year) with some very fishy techniques: Scratch 'n' Sniff leaflets; Whiff You Were Here postcards; and "Smelloons"—

---

[9]"Chocolate Bar Recalled as Allergy Precaution," *Columbia Daily Tribune*, April 22, 1995, 2A.

balloons filled with fresh-fishy odors. Scents include "Hint of Haddock," "Compressed Cod," and "Sentiment of Seaweed."[10] The center views this effort as a major success.

Required: Working with your class team, identify the changes in fixed and variable costs caused by this tourist campaign. Besides the increase in visitors, what criteria would you use to measure the success of this effort? What questions would you ask to help you determine its success?

11-43   Chico, Maria, Elaina, and Juan are taking a year off from school to form a partnership to develop the commercial potential of a cleaning solvent they accidentally discovered in their college chemistry lab. In liquid form, their product can remove grease stains from clothing and can launder shop towels so that they appear like new. In solid form (actually a buttery consistency), it can make a mechanic's hands look like those of a baby in two minutes, and as a foam, it can lift grease and oil spots out of concrete in a jiffy.

Chico's investigation of manufacturing requirements suggests that small-scale manufacturing is quite feasible and that output could be adjusted over a wide range of volumes with little change in efficiency. Potential suppliers of raw materials have been contacted, and a long list of friends who are eager for part-time work ensures the availability of direct labor.

Maria has located a steel farm building for rent near the interstate highway in an area recently rezoned for commercial and industrial use. Initial cleanup, some insulation, additional lighting, gravel for the drive and loading area, and a small amount of additional office space are all that seem to be necessary to make the building suitable for their needs. She also has acquired a delivery van, which was used by a local florist before his retirement.

Elaina has worked out all of the details for advertising and other promotional activities that would be needed for the three forms of the product. She and Juan have determined the range of possible sales volumes for each form, assuming primary customers to be commercial laundries for the liquid, auto service centers for the solid, and service stations for the foam. Juan has determined optimal delivery routes and schedules to cover the midstate area and is now considering the possibility of adding delivery to two major metropolitan areas.

After much planning and several meetings with local financial institutions, the four have concluded that the product can be produced and marketed in only one of the three forms—liquid, solid, or foam—until successful financial performance of the venture is established. This success would require at least $30,000 of profit to be earned during the first year.

The four asked for your advice recently at a party. Rather than talk business at the party, you suggested that they send you the information they had gathered. A few days after the party, you received the following information from them:

	Liquid	Solid	Foam
Selling price	$3.00	$3.50	$11.00
Variable costs:			
Direct materials	$0.40	$0.80	$3.00
Direct labor	0.10	0.20	0.60
Variable overhead	0.15	0.30	0.90
Variable selling and administrative	0.35	0.20	0.50
Total variable costs	$1.00	$1.50	$5.00
Fixed costs:			
Factory overhead	$ 5,000	$15,000	$85,000
Selling and administrative	10,000	15,000	5,000

Sales volume information (all possible sales volumes between the estimated minimum and maximum are equally likely):

	Liquid	Solid	Foam
Maximum sales volume	20,000 bottles	40,000 tubs	25,000 cans
Minimum sales volume	12,000	15,000	5,000
Expected sales volume	16,000	27,500	15,000

[10]"Something Fishy," *Fortune,* July 25, 1993, 14.

Required: (1) For each of the three forms of product, what is the
        (a) maximum possible profit?
        (b) minimum possible profit?
        (c) expected profit?
        (d) sales volume needed to earn $30,000?
      (2) Which form of product seems favored relative to each of the calculations you made in (1)?
      (3) Which form of the product would you recommend they introduce during the first year of operations? Why?

11-44 You talk on the phone a lot to your friends back home. Each phone call is long distance, and you have been using a "10 cents a minute" phone card to pay for each call. You are considering signing up for a long-distance calling plan with a phone company. The company has two plans from which to choose. Plan 1 requires a person to pay $4.95 per month plus $0.07 per minute. Plan 2 requires a person to pay $8.95 per month plus $0.05 per minute. You have reviewed your calls in previous months and, on the average you talk to your friends long distance about 240 minutes per month.

Required: (1) What is the "break-even" point (in minutes) where it makes no difference which plan you select?
      (2) Which of the two plans should you select? Why is it less expensive?

11-45 Yesterday, you received the following letter for your advice column in the local paper:

DR. DECISIVE

Dear Dr. Decisive:

I read your column every day, and I think you can help me with this problem I am having with my boyfriend. This morning, as we were walking to the gym, we started having a nice, normal conversation about my aerobics class. Then, before I knew it, we had a major parting of the ways.

This all started when I mentioned that I didn't think it was fair that we had to pay a fee for each aerobics class we attended and that the $150 facilities use fee that each student pays at the beginning of the semester ought to cover it. He looked at me like I was an idiot and informed me that I was WRONG and must not have been paying attention in my accounting class. According to him, it is obvious that the aerobics fee is assessed to cover the aerobics instructor's salary. Well, it's not so obvious to me, and I need for you to tell him how wrong he is (he reads your column too). Until we get this cleared up, we can't go to the gym together. (How could I work out with such a pompous FOOL?). Just sign me . . .

"Nobody's Fool"

Required: Meet with your Dr. Decisive team and write a response to "Nobody's Fool."

Federal Income Taxation— An Overview

CHAPTER LEARNING OBJECTIVES

- Discuss what constitutes a tax and the various types of tax rate structures that may be used to calculate a tax.

- Introduce the major types of taxes in the United States.

- Identify the primary sources of federal income tax law.

- Define *taxable income* and other commonly used tax terms.

- Introduce the calculation of taxable income for individual taxpayers and the unique personal deductions allowed to individuals.

- Develop a framework for tax planning and discuss the effect of marginal tax rates and the time value of money on tax planning.

- Make the distinction between tax avoidance and tax evasion.

- Introduce ethical considerations related to tax practice.

We have all heard the adage, "There's nothing certain but death and taxes." However, equating death and taxes is hardly a fair characterization of taxation. It is often stated that taxes are the price we pay for a civilized society. An early decision of the U.S. Supreme Court described a tax as "an extraction for the support of the government." Regardless of your personal view of taxation, society as we know it could not function without some system of taxation. People constantly demand that the government provide them with various services, such as defense, roads, schools, unemployment benefits, medical care, and environmental protection. The cost of providing the services that the residents of the United States demand is principally taxation. People are introduced to taxation at an early age. Remember the candy bar that had a price sticker of 25 cents yet actually cost 27 cents? The tax collector is all around us. Upon receiving their first paycheck, many are surprised that the $100 they earned resulted in a check of only $80 after taxes were deducted. The point is that taxes are a fact of life. Learning to deal with taxes, and perhaps using them to your advantage, is an essential element of success in today's world.

The federal income tax is a sophisticated and complex array of laws that imposes a tax on the income of individuals, corporations, estates, and trusts. Current tax law has developed over a period of more than 85 years through a dynamic process involving political, economic, and social forces. At this very minute, Congress is considering various changes in the tax law; the Internal Revenue Service (IRS) and the courts are issuing new interpretations of current tax law, and professional tax advisers are working to determine the meaning of all these changes.

The purpose of this book is to provide an introduction to the basic operation of the federal income tax system. However, before looking at some of the specifics, it is helpful to have a broad understanding of taxes and how the federal income tax fits into the overall scheme of revenue production. Toward this end, this chapter briefly discusses what constitutes a tax, how taxes are structured, and the major types of taxes in the United States before considering the federal income tax. Next, the primary sources of tax law authority are introduced. These sources provide the basis for calculating the tax and the unique terminology of federal income taxation. This chapter also introduces the tax calculation for individuals, the discussion of which serves as a reference for discussions in succeeding chapters. The next section of the chapter provides a framework for tax planning and a discussion of tax avoidance and tax evasion.

Because ethics is an important issue in the accounting profession, the chapter concludes with a brief discussion of the ethical considerations related to tax practice. The discussion provides the background that will help you detect ethical issues that you will face if you go on to practice in the tax area.

Because this is a tax text, one starting point is to define what is meant by the term *tax*. Particular types of taxes and tax rules are often criticized as being loopholes, unfair, or creating an excessive burden on a particular group of taxpayers. The discussion that follows presents the four criteria commonly used to evaluate these criticisms. In addition, three types of tax rate structure are presented as an aid in evaluating whether a particular tax is "good" or "bad."

Definition of a Tax

What is a tax? The Internal Revenue Service defines a tax as "an enforced contribution, exacted pursuant to legislative authority in the exercise of the taxing power, and imposed and collected for the purpose of raising revenue to be used for public or governmental purposes. Taxes are not payments for some special privilege granted or service rendered and are, therefore, distinguishable from various other charges imposed for particular purposes under particular powers or functions of government."[1]

A tax could be viewed as an involuntary contribution required by law to finance the functions of government. The amount of the contribution extracted from the taxpayer is unrelated to any privilege, benefit, or service received from the government agency imposing the tax. According to the IRS definition, a tax has the following characteristics:

1. The payment to the governmental authority is required by law.
2. The payment is required pursuant to the legislative power to tax.

3. The purpose of requiring the payment is to provide revenue to be used for public or governmental purposes.
4. Special benefits, services, or privileges are not received as a result of making the payment. The payment is not a fine or penalty that is imposed under other powers of government.

Although the IRS definition states that the payment of a tax does not provide the taxpayer with directly measurable benefits, the taxpayer does benefit from, among other things, military security, a legal system, and a relatively stable political, economic, and social environment. Payments to a government agency that relate to the receipt of a specific benefit—in privileges or services—are not considered taxes; they are payments for value received or are the result of a regulatory measure imposed by the government agency.

▶ **EXAMPLE 1**   Keith lives in Randal County, which enacted a law setting a 1% property tax to provide money for county schools. The 1% tax applies to all property owners in Randal County. All schoolchildren in the county will benefit from the tax, even if their parents do not own property or pay the tax. Is the 1% property tax a tax according to the definition?

Discussion: The property tax is a tax. The tax is a required payment to a government unit. The payment is imposed by a property tax law. The purpose of the payment is to finance public schools. The tax is levied without regard to whether the taxpayer receives a benefit from paying the tax.

▶ **EXAMPLE 2**   Assume that in example 1, the tax is imposed on a limited group of property owners to finance the construction of sewer lines to their properties. Is the 1% tax a tax as defined by the IRS?

Discussion: Each payer of the tax receives a direct benefit—a new sewer line. Therefore, the 1% tax payment is considered a payment to the government unit to reimburse it for improvements to the taxpayer's property. The taxpayers would treat the payment as an investment in their property and not as a tax. The 1% tax in this case is a special assessment for local benefits. An assessment differs from a tax in that an assessment is levied only on a specific group of taxpayers who receive the benefit of the assessment.

Certain payments that look like a tax are not considered a tax under the IRS definition. For example, an annual licensing fee paid to a state to engage in a specific occupation such as medicine, law, or accounting is not a tax, because it is a regulatory measure that provides a direct benefit to the payer of the fee. A fee paid for driving on a toll road, the quarter deposited in a parking meter, and payments to a city for water and sewer services are payments for value received and are not taxes according to the IRS's definition. Fines for violating public laws and penalties on tax returns are not taxes. Fines and penalties are generally imposed to discourage behavior that is harmful to the public interest and not to raise revenue to finance government operations.

Standards for Evaluating a Tax

In *The Wealth of Nations,* Adam Smith identified four basic requirements for a good tax system. Although other criteria can be used to evaluate a tax, Smith's four points are generally accepted as valid and provide a basis for discussion of the primary issues regarding taxes. These requirements are equality, certainty, convenience, and economy. Although Smith clearly stated the maxims, taxpayers have different opinions as to whether the federal income tax strictly satisfies the four requirements.

1. Equality—A tax should be based on the taxpayer's *ability to pay.* The payment of a tax in proportion to the taxpayer's level of income results in an equitable distribution of the cost of supporting the government.

The concept of equality requires consideration of both horizontal and vertical equity. **Horizontal equity** exists when two similarly situated taxpayers are taxed the same. **Vertical equity** exists when taxpayers with different situations are taxed differently but fairly in relation to each taxpayer's ability to pay the tax. This means that those taxpayers who have the greatest ability to pay the tax should pay the greatest proportion of the tax. These equity concepts are reflected to a great extent in the federal income tax. Certain low-income individuals pay no tax. As a person's taxable income level increases, the tax rate increases from 10 percent to 15 percent to 25 percent to 28 percent to 33 percent to 35 percent.

▶ EXAMPLE 3   Tom and Jerry each earn $15,000 a year and pay $1,500 in tax.

Discussion: The two taxpayers pay the same amount of tax on the same amount of income. Because they are treated the same, based on the facts given, horizontal equity exists.

Discussion: A slight change of facts provides a different result. If Tom is married and supports his wife and 3 children and Jerry is single with no one else to support, the tax appears unfair and not vertically equitable. The lack of vertical equity exists because the taxpayers' situations are no longer the same, yet they pay the same amount of tax on the same income.

▶ EXAMPLE 4   Assume that because of the size of his family, Tom (example 3) pays $500 in taxes. Jerry still pays $1,500.

Discussion: In this situation, vertical equity is considered to be present. Because he presumably has a greater ability to pay tax, Jerry pays a larger amount of tax than Tom—Jerry's income, although equal to Tom's, supports fewer people.

Some taxpayers consider inequitable the tax law provisions that treat similar income and deductions differently. For example, a person investing in bonds issued by a city does not have to pay tax on the interest income. In contrast, interest income earned on an investment in corporate bonds is taxed. People who operate proprietorships may deduct the cost of providing their employees with group term life insurance but may not deduct the cost of their own group insurance premium. If the proprietor incorporates, the cost of the insurance for both the shareholder-employee (owner) and employees can be deducted. Thus, the perception of equality often depends on the taxpayer's personal viewpoint. Because the concepts of equity are highly subjective, a tax rule considered equitable by one taxpayer is often considered unfair by a taxpayer who derives no benefit. Often, when evaluating the equality of a tax provision, taxpayers do not consider—or are not aware of—the economic, social, and administrative reasons for what may seem to be an inequity in the tax law.

▶ EXAMPLE 5   Karen is a single mother who earns $10,000 a year. Jane and her husband, Ben, earn $75,000 a year. Karen and Jane each pay Neighborhood Day Care $2,000 per year for taking care of one child while they work. Because the payment is for qualified child care, Karen is entitled to a $700 reduction in her income tax because of her low income level. Because of their high income level, Jane and Ben receive only a $400 reduction in their income tax. Who is more likely to view this treatment as being inequitable?

Discussion: Jane and Ben may view the tax rule as unfair, because Karen receives a larger reduction in tax for the same amount of payment for day care. However, there is increasing emphasis on tax relief for families. Congress has decided that it is important that children be adequately cared for while parents are at work. Thus, Karen's family is given a larger tax break to help provide child care. Without the larger tax reduction, Karen might not be able to afford to pay child-care costs. The difference in treatment could also be based on the ability to pay child-care costs. In addition, the difference in treatment depicts a situation of vertical equity. Because Jane and Ben have higher incomes, vertical equity requires that they pay a higher tax (through receiving a smaller tax credit).

    2. Certainty—A taxpayer should know when and how a tax is to be paid. In addition, the taxpayer should be able to determine the amount of tax to be paid.

Certainty in the tax law is necessary for tax planning. An individual's federal income tax return is due on the fifteenth day of the fourth month (usually April 15) after the close of the tax year. A corporation's return is due on the fifteenth day of the third month after the close of its tax year.[2] The balance of tax due with the return is usually paid by check to the IRS. However, determining the amount of tax due may not be so simple. When planning an investment that will extend over several tax years, the ability to predict with some degree of certainty how the results of the investment will be taxed is important to the investment decision.

Frequent changes in the tax law create uncertainty for the tax planner. For example, the Taxpayer Relief Act of 1997 amended approximately 800 code sections and added nearly 300 sections. In addition to these legislative amendments to the tax law, the IRS and the courts issue a constant stream of decisions and interpretations on tax issues, which results in a tax law that is in a continual state of refinement. However, for the average individual taxpayer, who has wages subject to withholding, receives some interest income, owns a home, pays state and local taxes, and perhaps donates to a church or other charities, there is little complexity and a great deal of certainty in the tax law despite the numerous changes to the tax system.

> **3. Convenience**—A tax should be levied at the time it is most likely to be convenient for the taxpayer to make the payment. The most convenient time for taxpayers to make the payment is as they receive income and have the money available to pay the tax.

Most taxpayers would argue that it is not convenient to keep records, determine the amount of tax due, and fill out complex forms. However, certain aspects of the income tax law make it more convenient than it might be otherwise. Based on the **pay-as-you-go concept,** taxes are paid as close to the time the income is earned as is reasonable. The pay-as-you-go system results in the collection of the tax when the taxpayer has the money to pay the tax. This tax payment system applies to all taxpayers, including the self-employed and those who earn their income from investing activities. This system is discussed in more detail later in this chapter.

The federal income tax is based on self-assessment and voluntary compliance with the tax law. Taxpayers determine in privacy the amount of their income, deductions, and tax due. The tax calculated by the taxpayer is considered correct unless the IRS detects an error and corrects it or selects the return for an audit. The federal income tax system relies on the honesty and integrity of taxpayers in determining their tax payments. This system of self-assessment and voluntary compliance promotes convenience for taxpayers.

> **4. Economy**—A tax should have minimum compliance and administrative costs. The costs of compliance and administration should be kept at a minimum so that the amount that goes to the U.S. Treasury is as large as possible.

The IRS operates on a budget of about one half of 1 percent of the total taxes collected. However, the IRS's budget does not reflect the full cost of administering the tax law. A taxpayer's personal cost of compliance can be substantial. Taxpayers often need to maintain accounting records for tax reporting in addition to those that are necessary for business decisions. A corporation, for example, might use different depreciation methods and asset lives for financial reporting and for income tax. The taxpayer's personal cost also includes fees paid to attorneys, accountants, and other tax advisers for tax-planning, compliance, and litigation services.

Tax Rates and Structures

Tax rates are often referred to as a *marginal rate,* an *average rate,* or an *effective rate.* In addition, a tax rate structure is frequently described as being *proportional, regressive,* or *progressive.* Because a tax rate structure indicates how the average tax rate varies with changes in its tax base, examining a rate's structure helps in understanding and evaluating the effect of a tax.

To compute a tax, it is necessary to know the tax base and the applicable tax rate. The tax is then computed by multiplying the tax base by the tax rate:

$$\text{Tax} = \text{Tax base} \times \text{Tax rate}$$

A **tax base** is the value that is subject to tax. The tax base for the federal income tax is called **taxable income.** Other common tax bases include the dollar amount of a purchase subject to sales tax, the dollars of an employee's wages subject to payroll tax, and the assessed value of property subject to property tax.

Tax Rate Definitions. When working with the federal income tax, different measures of the rate of tax paid from one year to the next are often compared to evaluate the effectiveness of tax planning and to help make decisions about future transactions. Three different rates are commonly used for these comparisons:

- The marginal tax rate

- The average tax rate
- The effective tax rate

The **marginal tax rate** is the rate of tax that will be paid on the next dollar of income or the rate of tax that will be saved by the next dollar of deduction. The marginal tax rate is used in tax planning to determine the effect of reporting additional income or deductions during a tax year. One objective of tax planning is to minimize the marginal rate and to keep the marginal rate relatively constant from one year to the next. The marginal tax rates for an individual taxpayer are 10 percent, 15 percent, 25 percent, 28 percent, 33 percent, and 35 percent.[3] If you know a person's taxable income (the tax base), you can find the marginal tax rate in the tax rate schedules in Appendix B.

EXAMPLE 6 Don has an asset he could sell this year at a $10,000 profit, which would increase his marginal tax rate from 15% to 28%. If he waits until next year to sell the asset, he is sure his other income will be less and the $10,000 gain will be taxed at 15%. Should Don sell the asset this year or wait until next year?

Discussion: By waiting until next year to sell the asset, Don's tax savings on the sale are $1,300 [$10,000 × (28% − 15%)]. In addition, he will postpone the payment of the tax interest-free for a year (a time value of money savings). Assuming that he can sell the asset early in the next year and does not need the proceeds from the sale before next year, he should wait until next year to sell the asset to take advantage of the lower marginal tax rate and the time value of money savings on the tax to be paid on the gain.

The **average tax rate** is the total federal income tax divided by taxable income (the tax base). This is the average rate of tax on each dollar of income that is taxable. The **effective tax rate** is the total federal income tax divided by the taxpayer's economic income (taxable income plus nontaxable income). Economic income is a broader base; it includes all the taxpayer's income, whether it is subject to tax or not. The effective tax rate is the average rate of tax on income from all taxable and nontaxable sources.

EXAMPLE 7 Assume that in example 6, Don sells the asset in 2004 and reports taxable income of $40,000. Also, Don collects $50,000 on a life insurance policy that is not taxable income. Don's tax on $40,000 is $6,738 (using the tax rate schedules in Appendix B). In addition, the only difference between Don's economic income and his taxable income is proceeds from the life insurance policy. What are Don's marginal, average, and effective tax rates?

Discussion: Based on the facts given, Don's marginal tax rate is 25% (from the tax rate schedules). His average tax rate is 16.8% ($6,738 ÷ $40,000). The effective tax rate on his economic income of $90,000 ($40,000 in taxable income + $50,000 in nontaxable income) is 7.49% ($6,738 ÷ $90,000) and is much less than both the marginal and average tax rates.

Tax Rate Structures. Tax rate structures are described as being proportional, regressive, or progressive. The structures explain how the tax rates vary with a change in the amount subject to the tax (the tax base).

PROPORTIONAL RATE STRUCTURE. A **proportional rate structure** is defined as a tax for which the average tax rate remains the same as the tax base increases. This rate structure is also referred to as a *flat tax*. If you charted a proportional tax rate structure on a graph, it would look like Chart 1 in Figure 1–1.

If a tax rate is proportional, the marginal tax rate and the average tax rate are the same at all levels of the tax base. As the tax base increases, the total tax paid will increase at a constant rate. Examples of proportional taxes are sales taxes, real estate and personal property taxes, and certain excise taxes, such as the tax on gasoline. The sales tax is a fixed percentage of the amount purchased, property tax is a constant rate multiplied by the assessed value of the property, and the gas tax is a constant rate per gallon purchased.

EXAMPLE 8 Betsy bought a new suit for $350. The sales tax at 7% totaled $24.50. Steve bought a new lawn tractor for $3,500. At 7%, the sales tax he paid came to $245. Is the sales tax proportional?

Discussion: Betsy's and Steve's marginal tax rate is 7%. In addition, Betsy's average tax rate is 7% ($24.50 ÷ $350), the same as Steve's (7% = $245 ÷ $3,500). The sales tax is proportional, because the marginal and average tax rates are equal at all levels of the tax base (the selling price).

REGRESSIVE RATE STRUCTURE. A **regressive rate structure** is defined as a tax in which the average tax rate decreases as the tax base increases. On a graph, a regressive tax rate structure would look like Chart 2 in Figure 1–1.

If a tax rate structure is regressive, the marginal tax rate will be less than the average tax rate as the tax base increases. Note that although the average tax rate and the marginal tax rate both decrease as the tax base increases, the total tax paid will increase. As a result, a person with a low tax base will pay a higher average and a higher marginal rate of tax than will a person with a high tax base. The person with the high tax base will still pay more dollars in total tax. Although a pure regressive tax rate structure (as defined earlier) does not exist in the United States, example 9 illustrates a regressive tax.

▶ **EXAMPLE 9**   Each year, Alan purchases $4,000-worth of egg rolls and Tranh purchases $17,000-worth of egg rolls. A tax is levied according to the dollar value of egg rolls purchased per the following tax schedule:

Tax Rate Schedule		Alan		Tranh	
Base	**Rate**	**Purchases**	**Tax**	**Purchases**	**Tax**
$-0- < $5,001	10%	$4,000	$400	$ 5,000	$ 500
$5,001 < $10,001	7%			5,000	350
More than $10,000	5%			7,000	350
Totals		$4,000	$400	$17,000	$1,200
Marginal tax rate			10%		5.0%
Average tax rate			10%		7.1%

Discussion: This tax rate schedule is regressive. The average tax rate applicable to Alan (10%) is greater than the average tax rate for Tranh (7.1%), even though Tranh's tax base is higher. Note that Tranh pays more total tax ($1,200) than Alan ($400).

If a different base is used to evaluate the tax rate structure, the same tax that may be viewed as proportional by one taxpayer may be considered regressive by another taxpayer. For example, using total wages as the tax base for evaluation, a person who spends part of her wages for items subject to sales tax would pay a lower average rate of tax than the person who spends all of his wages on taxable items.

▶ **EXAMPLE 10**   Judy earns $25,000 a year and spends it all on items subject to sales tax. Guillermo earns $30,000 a year and is able to save $5,000 of his earnings. He spends the remaining $25,000 on purchases subject to sales tax. If the sales tax rate is 10% of purchase price, is it a regressive tax?

Discussion: Judy and Guillermo pay the same total sales tax ($2,500). Thus, the tax is proportional when evaluated by using purchases as the tax base. However, Guillermo's average tax rate based on wages [8.3% = ($2,500 ÷ $30,000)] is less than Judy's [10% = ($2,500 ÷ $25,000)]. Thus, the sales tax is regressive when using wages to evaluate the tax.

Although property taxes are a proportional tax according to these definitions, an investor in property subject to property taxes might consider the effect of the tax on investments regressive compared with investments in stocks and bonds, which are not subject to property taxes. Similarly, low-income wage earners who pay Social Security tax on all their wages may consider this tax regressive compared with a person whose wages exceed the amount subject to the tax.

PROGRESSIVE RATE STRUCTURE. A **progressive rate structure** is defined as a tax in which the average tax rate increases as the tax base increases. On a graph, a progressive tax rate structure would look like Chart 3 in Figure 1–1.

If a tax rate structure is progressive, the marginal tax rate will be higher than the average tax rate as the tax base increases. The average tax rate, the marginal tax rate, and the total tax all increase with increases in the tax base. A person with

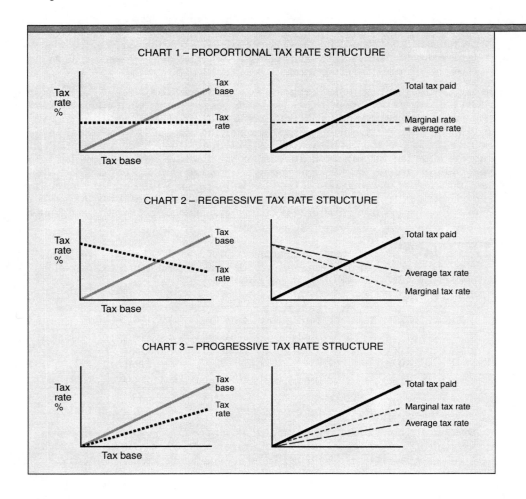

Figure 1–1
TAX RATE STRUCTURES

a low tax base will pay both lower average and marginal rates of tax than will a person with a high tax base.

The progressive tax rate structure reflects the embedding in the federal income tax rates of Adam Smith's equality criterion. Recall that according to this criterion, taxpayers should pay according to their ability to pay the tax. The use of progressive rate structures, wherein people with higher taxable income levels pay higher marginal tax rates, promotes equality.

▶ **EXAMPLE 11** Doug reports $16,000 a year in taxable income from wages he earns watering the greens at the Hot Water Golf Course. Shawana earns $35,000 in annual taxable income as a first grade teacher.

Discussion: Doug and Shawana's 2004 income taxes using the single taxpayer rates are as follows:

	Doug's Tax	**Shawana's Tax**
	(income: $16,000)	**(income: $35,000)**
Tax on income of $7,150 @ 10%	$ 715	$ 715
Tax on income from $7,150 to $29,050 @ 15%	1,328	3,285
Tax on income above $29,050 @ 25%	-0-	1,488
Total tax	$2,043	$5,488
Marginal tax rate	15%	25%
Average tax rate	12.8%	15.7%

Discussion: As a result of Shawana's larger tax base and the progressive tax rates, her marginal and average tax rates are higher than Doug's. Thus, the tax rate structure of the federal income tax promotes equality among taxpayers.

MAJOR TYPES OF U.S. TAXES

The federal, state, and local governments use a variety of taxes to fund their operations. Figure 1–2 depicts tax revenues generated by federal, state, and local government bodies by source for 1998. An examination of the sources of tax revenue shows that the bulk of the federal government's revenues is derived from the income tax and social insurance tax. State and local governments also receive a substantial portion of their revenues from a tax on income, with the sales tax and the property tax also providing significant revenue. In terms of overall taxes collected in the United States, the federal income tax produces almost as much revenue as all other forms of state and local taxes combined. Although this text covers the basic operation of the federal income tax, it is helpful to have a basic understanding of the other taxes levied by governments. As will be seen throughout the text, many taxes affect and interact with the rules for the federal income tax. Each major type of tax is discussed briefly in turn. Do not be concerned with the mechanics of the taxes at this point. Focus only on their general nature.

Income Taxes

The federal government levies a tax on the income of individuals, corporations, estates, and trusts. Most states also tax the income on these taxpayers, and a few local governments also impose an income tax on those who work or live within their boundaries. The income tax is levied on a *net* number—taxable income. In its simplest form, taxable income is the difference between the total income of a taxpayer and the deductions allowed that taxpayer. Thus, the study of income taxation is really the study of what must be reported as income and what is allowed as a deduction from that income to arrive at taxable income.

Each of the three government units that impose an income tax has its own set of rules for determining what is included in income and what is deductible from income to arrive at taxable income. However, because most state and local governments begin their taxable income calculations in relation to the federal income tax computation, an understanding of the federal income tax rules is essential for calculating most income taxes. This book makes no attempt to cover the myriad state and local income tax rules.

Income taxes are determined on an annual basis. However, the United States uses a pay-as-you-go collection system under which taxpayers pay an estimate of their tax as they earn their income. Employers must withhold income taxes from wages and salaries of their employees and remit them on a timely basis to the appropriate government body.[4] When taxpayers file their tax returns, these prepaid amounts are credited against their actual bill, resulting in either a refund of taxes, if the prepaid amount is greater than the actual tax, or an additional tax due, if the prepaid amount is deficient.[5] Self-employed taxpayers and those with other sources of income that are not subject to withholding (e.g., dividend and interest income)

Figure 1–2

1998 GOVERNMENT REVENUES BY SOURCE (ESTIMATED IN BILLIONS)

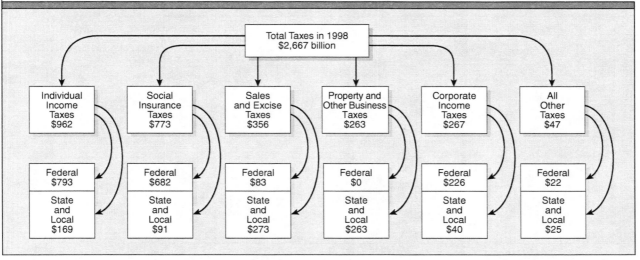

SOURCE: Tax Foundation, Inc., estimates based on National Income and Product Account definitions. Figures may not add up because of rounding.

must make quarterly estimated tax payments that are applied against their tax bills upon filing of the return.[6]

Employment Taxes

All employees and their employers pay taxes on the wages earned by employees. Employees pay **Social Security taxes** that are matched by their employers.[7] Self-employed individuals pay the equivalent of both halves of the Social Security tax by paying the **self-employment tax.**[8] In addition to the Social Security tax, employers pay unemployment compensation taxes to both the federal and state governments.

Social Security Taxes. Under the Federal Insurance Contribution Act (FICA), a tax is levied on wages and salaries earned. The Social Security system was originally designed to provide retirement benefits to all individuals who contributed to the system. This function has been expanded to include many other social programs, such as medical insurance, disability benefits, and survivor's benefits. The result of this expansion of coverage has been a great increase in the amount of Social Security taxes paid by workers and employers. It should be stressed that the Social Security system is not a "funded" system. Current payments into the system are used to pay current benefits; technically, any excess is placed in a fund. However, the federal government often borrows against this "fund" to pay general government expenses. Thus, there is no absolute guarantee that the amounts paid by current taxpayers will actually be available to them when they are eligible to receive their benefits.

The Social Security tax is imposed on employees and self-employed individuals. Employers are required to match employees' payments into the system.[9] Because a self-employed person is both an employee and an employer, the self-employment tax rate is twice the employee tax, resulting in an equivalent payment of tax by employee/employer and the self-employed.[10] The tax on employees and employers is a constant percentage of wages up to a maximum wage base. Both the percentage and the maximum wage base have been raised over time. As Table 1–1 shows, the tax has two components. In 2004, a tax of 6.2 percent is levied on the first $87,900 of wages for Old Age, Survivors, and Disability Insurance (OASDI). A tax of 1.45 percent on all wages pays for Medical Health Insurance (MHI). Before 1994, the MHI portion was subject to a maximum wage base. This established a maximum amount of Social Security tax that any taxpayer would pay. However, the abolition of the maximum MHI wage base for tax years after 1993 eliminated the ceiling on Social Security taxes.

▶ **EXAMPLE 12** Jenny earned $2,000 during February 2004 in her job as a carpenter for Acme Construction Company. How much Social Security tax must be paid by Jenny and Acme on her February earnings?

Year	OASDI[1]	MHI[2]	Total	Maximum Wage Base	Maximum Tax Paid
2000	6.20		6.20	$76,200	$4,724
		1.45	1.45	Wages earned	No maximum
			7.65		
2001	6.20		6.20	$80,400	$4,985
		1.45	1.45	Wages earned	No maximum
			7.65		
2002	6.20		6.20	$84,900	$5,264
		1.45	1.45	Wages earned	No maximum
			7.65		
2003	6.20		6.20	$87,000	$5,394
		1.45	1.45	Wages earned	No maximum
			7.65		
2004	6.20		6.20	$87,900	$5,450
		1.45	1.45	Wages earned	No maximum
			7.65		

Table 1–1
SOCIAL SECURITY TAX RATES FOR EMPLOYEES AND EMPLOYERS

[1]Old Age, Survivors, and Disability Insurance
[2]Medical Health Insurance

Discussion: Jenny must pay 6.2% (OASDI) and 1.45% (MHI) on the first $87,900 of income earned in 2004. Thus, Jenny must pay $153 [($2,000 × 6.2%) + ($2,000 × 1.45%)] in Social Security taxes on her wages. Acme must match the $153 in Social Security taxes Jenny paid on the wages.

▶ **EXAMPLE 13** Chandra earned $95,000 as the administrator of the Local Accounting Program in 2004. How much Social Security tax does Chandra pay in 2004?

Discussion: Chandra pays the maximum OASDI of $5,450 (6.2% × $87,900) and $1,378 (1.45% × $95,000) of MHI for a total Social Security payment of $6,828. Her employer is required to pay the same amount on Chandra's behalf.

As with income taxes, Social Security taxes are withheld from the employee's pay by the employer and remitted to the federal government with the employer's Social Security payment and other federal tax withholdings.

▶ **EXAMPLE 14** Assume that in example 12, Acme also withheld $312 in federal income tax and $87 in state income tax from Jenny's February earnings. What is Jenny's actual take-home pay for February?

Discussion: Jenny's February take-home pay is $1,448 after withholding for income tax and Social Security. Out of her earnings of $2,000, $153 is withheld for payment of Social Security tax, $312 for federal income tax, and $87 for state income tax. Acme must pay these taxes to the appropriate government units on a timely basis. Acme will also remit its $153 in Social Security taxes on Jenny's wages when it makes Jenny's payments.

Self-employed individuals pay a tax equal to the sum of the employee's and employer's payments. Thus in 2004, net self-employment income is subject to a tax of 12.4 percent (6.2% × 2) on the first $87,900 of income for OASDI and 2.9 percent (1.45% × 2) on all net self-employment income for MHI. Because employees are not taxed on the Social Security contribution made on their behalf by their employers, self-employed taxpayers are allowed to deduct one-half of their self-employment tax as a business expense to equalize the tax treatments of employees and the self-employed.

▶ **EXAMPLE 15** Assume that in example 13, Chandra's $95,000 in earnings constitutes net self-employment income rather than wages as an employee. How much self-employment tax must Chandra pay on her self-employment income?

Discussion: Chandra pays $10,900 (12.4% × $87,900) of OASDI and $2,755 (2.9% × $95,000) of MHI, for a total self-employment tax of $13,655. Note that this is equal to the total tax paid by Chandra and her employer ($6,828 × 2) in example 13. Because Chandra is self-employed, she must pay the equivalent of the employee's and employer's tax.

Unemployment Taxes. Employers must also pay state and federal unemployment taxes on wages paid to employees to fund unemployment benefits. The Federal Unemployment Tax (FUTA) is 6.2 percent of the first $7,000 in wages paid to each employee. Unemployment taxes do not have to be paid for employees who earn less than $1,500 per calendar quarter and certain classes of agricultural workers. Because each state also levies an unemployment tax, employers are allowed a credit of up to 5.4 percent for the state unemployment taxes they pay. Thus, the minimum FUTA tax rate is 0.8 percent (6.2% − 5.4%).

Sales Tax

Many state and local governments raise significant amounts of revenue from a sales tax. A sales tax is based on a flat percentage of the selling price of a product or service. In contrast to income and employment taxes, which are based on the income of taxpayers, a sales tax is based on a taxpayer's consumption of goods and services. The business that sells the goods or services subject to the tax collects the tax for the government. However, the tax is still paid by the taxpayer purchasing the goods or services. Each government unit that imposes a sales tax determines which goods and/or services are subject to the tax. Thus, not all goods and services are subject to a sales tax. For example, medical services are typically exempted from the tax. Other items that are often exempted from the sales tax are food, farm equipment, and sales to tax-exempt organizations.

Property Taxes

A tax on the value of property owned by taxpayers is called a *property tax.* In general, **real property** is land and any structures that are permanently attached to it, such as buildings. All other types of property are referred to as **personal property.** Because real property is immobile and difficult to conceal from tax assessors, local governments such as cities, counties, and school districts prefer it as a revenue source.

Property taxes are referred to as **ad valorem taxes,** because they are based on the value of the property being taxed. However, most property taxes are not based on the true fair market value of the property. Rather, the assessed value of the property is used to determine the tax. The *assessed value* of property varies widely but is typically 50 to 75 percent of the estimated market value of the property. Market values are determined by the designated assessment authority (e.g., the county assessor) based on various factors such as recent comparable sales, replacement cost per square foot, and other local market conditions. The assessed value is then computed as the predetermined percentage of the assessor's valuation.

▶ **EXAMPLE 16**   Maria Corporation owns a piece of land that it purchased for $6,000 in 1999. During the current year, the county tax assessor determines that the fair market value of the land is $8,000. In the county in which the land is located, assessed values are 50% of the fair market value. What is the assessed value of Maria Corporation's land?

Discussion: Maria Corporation's assessed value is $4,000 ($8,000 × 50%). Note that the local authority can increase or decrease property taxes on the land by varying the percentage of fair market value that is subject to tax. Thus, if the county raised the percentage to 75%, the corporation would pay property tax based on an assessed value of $6,000 ($8,000 × 75%).

Taxes on personal property are not as common as taxes on real property. The mobility and ease of concealment of personal property make the collection of a personal property tax administratively difficult. However, many local governments continue to selectively impose personal property taxes on types of property that are easier to track. Because of the relatively small number of establishments, property taxes on business property are still widely used. In addition, automobiles and boats are often assessed a personal property tax as part of their annual licensing fee.

▶ **EXAMPLE 17**   State A imposes an annual tag fee on automobiles. The licensing fee is $20. A personal property tax is also levied, based on the initial selling price of the automobile and its age. During the current year, Darla paid a $94 tag fee on her automobile. How much of the fee is a personal property tax?

Discussion: Darla's personal property tax on the automobile is $74 ($94 − $20). The $20 licensing fee is not a tax.

Other Taxes

Income taxes, employment taxes, sales taxes, and property taxes are the primary revenue producers for the various forms of government. However, businesses and individuals pay a number of other taxes. The most important of these are excise taxes and wealth transfer taxes. In addition, state and local governments impose taxes on certain occupations (e.g., liquor dealers) and franchise taxes for the privilege of doing business within their jurisdictions.

Excise Taxes. Excise taxes are imposed on various products and services. The federal government imposes excise taxes on a vast array of products and some services. Many states also levy excise taxes on the same products and services. An excise tax differs from a sales tax in that it is not based on the sales value of the product. Rather, an excise tax is typically imposed on a quantity, such as a gallon of gasoline or a pack of cigarettes. Some products subject to excise taxes include

alcohol	fishing equipment	shells	tobacco
coal	gasoline	telephone services	
diesel fuels	guns	tires	

Wealth Transfer Taxes. Transfers of wealth between taxpayers are taxed by the federal gift tax and the federal estate tax. Most states also impose taxes on the value of an estate. These taxes are essentially a tax on the right to transfer property to another. **Gift taxes** are paid by the donor of property—the person making the gift. The person who receives the gift, the donee, is not subject to either the gift tax or income tax on the gift. The **estate tax** is paid by the administrator (called the *executor*) of a deceased taxpayer's estate from the assets of the estate. Both the gift tax and the estate tax are based on the fair market value of the property being transferred. In addition, there are numerous exclusions from both taxes, the effect of which is to tax only relatively large gifts and estates. Although gift and estate taxes are vaguely familiar to many people, they are relatively minor revenue producers. However, a basic understanding of the operation of the two taxes will aid in understanding some of the income tax issues related to gifts and estates that are discussed later in the text.

FEDERAL GIFT TAX. A gift tax is imposed on the fair market value of gifts made between individuals.[11] Neither the donor nor the donee is subject to income tax on gifts. The donor of the gift property is responsible for reporting and paying the gift tax. The gift tax has several exclusions, the most basic of which is an annual exclusion of $11,000 per donee.[12] Under this provision, taxpayers can give as many individuals as they wish as much as $11,000 a year each and pay no gift tax. A married couple can use this exclusion to make tax-free gifts of up to $22,000 per person per year. The annual gift exclusion is indexed for inflation for gifts made after December 31, 1998. Taxpayers are also allowed to make unlimited gifts to their spouses and to charities without payment of the gift tax.

> **EXAMPLE 18** Ansel and Hanna gave their daughter a new car for graduation. The car cost $18,000. Is the gift subject to the gift tax?

Discussion: Ansel and Hannah each are entitled to give $11,000 to any person each year. Therefore, they may make gifts of up to $22,000 to an individual without incurring any gift tax. Because the fair market value of the car is less than $22,000, it is not subject to gift tax.

> **EXAMPLE 19** On their 25th wedding anniversary, Ansel gave Hannah a diamond ring that cost $30,000. Is the gift subject to the gift tax?

Discussion: Gifts to a spouse are not subject to gift tax, regardless of the value transferred. Therefore, the ring is not subject to the gift tax.

As these examples illustrate, the most common forms of gifts, such as those for birthdays, graduations, weddings, and anniversaries, are not subject to the gift tax. However, when a gift is made that is not totally excludable under one of these provisions, the taxpayer may use the unified donative-transfers credit to avoid payment of the gift tax.[13] The **unified donative-transfers credit** allows a lifetime credit against gift and estate taxes. The credit is equivalent to being able to exclude $1.5 million in property from the gift and/or the estate tax in 2004. In 2006, the estate tax exemption amount increases to $2 million, and it will continue to increase until the amount reaches $3.5 million in 2009. The estate tax is repealed in 2010. The gift tax exemption will remain at $1 million throughout the period.

FEDERAL ESTATE TAX. The estate tax is levied on the fair market value of the assets a taxpayer owned at death.[14] The executor of the estate is responsible for valuing the assets of the estate, administering the assets before their distribution to the heirs, paying the estate taxes, and distributing the assets to the estate's beneficiaries. As with the gift tax, several exclusions and the unified donative-transfers credit limit taxation of estates to those estates that are fairly substantial.[15] The fair market value of the estate's assets is reduced by funeral and administrative costs, debts owned by the taxpayer, amounts bequeathed to charities, and the marital deduction for property passing to the surviving spouse. The marital deduction is unlimited—all amounts that pass to a surviving spouse are exempt from the estate tax. Judicious use of the marital deduction and the donative-transfers credit lets the value of most estates go untaxed at the death of the first spouse. Because the unified donative-transfers credit is a cumulative lifetime amount that applies to both gifts and property passing through the estate, careful planning is required to minimize the lifetime tax on gifts and property held at death. Suffice it to say that the gift and estate tax provisions can be quite complex. Taxpayers with substantial assets should seek competent professional advice in planning their estates to minimize the liability for these taxes.

Although the transfer of property from an estate to the heirs of the decedent has no income tax effect, the estate itself is subject to income tax while it holds the assets of the decedent. The executor of the estate must file an income tax return that reports the income and the deductions related to the assets of the decedent for the period between the date of death and the final distribution of the estate's assets.

SOURCES OF FEDERAL INCOME TAX LAW

This text contains a general discussion of the federal income tax and by itself should not be considered a substitute for the original sources of the tax law. Before making a final decision about a tax issue, you should review the appropriate original source of the tax rule on which you are going to rely. Thus, it is important to be aware of the legislative, administrative, and judicial sources of tax law. These sources are frequently referred to as *primary sources* of tax law. The discussion that follows briefly outlines the primary sources. A more detailed discussion of the primary authorities is contained in Chapter 16, Tax Research. The remainder of this text generally will not make specific references to sources of tax law. Instead, this book makes generic reference to "tax law" to simplify the discussion.

The end of each chapter includes a list of applicable sources keyed to footnote numbers in the chapter and a brief summary of each source. For those who wish to read the primary sources, they are available in most university and public libraries. Briefly, our citations follow common tax practice, with deference to *The Bluebook: A Uniform System of Citation* (Harvard Law Review Association) and the *Chicago Manual of Style* (University of Chicago Press). For example, *Sec. 61* refers to Section 61 of the Internal Revenue Code of 1986 as amended. *Reg. Sec. 1.61-2* refers to the second Treasury regulation issued that interprets Section 61. *Helvering v. Gregory, 69 F.2d 809 (2d Cir. 1934)*, is a citation to a 1934 court case that was decided by the U.S. Court of Appeals for the Second Circuit. The case is located in volume 69 of the *Federal Reporter* case series, beginning at page 809. A complete explanation of all citations and how to locate the primary sources can be found in Chapter 16.

The federal income tax law dates to 1913 and has been amended, revised, and reworked numerous times since. The current statutory source of federal income tax law is the **Internal Revenue Code of 1986,** as amended (referred to as the *Code*). The tax law is laid out in the Code by section number. Thus, the basic reference to a particular tax law provision is to the section of the Code in which the law is stated. Often, particular tax treatments are referred to by their Code section number. For example, Section 179 lets a taxpayer deduct up to $102,000 of the cost of qualifying depreciable property in the year of acquisition (rather than depreciating it over its tax life). Tax practitioners refer to this election as the *Section 179 election.* Therefore, when appropriate, references to Code sections will include the popular terminology associated with that section.

The Internal Revenue Service is the branch of the Treasury Department that is responsible for interpreting and administering the tax law. The Treasury provides overall interpretive guidance on the Code by issuing **Treasury regulations.**[16] Regulations undergo an intensive review and public comment process before they are issued. Because of this intensive review, interpretations of regulations generally carry considerable authority, sometimes approaching that of the Code.

In fulfilling its administrative function, the IRS issues revenue rulings, revenue procedures, and a variety of other pronouncements that provide guidelines on the interpretation of the Code. Because the IRS issues several hundred rulings each year, they do not undergo the extensive review process accorded regulations. As such, they are given less weight as an authority than a Treasury regulation.

In addition to providing interpretive guidance, the IRS has responsibility for ensuring taxpayers' compliance with the tax law. During 2001, the IRS processed 228 million tax returns, provided tax preparation assistance to 10 million taxpayers, and audited 3.3 million tax returns filed by individual taxpayers. When audited by the IRS, taxpayers are allowed to present their reasoning for the items in question on their return. As might be expected, disputes often arise between taxpayers and the IRS concerning its interpretations and enforcement of the tax law. Most disputes are resolved through the IRS appeals process. However, taxpayers who are dissatisfied with the result of the appeals process are entitled to take their disputes to court for settlement.

Court decisions establish precedent in the interpretation of the tax law. Taxpayers and the IRS are generally bound by the interpretation of a court on a particular issue. However, the loser of an initial court case may appeal the decision to a U.S. Circuit Court of Appeals. A loss at the appellate level may be further appealed to the U.S. Supreme Court. However, the Supreme Court limits its review of tax cases to those of major importance (e.g., a constitutional issue) or to resolving conflicting decisions in the appellate courts. A Supreme Court decision is not subject to review—it is the final interpretation of the law. Only Congress can override an interpretation of the Supreme Court by amending the Code section in question.

Tax information is also published in a variety of secondary sources. These include tax reference services, professional tax journals, tax newsletters, and textbooks. Secondary sources are useful when researching an issue, and they are often helpful for understanding the primary sources. However, you should exercise care when using secondary sources, because their interpretations are not authoritative.

FEDERAL INCOME TAX TERMINOLOGY

Individuals, corporations, and certain estates and trusts are subject to tax on their federal taxable income. Federal taxable income is defined by the tax law and differs from both financial accounting and economic measures of income. The general computational framework for determining the taxable income of all taxpayers is shown in Exhibit 1–1. Both the terms used in the computations and the order of the computational framework are prescribed in the tax law.

Income

The term *income* is used in several ways. Therefore, always be sure you understand the context in which the term is used. As broadly defined, income includes both taxable and nontaxable types of income. This definition includes all income that belongs to the taxpayer. *Gross income* is a more restrictive term. As Exhibit 1–1 shows, **gross income** is income broadly defined minus income items that are excluded from taxation.[17] Items of gross income are included in the computation of taxable income. Generally, gross income is the starting point for reporting income items on a tax return. Chapter 3 discusses the most commonly encountered gross income items.

A fundamental rule in regard to income is that an item is included in gross income unless it is specifically excluded by the tax law. **Exclusions** represent increases in a taxpayer's wealth and recoveries of the taxpayer's capital investment that Congress has decided should not be subject to income tax. Thus, income exclusions are not counted as gross income. Common income exclusions include inheritances, gifts, and interest on certain municipal bonds. Exclusions are discussed in Chapter 4.

Although not an explicit part of the income tax computation, deferrals of income and deductions are also found in the tax law. A **deferral** is an item that does not affect the current period's taxable income but will affect taxable income in a future tax year. Thus, a deferral is like an exclusion in that it does not have a current tax effect. However, it differs in that an exclusion is *never* subject to tax, whereas a deferral *will be* subject to tax at some point in the future.

Exhibit 1–1
INCOME TAX COMPUTATIONAL FRAMEWORK

	Income "Broadly Defined" (includes income from all sources)
Minus:	Excluded income
Equals:	Gross income
Minus:	Deductions and exemptions
Equals:	Taxable income
×	Tax rate (schedule of rates)
Equals:	Income tax
Minus:	Tax prepayments
	Tax credits
Equals:	Tax (refund) due with return

Taxable income is a net number and is the tax base. Taxable income is determined by subtracting deductions and exemptions from gross income. Taxable income is the tax base that is multiplied by the applicable tax rate to compute the federal income tax. Taxable income is usually different from financial accounting income computed by using generally accepted accounting principles.

The differences between financial accounting income and taxable income generally arise because taxable income is computed according to the rules prescribed by the tax law. Tax accounting rules are not based on generally accepted accounting principles (GAAP). GAAP are concerned with determining the "true income" for an annual period. The income tax is geared to producing and collecting tax revenues and providing incentives for particular economic and social transactions. An important difference between the two objectives is that the income tax system attempts to collect the tax on income in the period during which the taxpayer has the resources to pay the tax. Under GAAP, having the resources to pay taxes is of no concern. As a result, specific income and deduction items may be accelerated, deferred, or permanently excluded from the current year's taxable income computation, as opposed to the GAAP treatment. For example, prepaid rental income may be amortized over the lease period for financial reporting but must be reported in full in the year it is collected for tax reporting. Another example is the treatment of depreciable property. For tax purposes, assets must be depreciated by using a statutorily determined recovery period, without regard for their actual useful life. For financial reporting, the same asset is depreciated over its useful life. These are but two examples of income and deduction items that are different for financial and taxable income and that will be discussed throughout the text.

Income is also referred to as *ordinary income*. **Ordinary income** is the recurring income earned by a taxpayer for a tax year.[18] It is the common type of income that people and businesses expect to earn. Ordinary income typically includes business profits, rent from property, interest on investments, dividend income, and wages. Ordinary income is subject to tax using regular tax rates and computations explained in later chapters. That is, ordinary income receives no special treatment under the laws.

Income also results from gains. A **gain** is the difference between the selling price of an asset and its tax cost and is the result of disposing of the asset.[19] Usually, a gain will be the result of a sale of a single asset. Most gains produce ordinary income. However, gains on the sale of certain types of assets receive special treatment in the determination of taxable income and the tax liability. These gains are called *capital gains* and result from the sale of capital assets.

Deductions

Deductions are amounts that the tax law specifically allows as subtractions from gross income. Deductions are a matter of legislative grace. The concept of legislative grace gives us a basic rule to follow to determine items that qualify for deduction. The rule is that an item may not be deducted unless the tax law specifically permits it. Deductions are characterized as *expenses, losses,* and *exemptions.*

An **expense** is a current period expenditure that is incurred to earn income. Deductions for expenses are limited to those incurred in a trade or business,[20] in an income-producing activity (investment activity),[21] and certain specifically allowed personal expenses of individuals. Trade or business expenses and income-producing expenses must be ordinary, necessary, and reasonable in amount to be deductible. Allowable personal expenses are deductible as itemized deductions and are subject to strict limitations.

The term **loss** refers to two distinctly different types of events. A loss occurs when an asset is disposed of for a selling price that is less than its tax cost. This type of loss is referred to as a **transaction loss** and represents a loss of capital invested in the asset. In later chapters, it will be necessary to apply limits to the amount of a loss that can be deducted in a tax year. To apply the limits, losses are characterized as personal, business, or capital. These limits deny deductions for most personal losses, place a cap on the amount of capital losses that may be deducted in the year of the loss, and allow business losses to be fully deducted as incurred.

The second type of loss is an annual loss. An **annual loss** results from an excess of allowable deductions for a tax year over the reported income for the year. The treatment of the annual loss depends on the activity in which the loss is incurred. Chapter 7 discusses the limitations on and treatment of all losses, transaction and annual.

Individuals, trusts, and estates may subtract predetermined amounts called **exemptions** to determine their taxable incomes. The exemption deduction for individuals is, in effect, Congress's recognition that people need a minimum amount of income to provide for their basic living expenses. Thus, this minimum amount of income is deducted as an exemption and is not subject to tax. The deduction for individual exemptions is reduced for high-income taxpayers. Apparently, the reduction is Congress's way of saying that these taxpayers have enough income to support themselves and that the ability-to-pay concept should prevail. Since the minimum basic costs of living increase each year because of inflation, the exemption amounts are indexed to inflation and increase each year to reflect the increased costs individuals incur.

Income Tax Rates

The 2004 tax rate schedules for two classes of individual taxpayers and corporations are reproduced in Table 1–2.[22] A full set of tax rates for individuals, corporations, estates, and trusts for 2003 and 2004 is reproduced in Appendix B. The income tax is calculated by multiplying taxable income by the applicable tax rates. Each year, the IRS publishes new tax rate schedules that are adjusted for cost-of-living increases. Adjusting the tax rate schedules for changes in the cost of living helps to minimize a hidden tax that results from inflation.

Assume the following information (shown in Exhibit 1–2): A single taxpayer's taxable income in 2003 was $28,000. The rate of inflation in 2004 was 2.2 percent, and the taxpayer was able to keep up with inflation by increasing her income. Her taxable income goes up by $616 to $28,616 ($28,000 × 1.022%) in 2004. At this point, the taxpayer is no better or worse off in 2004 than in 2003. Her income increase merely kept up with the rate of inflation. The top panel of Exhibit 1–2 shows that failure to adjust the 2004 tax rates for the 2.2 percent inflation rate results in $114 in additional tax. The increased tax is attributable to two sources. First, the increased income results in an additional $92 (15% × $616) in tax, even if the marginal rate stays the same from the first to the second year. Second, the problem worsens when the inflated income pushes the taxpayer into a higher marginal tax bracket (tax bracket creep) causing an additional $22 in tax [(25% − 15%) × ($28,616 − $28,400)]. Thus, the taxpayer is worse off, because she pays $114 more tax on the same deflated income when tax rates are not adjusted for inflation. The net result is an increase of after-tax income of only $502 ($616 − $114), which is less than the rate of inflation.

The bottom panel of Exhibit 1–2 calculates the tax using the actual 2004 rates, which are adjusted for the 2.2 percent inflation rate. The tax on a 2004 taxable income of $28,616 is $3,935. This is a reduction of $29 ($3,964 − $3,935) over the tax calculated using 2003 rates on the same income. The adjustment for inflation in the tax rate brackets leaves the taxpayer with the same inflation-adjusted after-tax income in 2004 [($28,000 − $3,850) × 1.022% = $24,681 ($28,616 − $3,935)] that the taxpayer had in 2003. Thus, the adjustment of the tax brackets for inflation each year ensures that taxpayers whose income merely keeps pace with inflation will not realize a decrease in real after-tax income.

Tax Prepayments

The pay-as-you-go system requires the payment of tax as the income is earned and when the taxpayer has the resources available to pay the tax. Tax prepayments are subtracted from the income tax liability to determine whether the taxpayer has underpaid and owes additional tax with the return (tax due) or is entitled to a refund of overpaid taxes (refund due). Employees prepay taxes on wages through payroll-tax withholding. Other types of income, such as pensions and some gambling winnings, are also subject to the withholding of tax by the payer. The employer or other person withholding the tax pays the tax withheld to the IRS, to be credited to the taxpayer's account with the government.

Self-employed people and taxpayers with income not subject to withholding (trade or business income, interest income, dividend income, gains from sales of assets, etc.) are required to make quarterly payments of their current-year estimated tax payments. An individual usually makes quarterly payments on April 15, June 15, and September 15 of the tax year and on January 15 of the next year. This corresponds to the fifteenth day of the fourth, sixth, and ninth months of the tax year and the fifteenth day of the first month of the following year. A corporation

Table 1–2

2004 Tax Rate Schedules

Single Taxpayers

If taxable income is over	But not over	The tax is	Of the amount over
$ -0-	$ 7,150	10%	$ -0-
7,150	29,050	$ 715.00 + 15%	7,150
29,050	70,350	4,000.00 + 25%	29,050
70,350	146,750	14,325.00 + 28%	70,350
146,750	319,100	35,717.00 + 33%	146,750
319,100	· · · · · ·	92,592.50 + 35%	319,100

Married Taxpayers Filing Jointly and Surviving Spouse

If taxable income is over	But not over	The tax is	Of the amount over
$ -0-	$ 14,300	10%	$ -0-
14,300	58,100	$ 1,430.00 + 15%	14,300
58,100	117,250	8,000.00 + 25%	58,100
117,250	178,650	22,787.50 + 28%	117,250
178,650	319,100	39,979.50 + 33%	178,650
319,100	· · · · · ·	86,328.00 + 35%	319,100

Corporate Tax Rate Schedule

If taxable income is over	But not over	The tax is	Of the amount over
$ -0-	$ 50,000	15%	$ -0-
50,000	75,000	$ 7,500 + 25%	50,000
75,000	100,000	13,750 + 34%	75,000
100,000	335,000	22,250 + 39%	100,000
335,000	10,000,000	113,900 + 34%	335,000
10,000,000	15,000,000	3,400,000 + 35%	10,000,000
15,000,000	18,333,333	5,150,000 + 38%	15,000,000
18,333,333	· · · · · ·	6,416,667 + 35%	18,333,333

Exhibit 1–2

The Hidden Inflation Tax

	Tax Year	
	2003	2004
Taxable income	$28,000	$28,000
Increase in taxable income due to 2.2% inflation	-0-	616
Inflation-adjusted taxable income	$28,000	$28,616
Tax using 2003 single taxpayer rates:		
Tax on base amount	$ 700	$ 3,910
Excess taxed at marginal rate		
15%	3,150	
25%		54
Total tax	$ 3,850	$ 3,964
Additional tax resulting from inflation		$ 114

Tax on $28,616 at 2004 tax rates		
Tax on $7,150		$ 715
Tax on income in excess of $7,150		
($28,616 − $7,150) × 15%		$ 3,220
Tax at 2004 rates		$ 3,935
2003 After-tax income $28,000 − $3,850		$24,150
2003 Inflation rate adjustment		× 1.022
2004 Real after-tax income		$24,681
Actual 2004 after-tax income $28,616 − $3,935		$24,681

makes its estimated tax payments on the fifteenth day of the fourth, sixth, ninth, and twelfth months of its tax year. Estates and trusts follow the estimated tax schedule used by individuals. Estimated tax payments, like withheld amounts, are subtracted as credits for the prepayment of tax.

Tax Credits

A **tax credit** is a direct reduction in the income tax liability. In effect, tax credits are treated like tax prepayments. As Exhibit 1–1 shows, a credit is not deducted to arrive at taxable income but is instead subtracted directly from the income tax liability. Thus, a tax credit is more valuable than a deduction of an equal amount, because the credit yields a larger reduction in the total tax due. Tax credits are often used as incentives to encourage taxpayers to enter into specific types of transactions that Congress feels will further some public purpose.

If a taxpayer's marginal tax rate is 28 percent, a $5,000 tax deduction has the same value as a $1,400 tax credit ($5,000 × 28%). Likewise, a $1,000 tax credit has the same value as a $3,571 deduction if the marginal rate is 28 percent ($1,000 ÷ 28%).

▶ **EXAMPLE 20**   Ron and Martha, whose marginal tax rate is 28%, paid $1,000 for child care.

Discussion: If the expenditure is treated as a credit, the tax they owe for the year will be reduced by the full $1,000. If the expenditure is treated as a deduction, their tax would be reduced by $280 ($1,000 × 28% marginal rate). Treatment of the expenditure as a credit would save them $720 more than treatment as a deduction.

The most common business tax credits are discussed in Chapter 15. Individuals are also allowed tax credits for certain circumstances and activities. For example, individuals with dependents are allowed a credit of $1,000 for each qualifying dependent. Restrictions and limitations associated with this tax credit and other common individual tax credits are discussed in Chapter 8.

Filing Returns

In general, all income tax entities must file an annual tax return. (See Chapter 8 for individual filing requirements.) Returns for individuals, estates, trusts, and partnerships must be filed on or before the fifteenth day of the fourth month following the close of the entity's tax year (April 15 for calendar-year taxpayers). Corporate tax returns are due on or before the fifteenth day of the third month following the close of a corporation's tax year (March 15 for calendar-year corporate taxpayers). Taxpayers who cannot complete and file their returns by the regular due date can apply for extensions for filing the return. Individuals are granted an automatic four-month extension by applying for the extension by the due date of the return. Corporations are allowed an automatic six-month extension; partnerships and trusts can automatically extend their filing date by three months. Filing an extension does not extend the time for paying the tax. Applications for automatic extensions must show and include payment of the estimated amount due with the final return.

▶ **EXAMPLE 21**   Thelma procrastinates about preparing her tax return and determines that she cannot complete the return by April 15. She has withholdings and estimated tax payments totaling $8,600 and estimates that her total tax liability for the year will be $8,950. What must Thelma do to extend the date for filing her return?

Discussion: Thelma can extend the period for filing her return to August 15 (four months from April 15) by filing the application for automatic extension by April 15. This only grants Thelma permission to delay the filing of the return. She must pay the $350 ($8,950 − $8,600) estimated tax she owes when she applies for the extension.

Taxpayers and the government can correct errors on returns within a limited time period called the **statute of limitations.** Generally, once the statute of limitations has expired, corrections cannot be made. The general statute of limitations is three years from the due date of the return, not including extensions. The three-year statute of limitations has several exceptions, the most important of which deal with fraudulently prepared returns. The statute of limitations runs for six years when a taxpayer omits gross income in excess of 25 percent of the gross income reported on the return. The government can bring charges of criminal fraud against a taxpayer at any time. That is, neither the three-year nor the six-year statute of limitations protects a taxpayer who willfully defrauds the government.

The government corrects errors on taxpayers' returns through its audit process. Taxpayers correct errors on prior year returns by filing amended returns. Amended returns are not used to adjust returns for previous years. (See discussion of the tax benefit rule in Chapter 2.) An amended return should be filed only if a taxpayer finds that an item of income that should have been included in gross income was omitted in the original filing or if the taxpayer improperly included an item of income in a prior year. Taxpayers also should file amended returns if they find that they failed to take an allowable deduction or if they find that they took an improper deduction on an earlier return.

▶ EXAMPLE 22 Geraldo Corporation incurred a net operating loss in 2003, its first year of operation. Because the controller knew that Geraldo was going to suffer a loss, he took no deductions for depreciation for 2003. Geraldo's independent auditor came upon the error in 2004, and advised Geraldo that it must take all allowable deductions in the proper year. Should Geraldo file an amended return for 2003?

Discussion: Because the depreciation was not treated properly on the 2003 tax return, Geraldo should file an amended return that takes the proper depreciation deduction for 2003.

▶ EXAMPLE 23 Walstad Corporation is an accrual basis taxpayer. In 2003, Walstad determined that one of its customers with an accounts receivable balance of $40,000 was in bankruptcy. After conferring with the customer's lawyers, Walstad determined that it would be able to collect only $15,000 of the account and deducted the $25,000 uncollectible amount as a bad debt expense. In 2004, the customer's bankruptcy was settled, and Walstad received $10,000 as a final settlement of the account it had written off. Should Walstad file an amended return for 2003 and correct the bad debt deduction?

Discussion: The actual bad debt is $30,000 ($40,000 − $10,000). The $25,000 bad debt deduction that Walstad took in 2003 was an estimate of the amount of the bad debt. Therefore, the deduction was not incorrect at the time the return was filed. Walstad should deduct the additional $5,000 ($30,000 − $25,000) of actual bad debt in 2004 to adjust the estimate. Amended returns are not filed to adjust estimates on prior year returns. Adjustments to estimates are made on the return for the year in which the actual amount of the deduction becomes known.

THE AUDIT AND APPEAL PROCESS WITHIN THE IRS

The federal income tax system is based on self-assessment, which requires taxpayers to report and pay their taxes correctly. IRS examinations, or audits, can vary from a letter that requests supporting information by mail to a full-scale, continuous examination of large corporations in which teams of IRS agents work at each taxpayer's office. Taxpayers who do not agree to changes suggested by the IRS during an audit can appeal the matter to a higher administrative level within the IRS. Generally, taxpayers cannot be charged with any additional taxes, interest, or penalties without first being formally notified. Whenever settlement cannot be reached with the IRS, the taxpayer can initiate litigation in one of the trial courts.

Tax Return Selection Processes

The IRS cannot possibly examine every return that is filed. It does examine as many returns as possible, given its staffing and facility levels. Currently, this amounts to only about 2 percent of all returns filed. The IRS uses five general methods to verify that taxpayers are properly self-assessing their taxes. One of the most important is a computerized return selection program called the **Discriminant Function System (DIF).** Through mathematical analysis of historical data, this program selects those returns with the highest probability of containing errors. Selected returns are typically examined only for specific items such as charitable contributions or employee business expenses. A related program is the **Taxpayer Compliance Measurement Program (TCMP).** Returns are randomly selected from different income levels, and every item on the return is comprehensively audited. The results are used to set the parameters for the DIF computer selection program. The IRS suspended the TCMP audits in 1996 because of reductions in its budget.

Virtually all returns are checked for mathematical, tax calculation, and clerical errors during the initial processing of the returns. If an error is discovered under this **document perfection program,** the IRS recalculates the amount of tax due and sends an explanation to the taxpayer. Another program of increasing importance is called the **information-matching program.** Information from banks, employers, and others on forms such as the W-2 for wages and withholding and the 1099 for miscellaneous income are matched to the taxpayer's return. For any omitted or incorrect items, the IRS recomputes the tax and sends an explanation to the taxpayer. Finally, a number of **special audit programs** are designed by the IRS and combine computer and manual selection based on various standards that are changed periodically. Some of the standards used include the size of the refund, the amount of adjusted gross income reported, and the amount or type of deduction claimed.

Types of Examinations

There are three basic types of IRS examinations. **Correspondence examinations** are those that can be routinely handled by mail. Most originate at the IRS service centers and involve routine requests for supporting documents such as canceled checks or some other written instruments. A written reply to the questions raised, along with copies of supporting documents, usually completes the examination.

Office examinations are conducted at the local district office of the IRS and usually involve middle-income, nonbusiness returns, and small sole proprietorships. The taxpayer is notified by letter of the date and time of the exam, as well as the items for which proof is requested. Most taxpayers appear for themselves, although some are represented by their return preparers or other tax advisers. The audit is relatively informal, and the IRS agent has considerable discretion in resolving factual questions such as substantiation of travel expenses. For questions of law, however, the agent must follow IRS policy as expressed in Treasury regulations, revenue rulings and procedures, and the like, even if court decisions indicate otherwise.

Field examinations are conducted at the taxpayer's place of business and can involve any item on the income tax return as well as any items on the payroll and excise tax returns. These examinations are handled by more-experienced IRS agents, and almost all taxpayers are represented by their tax advisers. As with office examinations, IRS agents must follow IRS policy on matters of law and are accorded a great deal of latitude in settling matters of fact.

Settlement Procedures

After the examination, the agent prepares a report, known as the revenue agent's report (RAR), describing how each issue was settled and the amount of any additional tax or refund due the taxpayer. The agent also prepares a waiver of restrictions on assessment (Form 870), which states that the taxpayer waives any restrictions against assessment and collection of the tax by the IRS. Both items are mailed to the taxpayer in a letter commonly called a 30-day letter, along with an IRS publication describing the taxpayer's appeal rights.

A signed Form 870 means that the taxpayer agrees to the proposed changes, but it is not binding on either the taxpayer or the IRS. The taxpayer merely agrees to pay the additional tax due while reserving the right to file for a refund in a subsequent court action. Generally, the IRS rejects a settlement reached by its agents only if there is fraud or a misrepresentation of a material fact.

Administrative Appeals

A taxpayer who does not agree with the agent's report may request a meeting with agents from the **IRS Appeals Division** within 30 days of the date of the letter. If the additional tax due exceeds $2,500, the taxpayer must include a written response to the agent's findings; the taxpayer's response is called a **protest letter.** When the amount is less than $2,500 or is the result of a correspondence or office examination, no written protest is required.

The administrative appeal process allows taxpayers one additional opportunity to reach a settlement before resorting to the courts. The appeals division has the authority to consider the hazards of litigation. For example, when the facts or the law are uncertain, or both, the appeals division may settle issues it does not want to litigate, even if the IRS position has some merit. After what may be lengthy negotiation, taxpayers who finally reach an agreement with the IRS, or who simply don't want to pursue the matter, sign the Form 870 (or Form 870-AD, if the IRS has conceded some issues) and pay the full amount of the deficiency plus any penalties and interest.

Taxpayers unable to reach an agreement in the appeals division, or who have bypassed the appeals division by failing to respond to the 30-day letter, are sent a statutory notice of deficiency. This letter is the official notification by the IRS that it intends to assess or charge the taxpayer for some additional taxes, and is commonly referred to as a 90-day letter.

Taxpayers who are not interested in going to court can simply wait 90 days to have the deficiency formally assessed and then pay any additional amounts due. Taxpayers who want to litigate in district court or the claims court first must pay the amounts due and file for a refund in the court of their choice. Taxpayers who do not want to pay first must file a petition with the U.S. Tax Court within 90 days of the date of the letter. The decision to take an unresolved issue to court involves a number of additional factors and typically is made only with the advice of legal counsel specializing in tax litigation.

INDIVIDUAL INCOME TAX CALCULATION

The general tax calculation presented in Exhibit 1–1 applies to all taxpayers. However, the tax law modifies this calculation for individuals to take into account the unique characteristics of individual taxpayers.

The calculation of an individual's taxable income is outlined in Exhibit 1–3. Note that the general flow remains the same—deductions are subtracted from gross income to arrive at taxable income. Gross income is determined under the general tax formula. The distinguishing feature of the individual taxable income calculation is that deductions are broken into two classes—deductions for adjusted gross income and deductions from adjusted gross income. This dichotomy of deductions results in an intermediate income number called the **adjusted gross income (AGI).**[23] As will become clear in the discussion that follows, this is a very important income number, because it is used to limit the deductions from adjusted gross income of an individual taxpayer. Deductions are discussed in more detail in later chapters. However, at this point, a general knowledge of the computational form and allowable deductions of individuals is necessary. Each type of deduction is discussed in turn.

Deductions for Adjusted Gross Income

Individuals are always allowed to deduct the qualified expenses they incur as **deductions for adjusted gross income.** In contrast to deductions from adjusted gross income, deductions in this class are not subject to reduction based on the income of the taxpayer. That is, once the allowable amount of an expenditure in this category has

	All sources of income (broadly defined)	$XXX
Minus:	Exclusions from income	(XXX)
Equals:	Gross income	$XXX
Minus:	Deductions *for* adjusted gross income	
	Trade or business expenses	
	Rental and royalty expenses	
	Other specifically allowable deductions	(XXX)
Equals:	**ADJUSTED GROSS INCOME**	$XXX
Minus:	Deductions *from* adjusted gross income	
	Personal deductions: the greater of	
	1. itemized deductions (allowable personal expenses and certain other allowable deductions)	
	OR	
	2. individual standard deduction	(XXX)
Minus:	Personal and dependency exemptions	(XXX)
Equals:	Taxable income	$XXX

Exhibit 1–3

INDIVIDUAL INCOME TAX FORMULA

been determined, it is not subject to further reduction based on the income of the taxpayer. The allowable deductions for adjusted gross income are generally those that are incurred in a trade or business of the taxpayer or that are related to the earning of other forms of income. In addition, several other specifically allowed items are deductible for adjusted gross income. Deductions for adjusted gross income include

> Trade or business expenses
> Rental and royalty expenses
> Capital loss deductions
> Alimony paid
> Contributions to individual retirement accounts (IRAs)
> Moving expenses
> Reimbursed employee business expenses
> Self-employment taxes paid
> Self-employed medical insurance premiums
> Up to $2,500 of interest on qualified student loans

Although these expenditures are not limited by the income of the taxpayer, other limitations in the tax law may reduce the current period's tax deduction. For example, the allowable deductions for rental properties may be limited by either the vacation home rules or the passive activity loss rules. Losses on the sale of capital assets are deductible but are first netted against capital gains. If the result is a net capital loss, the current year's deduction is limited to a maximum of $3,000.[24] These losses and other limits are covered in the chapters on deductions and losses. The important point to remember for now is that once the allowable amount of a deduction for adjusted gross income has been determined, it is not subject to further reduction. In addition, there is no preset minimum allowable amount of deductions for adjusted gross income.

Deductions from Adjusted Gross Income

Individuals are allowed to deduct certain personal expenditures and other specified nonpersonal expenditures as **deductions from adjusted gross income.** These deductions are commonly referred to as **itemized deductions.** Note in Exhibit 1–3 that individuals deduct the greater of their allowable itemized deductions or the standard deduction.[25] The **standard deduction** is an amount that Congress allows all taxpayers to deduct regardless of their actual qualifying itemized deduction expenditures. Thus, taxpayers itemize their deductions only if their total allowable itemized deductions exceed the standard deduction. For 2004, the standard deduction is $4,850 for a single individual and $9,700 for a married couple.

EXAMPLE 24 Festus is a single taxpayer with total allowable itemized deductions of $1,800 in 2004. What is Festus's allowable deduction from adjusted gross income?

Discussion: Festus deducts the larger of his $1,800 in itemized deductions or the $4,850 standard deduction for a single individual. In this case, Festus deducts the $4,850 standard deduction.

EXAMPLE 25 Assume that in example 24, Festus's total allowable itemized deductions are $6,700 in 2004. What is his allowable deduction from adjusted gross income?

Discussion: Festus would deduct the $6,700 in actual itemized deductions because it exceeds his $4,850 standard deduction.

As these examples illustrate, just because a particular expenditure is allowed as an itemized deduction does not necessarily mean that a taxpayer incurring the expense will actually deduct it. Itemized deductions reduce taxable income only when a taxpayer's total itemized deductions exceed the allowable standard deduction.

In addition to giving all taxpayers some minimum amount of deduction, the standard deduction eliminates the need for every taxpayer to list every qualifying personal expenditure. This makes it easier for taxpayers with small amounts of qualifying expenditures to comply with the tax law and relieves the government from having to verify millions of deductions that would have been claimed as a result of itemizing. Thus, the standard deduction is an important tool that the government uses to promote income tax law compliance by removing the burden of record-keeping and reporting for relatively small amounts of deductible items.

In the deduction classification scheme, specifically allowed personal expenditures are classified as itemized deductions.[26] In addition to personal expenditures, investment expenses and certain other employment-related expenses are deductible as itemized deductions. Many allowable itemized deductions are subject to an income limitation. That is, the amount of the qualifying expenditure must be reduced by a percentage of the taxpayer's adjusted gross income to determine the actual deduction. The effect of using this type of income limitation is to disallow deductions for amounts that are small in relation to the taxpayer's income.

EXAMPLE 26 Qualifying medical expenses are deductible to the extent that they exceed 7.5% of a taxpayer's adjusted gross income. During the current year, Li has an adjusted gross income of $40,000 and incurred $4,200 in qualified medical expenses. What is Li's itemized deduction for medical expenses?

Discussion: Li must reduce the $4,200 of qualified medical expenses by $3,000 ($40,000 × 7.5%), resulting in deductible medical expenses of $1,200.

Note that the effect of the limitation is to allow larger deductions for taxpayers with smaller incomes. Another taxpayer incurring the same $4,200 in expenses who had an adjusted gross income of only $25,000 would be allowed to deduct $2,325 [$4,200 − ($25,000 × 7.5% = $1,875)] of the medical expenses.

The following list is intended to acquaint you with the categories of itemized deductions available to individuals. At this point, you should note the types of personal expenses that are allowed as a deduction. Do not be concerned about the detailed deduction requirements and limitations. These issues are explained in more detail in Chapter 8.

MEDICAL EXPENSES—Unreimbursed medical expenses are deductible to the extent that they exceed 7.5 percent of adjusted gross income. Medical expenses include the cost of medical insurance, physicians, hospitals, glasses and contact lenses, and a multitude of other items. Because of the AGI limit, many taxpayers benefit from these deductions only when there is a major illness in the family.[27]

TAXES—State, local, and foreign income taxes, real estate taxes, and state and local personal property taxes may be deducted.[28]

INTEREST—An individual's itemized deduction for personal interest expense is limited to the following:[29]

- Home mortgage interest related to the acquisition of a home or to a home equity loan
- Investment interest expense

CHARITABLE CONTRIBUTIONS—Gifts to qualified charitable organizations may be deducted. Generally, the deductible contribution may not exceed 50 percent of the taxpayer's adjusted gross income.[30]

PERSONAL CASUALTY AND THEFT LOSSES—Deductions are allowed for losses of property from casualty or theft, subject to two limitations. Because of the limitations, most taxpayers must have a large total loss for the year to get a deduction for a personal casualty or theft loss.[31]

MISCELLANEOUS ITEMIZED DEDUCTIONS—This is a broad category of deductions that includes most expenses related to the production of investment income. The following list of miscellaneous deductions illustrates the types of items deducted in this category:

- Business expenses of an employee not reimbursed by an employer
- Investment-related expenses
- Expenses related to tax return preparation, planning, and examination

Generally, the deduction allowed for miscellaneous itemized deductions must be reduced by 2 percent of the taxpayer's adjusted gross income.[32]

Personal and Dependency Exemptions

Individuals are allowed to deduct a predetermined amount for each qualifying exemption.[33] In 2004, individuals deduct $3,100 for each qualifying personal and dependency exemption. The intention is to exempt from tax a minimum amount of income that is used to support the taxpayer and those who are dependent on that taxpayer. Because support costs increase with inflation, the exemption amounts are increased each year to account for the prior year's inflation. **Personal exemptions** are allowed for the taxpayer and the taxpayer's spouse. **Dependency exemptions** are granted for individuals who are dependent on the taxpayer for support. Although five technical tests (discussed in Chapter 8) must be met to qualify as a dependent, the underlying reasoning is that the dependent must rely on the taxpayer for basic living costs. Thus, children of a taxpayer and other relatives, such as parents and grandchildren who live with the taxpayer, are the most common dependents.

▶ **EXAMPLE 27**   John and Nancy are married and have 3 small children who live with them and depend on them for their support. What is John and Nancy's 2004 exemption deduction?

Discussion: John and Nancy are entitled to 2 personal exemptions and 3 dependency exemptions. Their deduction is $15,500 ($3,100 × 5 exemptions).

TAX PLANNING

The objective of tax planning is to maximize after-tax wealth. An effective tax plan results in a reduction of taxes for the planning period. Because a planning period may be two or more years, focusing on reducing tax for one year without considering any offsetting effects for other years can lead to excessive tax payments. The traditional planning technique of deferring income and accelerating deductions may not always be the best tax plan. The traditional technique considers only the time value of money savings that can be obtained from delaying tax payments on income or receiving tax savings from deductions sooner. Although the time value of money must always be considered, changes in marginal tax rates from one year to the next can have effects that offset the time value of money. Thus in many cases, changes in both the marginal tax rate and the time value of money must be considered when developing a tax plan. The mechanics of tax planning demonstrate basic techniques that can be used to help make tax-planning decisions. The planning discussion concludes by pointing out that tax avoidance is acceptable but tax evasion is not.

Mechanics of Tax Planning

The mechanics of tax planning focus on the issues of timing and income shifting. The timing question to be answered is when income and deductions should be claimed to save the most *real tax*. To make decisions involving timing, it is necessary to compare the tax effects of changes in marginal tax rates and the time value of money. To make the optimal choice among different alternatives, the calculations must be done to determine the *real* after-tax cost of each alternative. Income shifting involves moving income among related taxpayers to achieve the lowest marginal taxes (and lowest total tax) on the entire income of the related taxpayers. Shifting is commonly done by transferring income-producing property among family members and by using corporations that taxpayers control to shift income into the lowest marginal tax rates.

Timing Income and Deductions. A taxpayer's marginal tax rate and the time value of money must be considered in tax planning. The traditional technique of deferring income and accelerating deductions relies solely on the time value of money savings from delaying the tax payment or receiving the tax deduction savings earlier. For example, a taxpayer who expects to be in a 28-percent marginal tax bracket for the next several years might be indifferent about reporting $1,000 in extra income in 2004 or 2005. Regardless of which year the income is reported, the taxpayer pays $280 in tax and keeps $720 ($1,000 − $280) in after-tax income. When the present value of the tax payment is considered (see Table 1–3 for present values factors), it becomes clear that choice of years does make a difference. If the taxpayer's applicable interest rate is 10 percent and the marginal rate is expected to remain the same, deferring payment of the tax until 2005 results in an interest-free loan. The present value of the tax savings is $25:

Tax paid in 2005	$ 280
10% present value factor	× 0.909
Present value of tax paid in 2005	$ 255
Present value of tax paid in 2004	280
Real tax savings by deferring income	$ 25

			Present Value of a Single Payment				
Year	5%	6%	7%	8%	9%	10%	12%
1	0.952	0.943	0.935	0.926	0.917	0.909	0.893
2	0.907	0.890	0.873	0.857	0.842	0.826	0.797
3	0.864	0.840	0.816	0.794	0.722	0.751	0.712
4	0.823	0.792	0.793	0.735	0.708	0.683	0.636
5	0.784	0.747	0.713	0.681	0.650	0.621	0.567
6	0.746	0.705	0.666	0.630	0.596	0.564	0.507
7	0.711	0.665	0.623	0.583	0.547	0.513	0.452
8	0.677	0.627	0.582	0.540	0.502	0.467	0.404
9	0.645	0.592	0.544	0.500	0.460	0.424	0.361
10	0.614	0.558	0.508	0.463	0.422	0.386	0.322

Table 1–3
PRESENT VALUE TABLES

If the marginal rate is expected to decrease to 15 percent in 2005, the taxpayer has a greater incentive to defer the income. By deferring the income to 2005, the taxpayer receives the benefit of an interest-free loan for one year plus the benefit of the lower marginal tax rate. Deferring the income to 2005 would result in a real tax benefit of $144:

Tax paid in 2005 ($1,000 × 15%)	$ 150
10% present value factor	× 0.909
Present value of tax paid in 2005	$ 136
Present value of tax paid in 2004	280
Real tax savings by deferring income	$ 144

Table 1–3 shows how much $1 to be paid at a future date is worth today at the discount rate indicated.

If the taxpayer expects the marginal tax rate to increase to 35 percent next year, the income should be reported in 2004. Deferring the income to 2005 would have a real tax cost of $38:

Tax paid in 2005 ($1,000 × 35%)	$ 350
10% present value factor	× 0.909
Present value of tax paid in 2005	$ 318
Present value of tax paid in 2004	280
Real tax cost of deferring income	$ 38

The same approach can be used to determine the best timing for a deduction. However, keep in mind that deductions are the opposite of income—they reduce taxes paid. Therefore, the optimal choice for deductions is to maximize the real after-tax reduction in taxes paid. In many situations, it may be necessary to compare the offsetting effects of income and deduction items.

▶ **EXAMPLE 28**   Ann Corporation owes a $2,000 expense that may be paid and deducted on the cash basis of accounting in either 2004 or 2005. The applicable interest rate is 10%. In which year should Ann Corporation take the deduction if its 2004 marginal tax rate is 25%?

Discussion: The optimal year for taking the deduction depends on Ann Corporation's expected marginal tax rate in 2005. The following schedule calculates the real tax savings (real tax cost) of deducting the expenses in 2004 as compared with deferring the deduction until 2005 at different assumed marginal tax rates:

	Assumed 2005 Marginal Tax Rates		
	15%	25%	34%
Tax saved by 2005 deduction	$ 300	$ 500	$ 680
Present value @ 10%	× 0.909	× 0.909	× 0.909
Present value of tax savings	$ 273	$ 455	$ 618
Less: Tax savings of deduction in 2004 @ 25% marginal tax rate	(500)	(500)	(500)
Deduction in 2004 will result in:			
Tax savings	$(227)	$ (45)	
Tax cost			$ 118

Discussion: Ann Corporation should claim the deduction in 2004 if it expects the marginal tax rate to remain at 25% or decrease to 15%. If the corporation expects its marginal rate to increase to 34%, it should defer the deduction to 2005 to save $118.

▶ **EXAMPLE 29**   Lanny's marginal tax rate for 2004 is 28%. Lanny has $20,000 in income and $10,000 in deductions that could be reported in 2004 or deferred to 2005. Lanny expects his 2005 marginal tax rate to be 35% and the applicable interest rate to be 10%. When should the items be reported if both the income and deductions must be reported in the same year?

Discussion: The result of reporting both the income and the deductions in 2004 as compared with 2005 is as follows:

	2004	2005
Increase in income	$ 20,000	$ 20,000
Less: Increase in deductions	(10,000)	(10,000)
Net increase in taxable income	$ 10,000	$ 10,000
Marginal tax rate	× 28%	× 35%
Tax on net increase in income	$ 2,800	$ 3,500
Present value factor		× 0.909
Present value of tax in 2004	$ 2,800	$ 3,182

Discussion: Lanny should report the items in 2004 to save $382 in real tax cost.

▶ **EXAMPLE 30**   If Lanny could report the income or deductions separately, when should the income and the deductions be reported to maximize the tax savings?

Discussion: The tax cost of reporting each item must be considered separately and the total result compared with reporting both items in 2004 (which was previously determined to be the optimal same-year reporting).

Income

	Report income in	
	2004	2005
Increase in taxable income	$20,000	$20,000
Marginal tax rate	× 28%	× 35%
Increase in tax	$ 5,600	$ 7,000
Present value factor		× 0.909
Present value of tax in 2004	$ 5,600	$ 6,363
Net tax savings from reporting in 2004	763	

Deductions

	Report income in	
	2004	2005
Decrease in taxable income	$10,000	$10,000
Marginal tax rate	× 28%	× 35%
Tax savings from deduction	$ 2,800	$ 3,500
Present value factor		× 0.909
Present value of tax savings	$ 2,800	$ 3,182
Net tax savings from reporting in 2005	382	

Discussion: If Lanny reports the $20,000 of income in 2004, he has a real tax savings of $763. Deferring the reporting of the $10,000 in deductions until 2005 results in a real tax savings of $382. Thus, by reporting each item separately in the period that is optimal, he saves $1,145. This compares with a savings of $382 when both income and deductions are reported in the same tax year.

In summary, there are four general rules of thumb when planning the timing of income and deductions; two are based on time value of money propositions, and two are based on marginal tax rate considerations:

Time Value of Money

1. Defer recognition of income.

2. Accelerate recognition of deductions.

Marginal Tax Rate

3. Put income into the year with the lowest expected marginal tax rate.

4. Put deductions into the year with the highest expected marginal tax rate.

These general rules of thumb can be used in most situations. However, if there is a conflict between the time value rule and the marginal tax rate rule, the only way to determine the optimal strategy is to calculate the real tax cost of each. Table 1–4 summarizes the rules of thumb and indicates when calculation of the real tax cost is necessary.

Income Shifting. Income shifting is a method commonly used to reduce taxes. The basic idea behind income shifting is to split a single stream of income among two or more taxpayers to lower the total tax paid. The total tax paid is lower because of the progressive tax rate structure. For example, if a taxpayer in the 28-percent marginal tax rate bracket can shift $1,000 in income to another taxpayer who is in the 10-percent marginal tax rate bracket, $180 [$1,000 × (28% − 10%)] of tax will be saved on the $1,000 in income. Obviously, taxpayers shifting income will want the income to go to taxpayers whom they want to benefit, such as children or grandchildren.

> **EXAMPLE 31** A married taxpayer has $100,000 in taxable income in 2004. The taxpayer has 2 children who have no taxable income. What are the tax savings if the taxpayer can legally shift $5,000 in income to each of her children?

Discussion: The taxpayer saves $1,500 in tax by shifting $5,000 in taxable income to each child. Using the rates for married taxpayers, the tax on $100,000 in taxable income is $18,475:

$$\$8,000.00 + 25\% \ (\$100,000 - \$58,100) = \$18,475$$

By splitting the income into 3 streams, the taxpayer pays tax on $90,000, and each child pays tax (at single-taxpayer rates) on $5,000. This results in a tax of $16,975:

Tax on $90,000 for a Married Couple
$$\$8,000.00 + 25\% \ (\$90,000 - \$58,100) = \$15,975$$

Tax on $5,000 for a Single Person
$$\$5,000 \times 10\% = \$500 \times 2 = \underline{\quad 1,000}$$
$$\text{Total tax paid} \quad \underline{\$16,975}$$

The result of the income shift to the children is a reduction in the total tax paid on the $100,000 in taxable income of $1,500 ($18,475 − $16,975).

It should be noted that numerous provisions in the tax law make it difficult to get the full advantage of income shifting. For example, merely directing that some of your income be paid to your children will not shift the income for tax purposes. To shift income to family members, you will generally need to transfer ownership of income-producing property to the children in order to shift the income from the property. Unless the parents are willing to give up ownership of income-producing property, income shifting to children is difficult to achieve. Even if a valid transfer of property ownership is made, if the child is younger than 14, provisions exist to take away much of the marginal rate advantage of such a shift.

| Type of Item | Marginal Tax Rate | | |
	Increasing	Decreasing	Unchanged
Income	Calculate	Defer	Defer
Deduction	Calculate	Accelerate	Accelerate

Table 1–4
SUMMARY OF TAX-PLANNING RULES

Another popular income-shifting technique used by owners of a business is to incorporate the business and split income between themselves and the corporation. A review of the corporate tax rates (see Table 1–2) shows that the first $50,000 in taxable income of a corporation is taxed at 15 percent. The owners can split the income by paying themselves salaries, which are deductible by the corporation, and reduce the corporation's taxable income to a lower tax bracket.

▶ **EXAMPLE 32** Assume that the $100,000 in taxable income in example 31 comes from a business owned by the taxpayer. If the taxpayer incorporates the business and pays herself a salary of $50,000, what is the tax savings?

Discussion: Splitting the income between the taxpayer and a corporation results in a tax savings of $4,190. The taxpayer pays tax on $50,000, and the corporation pays tax on $50,000 ($100,000 income − $50,000 salary). This results in a tax of $14,285:

Tax on $50,000 for a Married Couple

$1,430.00 + 15% ($50,000 − $14,300) = $ 6,785

Tax on $50,000 for a Corporation

$50,000 × 15% =   7,500

Total tax paid $14,285

Before incorporation, the tax paid by the married couple was $18,475. The incorporation and split of the income saves $4,190 ($18,475 − $14,285) in tax.

Numerous other income-shifting techniques can be used by owners of a business. These include shifting income by employing children and using fringe-benefit packages to get tax-subsidized health care. It should be noted that careful planning is required to gain the optimal tax advantage from such shifting plans. The tax law contains many provisions designed to block blatant shifting schemes that lack economic substance. These provisions are discussed throughout the remainder of the text as they apply to the study of income and deductions.

Tax Evasion and Tax Avoidance

Taxpayers do not have to pay more income tax than is required by the tax law. In fact, taxpayers may plan transactions to make their tax bills as low as possible. In this regard, Judge Learned Hand stated: "[A] transaction, otherwise within an exception of the tax law, does not lose its immunity, because it is actuated by a desire to avoid, or, if one choose, to evade, taxation. Any one may so arrange his affairs that his taxes shall be as low as possible; he is not bound to choose that pattern which will best pay the Treasury; there is not even a patriotic duty to increase one's taxes."[34]

Tax evasion occurs when a taxpayer uses fraudulent methods or deceptive behavior to hide the actual tax liability. Tax evasion usually involves three elements:

- Willfulness on the part of the taxpayer
- An underpayment of tax
- An affirmative act by the taxpayer to evade the tax

Tax evasion often involves rearranging the facts about a transaction to receive a tax benefit. An intentional misrepresentation of facts on a tax return to avoid paying tax is not acceptable taxpayer behavior. Tax evasion is illegal and is subject to substantial penalties. Note that unintentional mathematical or clerical errors on the return are not generally considered tax evasion.

Tax planning uses tax avoidance methods. **Tax avoidance** is the use of legal methods allowed by the tax law to minimize a tax liability. Tax avoidance generally involves planning an intended transaction to obtain a specific tax treatment. Further, tax avoidance is based on disclosure of relevant facts concerning the tax treatment of a transaction.

▶ **EXAMPLE 33** Ted, an accountant, uses the cash method of accounting. To avoid reporting additional income in 2004, he does not send his December bills to clients until January 2, 2005.

Discussion: The income was properly reported when collected in 2005. Under the cash method of accounting, Ted properly reported income when his clients paid him. Ted's activity involves permissible tax avoidance.

> **EXAMPLE 34**   Ken, a painter, spent all the cash he received for his art work. He deposited payments he received by check to his business bank account. When he filed his tax return, he intentionally did not report the cash receipts as income.

Discussion: Ken is engaged in tax evasion. Ken's method of reducing his tax is illegal, and he is subject to substantial penalties.

At this point, you are probably wondering, "How will the IRS ever know?" Most people are aware that it is almost impossible for the government to track every cash receipt of income. In fact, the probability that the IRS will detect underreporting of cash income is quite low. This has led many taxpayers to play the "audit lottery," omitting cash income or overstating deductions, because they know that they probably will not be caught. The IRS estimates that this behavior results in a loss of more than $100 billion per year in tax revenue. This loss must be made up through higher taxes on honest taxpayers. It is clear that if taxpayers were more honest in their reporting of income and deductions, everyone's taxes could be lowered. There is no clear-cut, cost-efficient solution to the evasion problem. However, as future professionals and taxpayers, you should recognize your obligations to your profession and the country when it comes to tax evasion situations. Only through education and ethical taxpayer behavior will the tax evasion problem be resolved. Keep in mind that avoiding detection by the IRS does not somehow magically transform a fraudulent act into allowable behavior. The idea that something is not illegal unless one is caught is an idea that should have died ages ago.

ETHICAL CONSIDERATIONS IN TAX PRACTICE

The field of tax practice is virtually unregulated—anyone who wishes to can prepare tax returns for a fee. However, anyone who prepares tax returns for monetary considerations, or who is licensed to practice in the tax-related professions, is subject to various rules and codes of professional conduct. For example, the Internal Revenue Code contains provisions (see Exhibit 1–4 for a list of preparer penalties) that impose civil and criminal penalties on tax return preparers for various improprieties.

All tax practitioners are subject to the provisions of *IRS Circular 230,* "Regulations Governing the Practice of Attorneys, Certified Public Accountants, Enrolled Agents, and Enrolled Actuaries Before the Internal Revenue Service." Tax attorneys are subject to the ethical code of conduct adopted by the state(s) in which they are licensed to practice. Certified Public Accountants (CPAs) who are members of the American Institute of Certified Public Accountants (AICPA) are governed by the institute's Code of Professional Conduct. The AICPA's Statements on Standards for Tax Services provide eight advisory guidelines for CPAs who prepare tax returns. Although tax practitioners who are not members of the AICPA are not bound by the Code of Professional Conduct

---

Understatement of taxpayer's liability because of unrealistic positions
Understatement of taxpayer's liability because of willful or reckless conduct
Failure to furnish a copy of a return to the taxpayer
Failure to sign a return
Failure to furnish identifying information
Failure to retain a copy or a list of returns prepared
Failure to file correct information returns
Negotiation of tax refund check
Improper disclosure or use of information on taxpayer's return
Organizing (or assisting in doing so) or promoting and making or furnishing statements
 with respect to abusive tax shelters
Aiding and abetting an understatement of tax liability
Aiding or assisting in the preparation of a false return

Exhibit 1–4
I.R.C. VIOLATIONS WITH PENALTIES FOR TAX RETURN PREPARERS

and the Statements on Standards for Tax Services, the rules and guidelines contained in them provide useful guidance for all return preparers.

The AICPA Code of Professional Conduct is a set of rules that set enforceable ethical standards for members of the institute. The standards are broad and apply to all professional services that a CPA may render, including tax advice and tax return preparation. For example,

1. Rule 102 requires CPAs to perform professional services with objectivity and integrity, and to avoid any conflict of interest. CPAs should neither knowingly misrepresent facts nor subordinate their judgment to that of others in rendering professional advice.

2. Rule 202 requires compliance with all standards that have been promulgated by certain bodies designated by the AICPA's governing council.

3. Rule 301 states that CPAs will not disclose confidential client data without the specific consent of the client, except under certain specified conditions.

The eight Statements on Standards for Tax Services (SSTS) provide guidance on what constitutes appropriate standards of tax practice. The statements are intended to supplement, not replace, the Code of Professional Conduct. Because they specifically address the problems inherent in tax practice, each statement is briefly described here. The full text of the SSTS is reproduced in Appendix D.

SSTS No. 1: *Tax Return Positions.* CPAs should not recommend that a position be taken on a return unless they believe that, if the position is challenged, it is likely to be sustained, which is known as the *realistic possibility standard.* CPAs should not prepare a return or sign as preparer of a return if they know the return takes a position that could not be recommended because it does not meet the realistic possibility standard. However, a CPA may recommend any return position that is not frivolous, so long as the position is adequately disclosed on the return. SSTS Interpretation No. 1-1 (reproduced in Appendix D) contains the AICPA interpretation of the realistic possibility standard.

SSTS No. 2: *Answers to Questions on Returns.* A CPA should make a reasonable effort to obtain from the client and provide appropriate answers to all questions on a tax return before signing as preparer. Where reasonable grounds exist for omission of an answer, no explanation for the omission is required, and the CPA may sign the return unless the omission would cause the return to be considered incomplete.

SSTS No. 3: *Procedural Aspects of Preparing Returns.* A CPA may in good faith rely upon, without verification, information furnished by the client or third parties. Reasonable inquiries should be made if the information furnished appears to be incorrect, incomplete, or inconsistent. The CPA should use previous years' returns whenever possible to avoid omissions. In addition, the CPA may appropriately use information from the tax return of another client if the information would not violate the confidentiality of the CPA-client relationship and is relevant to and necessary for proper preparation of the return.

SSTS No. 4: *Use of Estimates.* A CPA may prepare returns using estimates provided by the taxpayer if it is impracticable to obtain exact data and the estimates are reasonable, given the facts and circumstances.

SSTS No. 5: *Departure from Previous Position.* If a CPA follows the standards in SSTS No. 1, the result of an administrative proceeding or court decision with respect to a prior return of the taxpayer does not bind the CPA as to how the item should be treated in a subsequent year's return.

SSTS No. 6: *Knowledge of Error: Return Preparation.* A CPA who becomes aware of an error in a previous year's return—or of the client's failure to file a required return—should promptly inform the client and recommend measures to correct the error. The CPA may not inform the IRS of the error except when required to do so by law. If the client does not correct the error, the CPA should consider whether to continue the professional relationship and must take reasonable steps to ensure that the error is not repeated if the relationship is continued.

SSTS No. 7: *Knowledge of Error: Administrative Proceedings.* When a CPA becomes aware of an error in a return that is the subject of an administrative proceeding, the CPA should promptly inform the client of the error and recommend measures to be taken. The CPA should request the client's consent to disclose the error to the IRS but should not disclose the error without consent unless required to do so by law. If the client refuses disclosure, the CPA should consider whether to withdraw from representing the client in the administrative proceeding and whether to continue a professional relationship with the client.

SSTS No. 8: *Form and Content of Advice to Clients.* A CPA should use judgment to ensure that advice given to a client reflects professional competence and appropriately serves the client's needs. For all tax advice given to a client, the CPA should adhere to the standards of SSTS No. 1, pertaining to tax return positions. A CPA may choose to notify a client when subsequent developments affect advice previously given on significant tax matters but is under no strict obligation to do so.

CONCEPT CHALLENGE

Reinforce the concepts covered in this chapter by completing the on-line tutorials located at the *Concepts in Federal Taxation* website.

http://murphy.swlearning.com

SUMMARY

Taxes are a fact of everyday life. Taxes are levied on income, products, property holdings, and transfers of wealth. The federal income tax is the largest revenue producer of all the taxes in use in the United States. Therefore, a solid understanding of the basic rules of the income tax system is essential to maximize your after-tax income.

The term *tax* has been defined, and concepts have been examined that will help you reach your own conclusions about whether a tax is "good" or "bad." Keep these evaluations in mind as you continue through the text and as you read articles on proposed tax legislation.

The income tax law is a complex body of constantly changing information that is issued by legislative, administrative, and judicial sources. When evaluating a particular tax rule, it may be necessary to consult resources in all three areas.

Tax terms used in income tax computation have been defined in this chapter. Subsequent chapters explain the terms and build on the basic information. When you encounter a new term in later chapters, do not hesitate to refer to this chapter to see how the new term fits into the computational framework.

The study of federal income taxation will help you evaluate how business and personal financial decisions influence the amount of income tax you will have to pay. Awareness of basic income tax concepts will help you recognize opportunities to minimize compliance costs, save taxes, avoid IRS penalties, and make more informed business decisions.

The practical approach to tax planning discussed in this chapter does not require you to be a tax specialist to become an effective tax planner. In later chapters, you will be asked to solve tax-planning problems that require you to make decisions about when an item of income or deduction should be reported. When solving these problems, you will need to consider the effects of changes in the marginal tax rate and the time value of money.

Finally, always be aware of the difference between tax evasion and tax avoidance. Avoid tax evasion—it is illegal. Tax avoidance is legal and is expected of taxpayers.

KEY TERMS

adjusted gross income (AGI) (p. 23)
ad valorem tax (p. 13)
annual loss (p. 17)
average tax rate (p. 7)
certainty (p. 5)
convenience (p. 6)
correspondence examinations (p. 22)
deduction (p. 17)
deductions for adjusted gross income (p. 23)
deductions from adjusted gross income (p. 24)

deferral (p. 16)
dependency exemption (p. 26)
Discriminant Function System (DIF) (p. 21)
document perfection program (p. 22)
economy (p. 6)
effective tax rate (p. 7)
equality (p. 4)
estate tax (p. 14)
exclusion (p. 16)
exemption (p. 18)
expense (p. 17)

field examinations (p. 22)
gain (p. 17)
gift tax (p. 14)
gross income (p. 16)
horizontal equity (p. 4)
information-matching program (p. 22)
Internal Revenue Code of 1986 (p. 15)
IRS Appeals Division (p. 22)
itemized deduction (p. 24)
loss (p. 17)
marginal tax rate (p. 7)
office examinations (p. 22)

ordinary income (p. 17)
pay-as-you-go concept (p. 6)
personal exemption (p. 26)
personal property (p. 13)
progressive rate structure (p. 8)
proportional rate structure (p. 7)
protest letter (p. 22)
real property (p. 13)
regressive rate structure (p. 8)

self-employment tax (p. 11)
Social Security taxes (p. 11)
special audit programs (p. 22)
standard deduction (p. 24)
statute of limitations (p. 20)
taxable income (p. 6)
tax avoidance (p. 30)
tax base (p. 6)
tax credit (p. 20)

tax evasion (p. 30)
Taxpayer Compliance Measurement
  Program (TCMP) (p. 22)
transaction loss (p. 17)
Treasury regulation (p. 15)
unified donative-transfers credit
  (p. 14)
vertical equity (p. 4)

PRIMARY TAX LAW SOURCES

[1]Rev. Rul. 77-29.

[2]Sec. 6072—Specifies the general rules for due dates of tax returns.

[3]Sec. 1—Imposes a tax on the taxable income of different classes of individual taxpayers; provides tax rates by class of taxpayer and requires adjustment of rate schedules each year for inflation; limits the tax rate on net long-term capital gains to 15%.

[4]Sec. 3402—Requires employers to withhold estimates of taxes on wages and salaries paid to employees.

[5]Sec. 31—Provides that amounts withheld as tax from salaries and wages are allowed as credits against that year's tax liability.

[6]Sec. 6654—Provides that all individuals must pay estimated taxes when their tax liability is expected to be greater than $1,000; imposes a penalty for not paying the proper amount of estimated tax.

[7]Sec. 3101—Imposes the Social Security tax on employees; provides rates of tax to be paid.

[8]Sec. 1402—Defines *self-employment income* and provides for the tax to be paid on base amounts as specified in the Social Security Act for each tax year.

[9]Sec. 3111—Imposes the Social Security tax on employers for wages paid to employees.

[10]Sec. 1401—Provides the tax rates for self-employment taxes.

[11]Sec. 2501—Imposes a tax on transfers of property by gift.

[12]Sec. 2503—Allows exclusion from gift tax of gifts up to $11,000.

[13]Sec. 2505—Allows unified credit against taxable gifts.

[14]Sec. 2001—Imposes a tax on the assets of an estate. Provides tax rates on estate assets and for unlimited marital exclusion.

[15]Sec. 2010—Provides for unified tax credit against tax liability of an estate.

[16]Sec. 7801—Directs the secretary of the Treasury to issue the regulations necessary to implement and interpret the tax law.

[17]Sec. 61—Provides the general definition of *gross income* as all income from whatever source derived.

[18]Sec. 64—Defines *ordinary income* as income that does not result from the sale or exchange of property that is not a capital asset or an asset described in Sec. 1231.

[19]Sec. 1001—Prescribes the calculation of gains and losses for dispositions of property; defines *amount realized* for purposes of determining gain or loss for dispositions.

[20]Sec. 162—Allows the deduction of all ordinary and necessary expenses incurred in a trade or business of the taxpayer.

[21]Sec. 212—Allows the deduction of all ordinary and necessary expenses incurred in a production-of-income activity of the taxpayer.

[22]Sec. 11—Imposes an income tax on corporations and provides the applicable tax rate schedules.

[23]Sec. 62—Defines *adjusted gross income* for individual taxpayers and specifies the deductions allowed as deductions for adjusted gross income.

[24]Sec. 1211—Sets forth the limit on deductions of capital losses of corporations and individuals.

[25]Sec. 63—Defines *taxable income*. Allows individual taxpayers to deduct the greater of their allowable itemized deductions or the standard deduction. Standard deduction amounts are specified and are required to be adjusted annually for inflation.

[26]Sec. 211—Generally allows specific personal expenditures as itemized deductions of individuals.

[27]Sec. 213—Allows the deduction of medical expenses as an itemized deduction for individual taxpayers; defines *medical expenses* and prescribes limitations on the amount of the deduction.

[28]Sec. 164—Specifies the allowable deductions for taxes.

[29]Sec. 163—Specifies the allowable deductions for interest.

[30]Sec. 170—Allows the deduction of contributions to qualified charitable organizations.

[31]Sec. 165—Specifies the allowable deductions for losses.

[32]Sec. 67—Limits the allowable deduction for miscellaneous itemized deductions to the excess of 2% of adjusted gross income.

[33]Sec. 151—Allows an exemption deduction for the taxpayer, the taxpayer's spouse, and for each qualifying dependent.

[34]Helvering v. Gregory, 69 F.2d 809 at 810 (2d Cir. 1934).

DISCUSSION QUESTIONS

1. Briefly state Adam Smith's four requirements for a good tax system.
2. Based on the discussion in the chapter, evaluate how well each of these taxes meets Adam Smith's four requirements:
   a. Income tax
   b. Employment taxes
3. Based solely on the definitions in the chapter, is the Social Security tax a proportional, regressive, or progressive tax? Explain, and state how the tax might be viewed differently.
4. Based solely on the definitions in the chapter, is the sales tax a proportional, regressive, or progressive tax? Explain, and state how the tax might be viewed differently.
5. As stated in the text, the federal income tax is the largest revenue-producing tax in use in the United States. Why do you think the income tax produces more revenue than any other tax?
6. How are federal, state, and local income taxes collected by the government? Consider the cases of an employee and a self-employed taxpayer.
7. How is a sales tax different from an excise tax?
8. Who is responsible for collecting sales and excise taxes? Who actually pays the tax?
9. Why is a tax on real property used more often than a tax on personal property?
10. The gift tax is supposed to tax the transfer of wealth from one taxpayer to another. However, the payment of gift tax on a transfer of property is relatively rare. Why is gift tax not paid on most gifts?
11. The estate tax is a tax on the value of property transferred at death. Why is payment of the estate tax not a common event?
12. What is the basis for valuing assets transferred by gift and at death?
13. Who is responsible for reporting and paying gift taxes? estate taxes?
14. Identify three primary sources of tax law.
15. Explain why the following statement is not necessarily true: "If the IRS disagrees, I'll take my case all the way to the Supreme Court."
16. What is the federal income tax base?
17. What is an exclusion?
18. How is a deferral different from an exclusion?
19. How is gross income different from income?
20. What are the three basic tests that an expense must satisfy to be deductible?
21. What is the difference between an expense and a loss?
22. How is a transaction loss different from an annual loss?
23. How does the legislative grace concept help identify amounts that qualify for deduction?
24. What is the purpose of the exemption deduction?
25. Based on the example in Exhibit 1–2, explain how inflation can have two effects that result in a hidden tax.
26. Explain the pay-as-you-go system.
27. What is a tax credit?

28. How is a tax credit different from a tax deduction?

29. If you were in the 28% marginal tax bracket and you could choose either a $1,000 tax credit or a $3,000 tax deduction, which would give you the most tax saving? Why?

30. What is the statute of limitations, and what role does it play in the filing of tax returns?

31. Briefly describe the types of programs used by the IRS to select a return for audit.

32. What are the three types of IRS examinations?

33. What is included in the 30-day letter, and what options does the taxpayer have after receiving one?

34. What does the 90-day letter represent, and what are the choices the taxpayer has after receiving one?

35. How is the calculation of taxable income for an individual different from the calculation of a corporation's taxable income?

36. How do deductions for adjusted gross income and deductions from adjusted gross income of an individual differ?

37. What is the purpose of the standard deduction for individuals?

38. Randy is studying finance at State University. To complete the finance major, he has to take a basic income tax course. Because Randy does not intend to be a tax expert, he considers the course a waste of his time. Explain to Randy how he can benefit from the tax course.

39. Evaluate the following statement: "The goal of good tax planning is to pay the minimum amount of tax."

40. It has often been said that only the rich can benefit from professional tax planning. Based on the information presented in this chapter, why is this statement at least partially true?

PROBLEMS

41. State whether each of the following payments is a tax. Explain your answers.

    a. To incorporate his business, Alex pays the state of Texas a $2,000 incorporation fee.

    b. The city paves a road and assesses each property owner on the road $4,000 for his or her share of the cost.

    c. The city of Asheville charges each residence in the city $10 per month to pick up the trash.

    d. Rory pays $450 of income tax to the state of California.

    e. Lanny is fined $45 for exceeding the speed limit.

42. Explain why each of the following payments does or does not meet the IRS's definition of a tax:

    a. Jack is a licensed beautician. He pays the state $45 each year to renew his license to practice as a beautician.

    b. Polly Corporation pays state income taxes of $40,000 on its $500,000 of taxable income.

    c. Winona pays $15 annually for a safety inspection of her automobile that is required by the state.

    d. The Judd Partnership owns land that is valued by the county assessor at $30,000. Based on this valuation, the partnership pays county property taxes of $800.

    e. Andrea fails to file her income tax return on time. She files the return late, and the IRS assesses her $25 for the late filing and $5 for interest on the tax due from the due date of the return until the filing date.

43. Susan is single with a gross income of $90,000 and a taxable income of $78,000. In calculating gross income, she properly excluded $10,000 of tax-exempt interest income. Using the tax rate schedules in the chapter, calculate Susan's

    a. Total tax   c. Average tax rate
    b. Marginal tax rate   d. Effective tax rate

44. A taxpayer has $95,000 of taxable income for the current year. Determine the total tax, the marginal tax rate, and the average tax rate if the taxpayer is a

    a. Single individual
    b. Married couple
    c. Corporation

45. Rory earns $60,000 per year as a college professor. Latesia is a marketing executive with a salary of $120,000. With respect to the Social Security tax, what are Rory's and Latesia's

a. Total taxes? **c.** Average tax rates?

b. Marginal tax rates? **d.** Effective tax rates?

46. For each of the following, explain whether the rate structure is progressive, proportional, or regressive:

a. Plymouth County imposes a 5% tax on all retail sales in the county. Taxpayers with incomes less than $12,000 receive a refund of the tax they pay.

b. The country of Zambonia imposes a 10% tax on the taxable income of all individuals.

c. Regan County imposes a property tax using the following schedule:

Assessed Value	Tax
$ 0 to $10,000	$ 40
$10,001 to $40,000	$ 40 + 1% of the value in excess of $10,000
$40,001 to $80,000	$ 340 + 2% of the value in excess of $40,000
$80,001 and above	$1,140 + 3% of the value in excess of $80,000

d. The city of Thomasville bases its dog licensing fee on the weight of the dog per the following schedule:

Weight (in pounds)	Tax Rate
0 to 40	$ 2 + 50% of weight
41 to 80	$22 + 40% of weight in excess of 40 lbs.
81 and above	$36 + 30% of weight in excess of 80 lbs.

47. The country of Boodang is the leading producer of sausage. Boodang imposes three taxes on its residents and companies to encourage production of sausage and discourage its consumption. Each tax applies as follows:

• Income tax—Rates apply to each taxpayer's total income:

$ -0- –$ 50,000	5% of total income
$ 50,001–$200,000	$ 2,500 + 10% of income in excess of $ 50,000
$200,001–$500,000	$17,500 + 20% of income in excess of $200,000
$500,001 or more	40% of total income

In calculating total income, sausage workers are allowed to deduct 25% of their salaries. Companies that produce sausage are allowed to deduct 50% of their sales. No other deductions are allowed.

• Sausage tax—All sausage purchases are subject to a 100% of purchase price tax. Residents who consume less than 10 pounds of sausage per year are given a 50% rebate of the sausage tax they paid.

• Property tax—Taxes are based on the distance of a taxpayer's residence from state-owned sausage shops per the following schedule:

0–2 miles	$15,000 per mile
2 miles–5 miles	$ 5,000 per mile
5 miles or more	$ 2,000 per mile

Given the definitions in the chapter, are Boodang's taxes progressive, proportional, or regressive? Evaluate and discuss each tax and the aspect(s) of the tax that you considered in making your evaluation.

48. Joe Bob is an employee of Rollo Corporation who receives a salary of $9,000 per month. How much Social Security tax will be withheld from Joe Bob's salary in

a. March?

b. November?

49. Return to the facts of problem 48. Assume that each month, Joe Bob has $2,400 in federal income tax and $800 in state income tax withheld from his salary. What is Joe Bob's take-home pay in

a. March?

b. November?

50. Gosney Corporation has 2 employees. During the current year, Clinton earns $64,000 and Trahn earns $94,000. How much Social Security tax does Gosney have to pay on the salaries earned by Clinton and Trahn?

51. Eric is a self-employed financial consultant. During the current year, Eric's net self-employment income is $98,000. What is Eric's self-employment tax?

52. Darrell is an employee of Whitney's. During the current year, Darrell's salary is $100,000. Whitney's net self-employment income is also $100,000. Calculate the Social Security and self-employment taxes paid by Darrell and Whitney. Write a letter to Whitney in which you state how much she will have to pay in Social Security and self-employment taxes and why she owes those amounts.

53. Classify the following items as ordinary income, a gain, or an exclusion:
   a. The gross revenues of $160,000 and deductible expenses of $65,000 of an individual's consulting business
   b. Interest received on a checking account
   c. Sale for $8,000 of Kummel Corporation stock that cost $3,000
   d. Receipt of $1,000 as a graduation present from grandfather
   e. Royalty income from an interest in a gold mine

54. Classify the following items as ordinary income, a gain, or an exclusion:
   a. The salary received by an employee
   b. Dividends of $400 received on 100 shares of corporate stock
   c. Sale for $10,000 of an antique chair that cost $3,500
   d. Rental income from an apartment building
   e. Receipt of an automobile worth $20,000 as an inheritance from Aunt Ruby's estate

55. Explain why each of the following expenditures is or is not deductible:
   a. Lumbar, Inc., pays $12,000 as its share of its employees' Social Security tax. The $12,000 is deductible.
   b. Leroy pays a cleaning service $250 per month to clean his real estate office. The $250 is deductible.
   c. Janice pays a cleaning service $75 per month to clean her personal residence. The $75 is not deductible.
   d. Leyh Corporation purchases land to use as a parking lot for $35,000. The $35,000 is not deductible.
   e. Martin spends $50 per month on gasoline for the car he uses to drive to his job as a disc jockey. The $50 is not deductible.

56. Classify each of the following transactions as a deductible expense, a nondeductible expense, or a loss:
   a. Nira sells for $4,300 stock that cost $6,000.
   b. Chiro Medical, Inc., pays $2,200 for subscriptions to popular magazines that it places in its waiting room.
   c. Lawrence pays $200 for subscriptions to fly-fishing magazines.
   d. The Mendota Partnership pays $200,000 to install an elevator in one of its rental properties.
   e. Sterling Corporation pays $6,000 for lawn maintenance at its headquarters.

57. Based on the following information, what are the taxable income and tax liability for a single individual?

Total income	$91,000
Excludable income	2,000
Deductions for adjusted gross income	2,500
Deductions from adjusted gross income	6,850

58. Based on the facts in problem 57, calculate the taxable income and the tax liability for a married couple.

59. Reba's 2004 income tax calculation is as follows:

Gross income	$120,000
Deductions for adjusted gross income	(3,000)
Adjusted gross income	$117,000
Deductions from adjusted gross income:	
Standard deduction	(4,850)
(Total itemized deductions are $2,100)	
Personal exemption	(3,100)
Taxable income	$109,050

Before filing her return, Reba finds an $8,000 deduction that she omitted from these calculations. Although the item is clearly deductible, she is unsure whether she should deduct it for or from adjusted gross income. Reba doesn't think it matters where she deducts the item, because her taxable income will decrease by $8,000 regardless of how the item is deducted. Is Reba correct? Calculate her taxable income both ways. Write a letter to Reba explaining any difference in her taxable income arising from whether the $8,000 is deducted for or from adjusted gross income.

60. Since graduating from college, Mabel has used the firm of R&P to prepare her tax returns. Each January, Mabel receives a summary information sheet, which she fills out and sends to R&P along with the appropriate documentation. Because she has always received a refund, Mabel feels that R&P is giving her good tax advice. Write a letter to Mabel explaining why she may not be getting good tax advice from R&P.

Communication

61. Michiko and Saul are planning to attend the same university next year. The university estimates tuition, books, fees, and living costs to be $9,000 per year. Michiko's father has agreed to give her the $9,000 she needs to attend the university. Saul has obtained a job at the university that will pay him $11,000 per year. After discussing their respective arrangements, Michiko figures that Saul will be better off than she will. What, if anything, is wrong with Michiko's thinking?

62. Inga, an attorney, completed a job for a client in November 2004. If she bills the client immediately, she will receive her $10,000 fee before the end of the year. By delaying the billing for a month, she will not receive the $10,000 until 2005. What factors should Inga consider in deciding whether she should delay sending the bill to the client?

63. Art is in the 28% marginal tax bracket for 2004. He owes a $10,000 bill for business expenses. Because he reports taxable income on a cash basis, he can deduct the $10,000 in either 2004 or 2005, depending on when he makes the payment. He can pay the bill at any time before January 31, 2005, without incurring the normal 8% interest charge. If he expects to be in a 33% marginal tax bracket for 2005, should he pay the bill and claim the deduction in 2004 or 2005?

64. Elki would like to invest $50,000 in tax-exempt securities. He now has the money invested in a certificate of deposit that pays 5.75% annually. What rate of interest would the tax-exempt security have to pay to result in a greater return on Elki's investment than the certificate of deposit? Work the problem assuming that Elki's marginal tax rate is 15%, 25%, 28%, and 33%.

65. Leroy and Amanda are married and have three dependent children. During the current year, they have the following income and expenses:

Salaries	$96,000
Interest income	45,000
Royalty income	27,000
Deductions for AGI	3,000
Deductions from AGI	9,000

   a. What is Leroy and Amanda's current year taxable income and income tax liability?

   b. Leroy and Amanda would like to lower their income tax. How much income tax will they save if they validly transfer $5,000 of the interest income to each of their children? Assume that the children have no other income and that they are entitled to a $800 standard deduction but are not allowed a personal exemption deduction.

66. Tina owns and operates Timely Turn Tables (TTT) as a sole proprietorship. TTT's taxable income during the current year is $80,000. In addition to the TTT income, Tina has the following income and expenses during the current year:

Interest income	$ 11,000
Royalty income	28,000
Deductions for AGI	2,500
Deductions from AGI	12,000

   a. What is Tina's current year taxable income and income tax liability?

   b. Tina would like to lower her tax by incorporating Timely Turn Tables. How much income tax will she save if she incorporates TTT and pays herself a salary of $40,000?

67. For each of the following situations, state whether the taxpayer's action is tax evasion or tax avoidance.

   a. Tom knows that farm rent received in cash or farm produce is income subject to tax. To avoid showing a cash receipt on his records, he rented 50 acres for 5 steers to be raised by the tenant. He used 2 of the steers for food for his family and gave 3 to relatives. Because he did not sell the livestock, he did not report taxable income.

   b. Betty applied for and received a Social Security number for Kate, her pet cat. Surprised by how easy it was to get a Social Security number, she decided to claim a dependent exemption on her tax return for Kate. Other than being a cat, Kate met all the tests for a dependent.

   c. Glen has put money in savings accounts in 50 banks. He knows a bank is not required to report to the IRS interest it pays him that totals less than $10. Because the banks do not report the payments to the IRS, Glen does not show the interest he receives as taxable income. Although Glen's accountant has told him all interest he receives is taxable, Glen insists that the IRS will never know the difference.

   d. Bob entered a contract to sell a parcel of land at a $25,000 gain in 2003. To avoid reporting the gain in 2003, he closed the sale and delivered title to the land to the buyers on January 2, 2004.

   e. Asha's taxable income for 2004 puts her in the 33% marginal tax bracket. She has decided to purchase new equipment for her business during 2005. A special election allows Asha to treat the $25,000 of the cost of the equipment as a current period expense. Because she expects to be in a lower tax bracket next year, Asha buys and begins using $25,000 worth of the equipment during December 2004. She claims a $25,000 expense deduction under the special election for 2004.

68. In each of the following situations, explain why the taxpayer's action is or is not tax evasion:

   a. Jamal owns an electrical appliance repair service. When a client pays him in cash, he gives the cash to his daughter Tasha. Jamal does not report the cash he gives to Tasha in his business income. Tasha has no other income, and the amount of cash that she receives from Jamal is small enough that she is not required to file a tax return.

   b. Roberta and Dudley are married. Roberta usually prepares their tax return. However, she was in the hospital and unable to prepare the return for 2003, so Dudley did it. In preparing their 2004 return, Roberta notices that Dudley included $1,000 of tax-exempt municipal bond interest in their 2003 gross income. To correct this mistake, Roberta takes a $1,000 deduction on the 2004 return.

   c. In 2004, Hearthome Corporation receives notice that the IRS is auditing its 2002 return. In preparing for the audit, Hearthome's controller, Monique, finds a mistake in the total for the 2002 depreciation schedule that resulted in a $5,000 overstatement of depreciation expense.

   d. While preparing his tax return, Will becomes unsure of the treatment of a deduction item. He researches the issue and can find no concrete tax law authority pertaining to the particular item. Will calls his buddy Dan, an accounting professor, for advice. Dan tells Will that if the law is unclear, he should treat the deduction in the most advantageous manner. Accordingly, Will deducts the full amount of the item, rather than capitalizing and amortizing it over 5 years.

   e. Sonja is a freelance book editor. Most companies for which she works pay her by check. In working out the terms of a job, a new client agrees to pay her by giving her a new computer valued at $3,600. In preparing her tax return, Sonja notes that the client failed to report to the IRS the value of the computer as income for Sonja. Aware that her chances of getting caught are small, Sonja does not include the $3,600 value of the computer in her gross income.

ISSUE IDENTIFICATION PROBLEMS

In each of the following problems, identify the tax issue(s) posed by the facts presented. Determine the possible tax consequences of each issue that you identify.

69. Marla had $2,100 in state income taxes withheld from her 2004 salary. When she files her 2004 state income tax return, her actual state tax liability is $2,300.

70. While reading a State College alumni newsletter, Linh is surprised to learn that interest paid on student loans is deductible. Linh graduated from college 2 years ago and paid $1,200 in interest during the current year on loans that he took out to pay his college tuition.

71. Victoria's son needs $5,000 for tuition at the Motown School of Dance. Victoria, who is in the 35% marginal tax rate bracket, intends to pay the tuition by selling stock worth $5,000 that she paid $2,000 for several years ago.

72. Joey and Camilla are married and have three children, ages 8, 16, and 18. They own a commercial cleaning business that is organized as a sole proprietorship and makes $120,000 annually. They have $30,000 of other taxable income (net of allowable deductions).

73. **INTERNET ASSIGNMENT** The purpose of this assignment is to introduce you to the tax information provided by the Internal Revenue Service on its World Wide Web site (http://www.irs.ustreas.gov/). Go to this site and look at the various types of information provided and write a short summary of what the IRS offers at its site. Chapter 1 discusses the audit and appeals process. Locate Publication 17, Tax Information for Individuals, and find the discussion of the examination and appeals process. Print out the text of this discussion.

74. **INTERNET ASSIGNMENT** Many legislative, administrative, and judicial resources are available on the Internet. These can be located using a search engine or a tax directory site on the Internet. This assignment is designed to acquaint you with some of the tax directory sites. Go to one of the tax directory sites provided in Exhibit 16–6 (Chapter 16) and describe the types of information you can access from the site. Use at least three links to other sites and describe the information at each of the sites.

75. **RESEARCH PROBLEM** Audrey opened Hardy Consulting Services during the current year. She has one employee, Deng, who is paid a salary of $30,000. Audrey is confused about the amount of federal unemployment tax she is required to pay on Deng's salary. The state unemployment tax rate is 4%. Audrey has asked you to determine how much federal unemployment tax she is required to pay on Deng's salary. Write Audrey a letter explaining the amount of federal unemployment tax she must pay.

Communication

76. **RESEARCH PROBLEM** Shawna earns $70,000 as a biologist for Berto Corporation. She also consults with other businesses on compliance with environmental regulations. During the current year, she earns $25,000 in consulting fees. Determine the amount of self-employment tax Shawna owes on her consulting income.

77. **SPREADSHEET PROBLEM** Using the information below, prepare a spreadsheet that will calculate an individual's taxable income. The spreadsheet should be flexible enough to accommodate single and married taxpayers as well as changes in the information provided below.

Number of dependents	2
Salary	$75,000
Interest	8,000
Deductions for adjusted gross income	2,800
Deductions from adjusted gross income	12,100

78. A value-added tax has been the subject of much debate in recent years as a tax to use to help reduce the deficit. Various forms of value-added taxes are used throughout Europe, Canada, and in many other countries. To acquaint yourself with the basic operation of a value-added tax, read the following article:

Peter Chin and Joel G. Siegel, "What the Value-Added Tax Is All About," *TAXES— The Tax Magazine*, January 1989, pp. 3–13.

After reading the article, consider the following circumstances:

Joe is married and has 2 children. A brain surgeon, he earns about $300,000 annually from his medical practice and averages about $250,000 in investment income. Jane, Joe's wife, spends most of her time doing volunteer work for charitable organizations. Tom is also married and has 5 children. He earns $30,000 per year working as a maintenance man for Joe.

While Joe was working late one night, he and Tom had a serious disagreement about two new tax bills recently introduced to help reduce the deficit. The first bill would levy a 10% value-added tax on all goods and services. A second bill introduced at the same time would add an additional 10% tax to each of the six current tax rate brackets (i.e., 10% would become 20%, 15% would become 25%, 25% would become 35%, 28% would become 38%, 33% would become 43%, and 35% would become 45%).

Joe is concerned that the imposition of a value-added tax would mean that fewer people could afford medical treatment. Both his patients and his practice would suffer from the tax. Tom strongly disagrees with Joe. He thinks that Joe does not want to pay his fair share of taxes. Tom charges that Joe can afford to hire tax accountants to help him avoid paying higher income taxes, even with the higher tax rates. By enacting a value-added tax, Tom believes, high-income taxpayers like Joe will have to pay up. He thinks it is the only fair way to raise taxes to bring down the deficit.

After several hours of arguing, neither could convince the other that he was wrong. Joe finally ended the discussion by saying that he would get an independent person knowledgeable in tax law to decide who is right.

You work for the firm that prepares Joe's tax return and advises him on managing his finances. The tax partner of your firm asks you to prepare a memorandum discussing the merits and deficiencies of the two proposals as they apply to Joe and Tom. In your memorandum, you are directed to specifically consider the following and provide a response:

a. What is a value-added tax, and how does it work?

b. Evaluate the rate structures of the two proposed taxes. Are they proportional, progressive, or regressive?

c. What, if anything, is wrong with Tom's and/or Joe's point of view? Be sure to explain this part in depth.

79. Norman and Vanessa are married and have 2 dependent children. This is a summary of their 2003 tax return:

Adjusted gross income	$78,200
Deductions from adjusted gross income:	
Standard deduction	(9,500)
Exemptions ($3,050 × 4)	(12,200)
Taxable income	$56,500
Tax liability	$ 7,775

a. Assuming that Norman and Vanessa's 2004 adjusted gross income will increase at the 2.2% rate of inflation and that the standard deduction and exemption amounts do not change, calculate their 2004 taxable income. Calculate the tax liability on this income using the 2003 tax rate schedules (Appendix A).

b. Calculate Norman and Vanessa's projected 2004 taxable income and tax liability, assuming that their adjusted gross income will increase by 2.2% and that all other inflation adjustments are made. Compare these calculations with those in part a, and explain how the inflation adjustments preserve Norman and Vanessa's after-tax income.

TAX PLANNING CASES

Communication

80. Bonnie is married and has 1 child. She owns Bonnie's Rib Joint, which produces a taxable income of approximately $100,000 per year.

a. Assume that Bonnie's taxable income is $40,000 without considering the income from the rib joint. How much tax will she pay on the $100,000 of income from the rib joint?

b. You work for the firm that prepares Bonnie's tax return. Bonnie has asked the partner for whom you work to advise her on how she might lower her taxes. The partner has assigned you this task. Draft a memorandum to the partner that contains at least two options Bonnie could use to lower her taxes. For each option, explain the calculations that support the tax savings from your recommendation.

81. Barbara is going to purchase a car for $20,000. She has two financing options: She can finance the purchase through the dealer at 1 percent for 48 months, with monthly loan payments of $425, or she can take a $2,000 rebate on the purchase price and finance the remaining $18,000 with a 7.5 percent home equity loan whose monthly payment will be $435. The interest on the home equity loan is deductible; the interest on the dealer loan is not. Barbara is in the 33% marginal tax rate bracket. Determine her best course of action in financing the purchase of the car.

ETHICS DISCUSSION CASE

82. Return to the facts of problem 67. Assume that you are the CPA in charge of preparing the tax return for each of the taxpayers in the problem. Based on the Statements on Standards for Tax Services (Appendix D), explain what you should do in each case. Your discussion should indicate which, if any, of the eight statements is applicable and your obligations with regard to each applicable statement. If the facts are not sufficient to determine whether a statement applies to a situation, discuss the circumstances in which the statement would apply.